Building Resilience of Human-Natural Systems of Pastoralism in the Developing World: Interdisciplinary Perspectives

Shikui Dong • Karim-Aly S. Kassam
Jean François Tourrand • Randall B. Boone
Editors

Building Resilience of Human-Natural Systems of Pastoralism in the Developing World: Interdisciplinary Perspectives

Springer

Editors
Shikui Dong
Beijing, China

Karim-Aly S. Kassam
Ithaca, NY, USA

Jean François Tourrand
Montpellier, France

Randall B. Boone
Fort Collins, CO, USA

ISBN 978-3-319-80877-2 ISBN 978-3-319-30732-9 (eBook)
DOI 10.1007/978-3-319-30732-9

Printed on acid-free paper

This Springer imprint is published by Springer Nature
The registered company is Springer International Publishing AG Switzerland

To marginalized pastoralists in the world

Acknowledgements

We express our deep thanks to esteemed Prof. Jizhou Ren, Academician, Chinese Academy of Engineering, for writing the foreword for this book. We also offer our heartful thanks to Dr. James Lassoie, International Professor of Conservation at Cornell University, Dr. Dennis Ojima, Professor of Natural Resource Ecology Laboratory at Colorado State University, Dr. Xinquan Zhao, Professor and Director of Chengdu Institute of Biology at Chinese Academy of Sciences to present the insightful commentaries for this book. We acknowledge the substantial efforts of Dr. Saini Sherestha, Mr. Zachary Romano and Ms. Sasireka Kuppan, the editors from Springer to publish the manuscript. We thank all our collaborators, including farmers/herders, professionals, practitioners, and students, who have kindly provided assistances in our field trips and book writing. We would also like to thank funding organizations, such as Ministry of Science and Technology of the People's Republic of China (Grant No. 2016YFC0501906), National Natural Science Foundation of China, and Asian Scholar Foundation for their financial support. We are most grateful to our family members for their support, love and patience when we were conducting our field studies and writing the book.

Preface

Pastoralism is a production system and livelihood strategy that is based on extensive livestock grazing on rangelands/grasslands and often some form of herd mobility, which has been practiced in many regions of the world for centuries. Currently, extensive pastoralism occurs on about 25 % of Earth's land area, mostly in the developing world, from the drylands of Africa and the Arabian Peninsula to the highlands of Asia and Latin America where intensive crop cultivation is physically not possible because of a harsh environment and poor access. In addition, cattle and sheep ranchers in Western North America, Australia, New Zealand, and a few other regions of the world presently practice a modern form of pastoralism. Worldwide, pastoralism supports about 200 million households and herds of nearly a billion animals including camels, cattle, and smaller livestock that account for about 10 % of the world's meat production.

Pastoralism is globally important for the human population it supports, the food and ecological services it provides, the economic contributions it makes to some of the world's poorest regions, and the long-standing civilizations it helps to maintain. However, threats and pressures associated with human population growth, economic development, land use changes, and climate change, etc., at a global scale are challenging the sustainability of these invaluable social, cultural, economic, and ecological assets. Key services and functions such as food production and biodiversity conservation provided by pastoral ecosystems may be vulnerable to both natural stresses and human disturbances. These problems are widely recognized by professionals and practitioners in the field of pastoral sciences. Furthermore, it is commonly agreed that these problems cannot be addressed solely through technical innovations, political reform, or economic development. The newly developed portfolios of coupled human and natural systems may provide important insights into diverse complex systems of pastoralism that cannot be well understood or effectively managed within a single dimension. New research and monitoring programs for pastoral areas will need to be designed that can address ecological and socioeconomic interrelationships within a framework of coupled human and natural systems by necessitating effective collaborations among social scientists, biophysical scientists, and management practitioners, as well as forming an international interdisciplinary research network capable of investigating pastoralism on various scales, from local to global.

As a veteran researcher working in the field of grassland and pastoral sciences for more than 70 years, I am delighted to see the timely publication of *Resilience of Coupled Human–Natural Systems: Interdisciplinary Strategies*. To my knowledge, this is the first book to address the issue of resilience of human–natural systems of pastoralism. I congratulate the editors - leading scholars in resilience or pastoral studies - for presenting their new research findings from diverse pastoral systems in the world and their synthesis of other investigations of pastoralism across a vast range of pastoral landscapes on earth. This book provides a compendium of information and insights that will prove valuable for the design of research/monitoring projects and planning policy programs in pastoral areas across the world. I highly recommend it to scientists, planners, government officials, and public organizations concerned about the protection and sustainable development of pastoralism.

任继周
Jizhou Ren
Academician, Chinese Academy of Engineering
Professor, College of Pastoral Agriculture,
Science and Technology,
Lanzhou University
Lanzhou, China

Contents

Chapter 1
Overview: Pastoralism in the World

Shikui Dong

Abstract This chapter provides an overview of global pastoralism, including the definition, forms, structures, origin, development, distribution, value, and future of pastoralism. Pastoralism can be defined as mobile livestock herding in the dimension of either production or livelihood. Nomadic and transhumant rearing of domesticated animals are generally two essential forms of pastoralism, with pastoral farming/enclosed ranching as the third form of pastoralism in the broad meaning. A clan is generally the basis of pastoral organization, which is responsible for the control of the optimum territory and management of the livestock species herded in every corner of the world. Most of the burden of pastoral activities is borne by women, and empowering women remains a challenge in most of the pastoral regions across the world. Although the emergence of pastoralism was a complex and multifaceted phenomenon, primitive hunting has been commonly accepted as the primary source. The origin of pastoralism can be dated to 6000 B.P. in the Andes of South America, and even as early as 9000 B.P. in Northeast Africa. A multiple-center origination is more probable than a single-center origination for explaining the spread of pastoralism worldwide. Currently, extensive pastoralism occurs on about 25 % of Earth's land area, mostly in the developing world, from the drylands of Africa and the Arabian Peninsula to the highlands of Asia and Latin America. Globally, pastoralism is critically important in supporting huge human populations, providing tremendous ecological services, maintaining long-standing civilizations, and making significant contributions to subsistence economy in some of the world's poorest regions. However, the practices of pastoralism have been overwhelmed by agricultural expansion, industrial development, and sedentary livestock farming in recent decades. Pastoral societies across the world will have more unpleasant fates with the stress of global change in the future.

S. Dong (✉)
School of Environment, Beijing Normal University, Beijing 100875, China
e-mail: dongshikui@sina.com

S. Dong et al. (eds.), *Building Resilience of Human-Natural Systems of Pastoralism in the Developing World: Interdisciplinary Perspectives*,
DOI 10.1007/978-3-319-30732-9_1

1.1 Introduction

Extensive pastoralism exists on all continents except Antarctica, mostly in the drylands or highlands, where intensive crop cultivation is physically not possible (FAO 2001). Currently, pastoralism occurs in more than 100 countries on about 25 % of Earth's land area and supports about 200 million households and herds of nearly a billion animals, including camels, cattle, and smaller livestock that account for about 10 % of the world's meat production (FAO 2001). Pastoralism provides very important ecological services, such as primary production, biodiversity conservation, and erosion control. However, the social, economic, and environmental importance of worldwide pastoralism has been overlooked in the modern era. It is necessary to review the history, distribution, and importance of global pastoralism, especially in the developing world. Here, we provide an overview of global pastoralism and its human–natural systems.

1.2 Definition and Forms of Pastoralism

1.2.1 Definition of Pastoralism

The definition of pastoralism varies greatly in terms of purposes and focuses (e.g., intensional, extensional, descriptive, stipulative, etc.). Basically, two common definitions derived from either the production perspective or the livelihood perspective are broadly used for "pastoralism." In the dimension of production, pastoralism is animal husbandry, the branch of agriculture concerned with the care, tending, and use of grazing livestock in dry or cold rangeland areas. In the dimension of livelihood, pastoralism is a subsistence living pattern of tending herds of large animals (Blench 2001) or a successful livelihood strategy on less productive lands through livestock herding (IFAD 2008). As summarized by the International Fund for Agricultural Development in Fig. 1.1, pastoralism, with the features of mobility, adaptation, flexibility, diversification, conservation, and mutual support, is "the finely-honed symbiotic relationship between local ecology, domesticated livestock and people." As the traditional rangeland management strategy, pastoralism represents a complex form of natural resource management, involving the direct interaction between natural resources and their users done within a larger geopolitical context (Pratt et al. 1997). Therefore, pastoralism can be understood as one of the coupled human–natural systems in the developing world (including remote and marginalized areas of developed countries).

In the literature, "pastoral system" is often used as an alternative term for "pastoralism." A pastoral system is defined as a system occurring in rangeland areas, where livestock grazing is the predominant form of land use (FAO 2002). A pastoral system is a system that is adaptive to particular natural, political, and economic environments. In a pastoral system people who herd or raise livestock are called "pastoralists," and they currently live in more than 100 countries (Fig. 1.2).

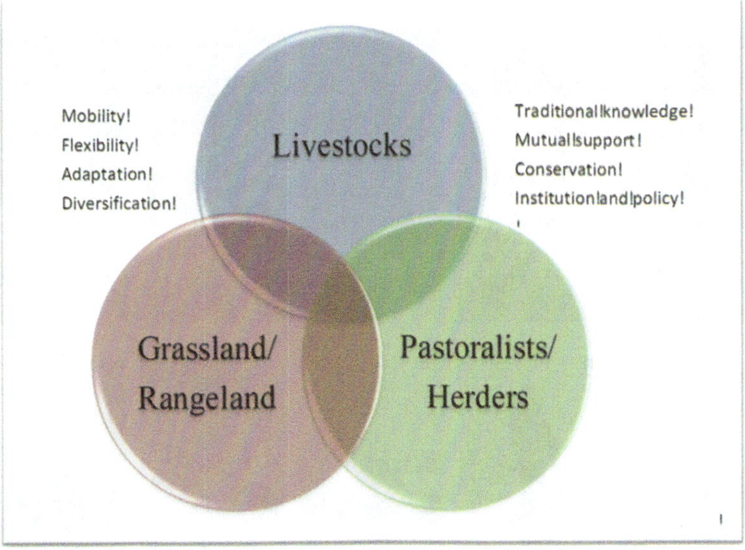

Fig. 1.1 Pastoralism: a sustainable natural resource management system. (Adapted from IFAD 2008 http://www.ifad.org/lrkm/factsheet/pastoralists.pdf)

Pastoralist groups can take many forms to adapt the particular natural, political, and economic environments across the world. The types of livestock kept by these pastoralists differ according to variations of the climate, environment, water and other natural resources, and geographical areas. As summarized by Blench (2001), cattle and sheep are broadly raised as pastoral herds, by ranchers in North America and Australia and by herders in Africa, Europe, and Asia. llamas and alpacas are mainly raised as the key pastoral herds in the Andes of South America. Horses are mostly herded in Central Asia. Donkeys and dromedaries are predominantly grazed in North Africa and West Asia. Bactrian camels are mostly raised in East and Central Asia. Goats are predominantly raised in West Africa and West and Central Asia. Yaks are mainly grazed in the highlands of Central Asia. Reindeers are grazed in circumpolar Eurasia (Table 1.1, Fig. 1.3).

A pastoral system is characterized by relatively large herd or flock sizes, a high proportion of females, and more steers than oxen in the case of cattle (FAO 2002). The management of livestock in a pastoral system is aimed at ensuring subsistence, averting risk, and adapting to the institutional environment, which consists mostly of communal grazing (FAO 2002). In addition, there is agropastoralism in the transition zone between pastoral areas and agricultural areas. People who live on agropastoralism are called "agropastoralists." Agropastoralism is defined as a set of practices that combine pastoral livelihoods with production of millet, sorghum, maize, vegetables, and pulses (annual legumes). This system is extremely important and is the most prevalent land use in arid and semiarid environments

Fig. 1.2 Some pastoral groups in the world: (**a**) Tibetan in Qinghai, China; (**b**) Kirghiz in Badakhshan, Afghanistan; (**c**) Boran in Borana, Ethiopia; (**d**) Massai in Kenya; (**e**) Mongol in Inner Mongolia, China; (**f**) Tajik in Yangi Qala, Afghanistan; (**g**) Bedouin in Negev, Israel; (**h**) Baloch in northern Pakistan;

Table 1.1 Pastoral species and their management strategies in the world

Species	Scientific name	Main regions	Nomadic	Transhumant	Agropastoral	Enclosed
Alpaca	*Lama pacos*	Andes	–	+	+	–
Bactrian camel	*Camelus bactianus*	East and Central Asia	+	+	+	–
Buffalo	*Bubalus bubalis*	Iran, India	+	+	+	?
Cattle (taurine)	*Bos taurus*	Europe, West Asia, West Africa	–	+	+	+
Cattle (zebu)	*Bos indicus*	Africa, Central Asia	+	+	+	+
Donkey	*Equus asinus*	Africa, Asia	+	+	+	–
Dromedary	*Camelus dromedarius*	Africa, West Asia	+	+	+	–
Goat	*Capra hircus*	Africa, Europe, Asia	+	+	+	+
Horse	*Equus caballus*	Central Asia	+	+	+	–
Llama	*Lama lama*	Andes	–	+	+	–
Reindeer	*Rangifer tarandus*	Circumpolar Eurasia	+	+	–	?
Sheep	*Ovis aries*	Africa, Europe, Asia	+	+	+	+
Yak	*Poephagus grunniens*	Highland Central Asia	–	+	–	–

Adapted from Blench (2001)

(USAID 2011). By definition, the difference between pastoralism and agropastoralism is that pastoralists derive most of their family incomes (more than 50 %) from livestock and livestock products, whereas agropastoralists derive most of their family income from cultivation and only a small amount from livestock production (IFAD 2008).

Fig. 1.2 (continued) (**i**) Kanets in Himachal Pradesh, India; (**j**) Gaddi in Himachal Pradesh, India; (**k**) Bedouin in Egypt; (**l**) Aymara herder in the Bolivian Andes; (**m**) Tamang in Rasuwa, Nepal; (**n**) Yugur in Gansu, China; (**o**) Kazak in Xinjiang, China; (**p**) Sami in Kola Peninsula, Russia; (**q**) Ngalop herder in Paro, Bhutan; (**r**) Wakhi in Wakhan, Afghanistan; (**s**) Pashtun in Pakistan; (**t**) Gujjar in Pakistan. (Photos by (**a**) Xukun Su, 2014; by Aziz Ali, 2012; (**c**) Allan Degen, 2010; (**d**) Jianchu Xu, 2012; (**e**) Wei Sha, 2014; (**f**) Shaoliang Yi, 2012; (**g**) Allan Degen; (**h**) Abdul Wahid Jasra, 2010; (**i**) Shikui Dong, 2010; (**j**) Shikui Dong, 2010; (**k**) Jean François Tourrand, 2012; (**l**) Jean François Tourrand; (**m**) Shikui Dong, 2007; (**n**) Kiran Elana, 2010; (**o**) Xi Wang, 2012; (**p**) Karim-Aly Kassam, 1996; (**q**) Shaoliang Yi, 2010; (**r**) Shaoliang Yi, 2010; (**s**) Abdul Wahid Jasra, 2010; (**t**) Abdul Wahid Jasra, 2010)

Fig. 1.3 Pastoral livestock in the world: (**a**) donkey; (**b**) reindeer; (**c**) buffalo; (**d**) dromedary; (**e**) Bactrian camel; (**f**) zebu cattle; (**g**) taurine cattle; (**h**) yak; (**i**) horse; (**j**) sheep; (**k**) goat; (**l**) alpaca; (**m**) llama. (Photos from (**a**) Afghanistan by Shaoliang Yi, 2010; (**b**) Russia by Karim Kassam, 1996;

1.2.2 *Forms of Pastoralism*

The forms of pastoralism are often classified by the method of mobility, a key feature qualifying pastoralism (Blench 2001). There are generally two essential forms of pastoralism: nomadic and transhumant (O'Neil 2011). Pastoral farming/enclosed ranching is considered the third form of pastoralism in the broad meaning (Blench 2001). Sometimes agropastoralism is also defined as one of the forms. The term "nomadic pastoralism" is used for pastoral mobility in highly irregular patterns. The term "transhumant pastoralism" is used for pastoral mobility of regular back-and-forth movements between relatively fixed locations. The term "pastoral farming" is used for pastoral mobility with little or no long-distance movement (i.e., enclosed ranching). Different forms of pastoralism have been practiced on different types of livestock by different indigenous pastoralists across different regions of the world (Table 1.1).

1.2.2.1 Nomadic Pastoralism

Nomadic pastoralism is the common practice in regions with little arable land, typically in the drylands and highlands of the world. It exists in areas of low rainfall, such as the Arabian Peninsula and Northeast Africa inhabited by Bedouins. It is also found in areas of harsh climate, such as northern Europe and arctic areas of Russia inhabited by Sami. In this pastoral system, the pastoralists follow a seasonal migratory pattern of a nomadic cycle varying from year to year with the grazing needs in the northern hemisphere, as exampled as follows for the Aertai region of Xinjiang:

- Spring (April to June)—transition
- Summer (July to late September)—high mountains
- Autumn (October to end of November)—transition
- Winter (from December to March)—dry plains

The migration routes of this nomadic cycle range from tens to hundreds of kilometers, even 1000 km (e.g., in Central Asia). Sometimes pastoralists have to find the ways for mobile livestock among the heavy traffic on the highway during the long migration (Fig. 1.4). Nomadic pastoralists live in the tents or other movable dwellings all year round (Fig. 1.5). Camps or semipermanent shelters are usually established in the same place along the yearly migration route.

There are about 30 million to 40 million pastoral nomads, the people who practice nomadic pastoralism. The Mongols in Mongolia, Russia, and China, the Tatars and Turkic people of eastern Europe and the Kazaks in Central Asia practiced nomadic pastoralism along Asian–European steppes in the past. Some of these populations

Fig. 1.3 (continued) (**c**) Pakistan by Abdul Wahid Jasra, 2010; (**d**) Egypt by Jean François Tourrand, 2010; (**f**) China by Wei Sha, 2012; (**f**) Kenya by Jianchu Xu, 2013; (**g**) China by Mingjiu Wang, 2010; **h** China by Shikui Dong, 2012; **i** China by Xukun Su, 2014; (**j**) Afghanistan by Shaoliang Yi, 2012; (**k**) India by Shikui Dong, 2011; (**l**) Bolivia by Jean François Tourrand, 2010; (**m**) Bolivia by Jean François Tourrand, 2010)

Fig. 1.4 Tibetan pastoralists transferring their flocks of sheep from one pasture to another along the highway at Daotanghe, Qinghai Province, China. (Photo by Shikui Dong, 2014)

still practice nomadic pastoralism. Presently, about 40% of the populations in Mongolia are nomadic pastoralists.

In the high HImalayas, where the average elevation is more than 4000 m above seal level, some people of Tibetan origin practice nomadic pastoralism as the dominant livelihood as sedentary agriculture is impossible because of low temperatures and limited irrigation. In arid and semiarid regions of Central Asia, pastoral people such as Kazaks, Kyrgyzs, and Tajiks migrate hundreds of miles in a year even in winter to herd their livestock (Fig. 1.6). In arctic regions, including northern Finland, Sweden, Norway, and the Kola Peninsula of Russia, the indigenous Sami practice nomadic pastoralism of reindeer raising in cold and harsh environments. In north Africa, the pastoralists include the Zaghawa, Kreda, and Mimi, and Bedouins also practice nomadic pastoralism in dry and infertile lands. Even in Europe, sheepherders such as the Mesta in Spain are keeping the traditional way of nomadic pastoralism in the grassland areas against small peasants through *cañadas* (Caballero et al. 2009).

1.2.2.2 Transhumant Pastoralism

Transhumant pastoralism has traditionally occurred throughout the pastoral world, particularly in the pastoral regions of Europe, Asia, Africa, and South America. It is a common practice characterized by the fact that the pastoralists move their

Fig. 1.5 Different types of camps used by nomadic pastoralists: (**a**) Mongolian herder's yurt in Inner Mongolia, China; (**b**) Massai herder's hut in Arusha, Tanzania; (**c**) Kuchi herder's tent in Pashtoon, Afghanistan; (**d**) Kyrgyz herder's yurt in Alichure, Tajikistan; (**e**) Tamang herder's camp in Rasuwa, Nepal; (**f**) Gaddi herder's camp in Himachal, India; (**g**) Boran herder's hut in Borana, Ethiopia; (**h**) Tibetan herder's tent in Qinghai, China. (Photos by (**a**) Mingjiu Wang, 2009; (**b**, **e**) Shikui Dong, 2007; (**c**, **d**) Shaoliang Yi, 2011; (**f**) Shikui Dong, 2011; (**g**) Chuan Liao, 2012; (**h**) Shikui Dong, 2010)

Fig. 1.6 Kazak pastoralists travel to find grazing pastures for their livestock in winter in Aertai, Xinjiang Autonomous Region, China. (Photo by Yining Lai, 2011)

livestock between fixed summer and winter pastures on a yearly basis. The pastoralists have a permanent home or dwelling typically in valleys or low-elevation areas. There are two types of transhumance: vertical and horizontal. Vertical transhumance occurs typically in mountain regions, where the movements shift between high-altitude pastures in summer and low-altitude pastures in winter (Fig. 1.7). Horizontal transhumance exists mostly in plain or plateau regions such as Mongolia, where the movement occurs between the summer pastures far from the home to the winter pastures close to the home (Fig. 1.8). In contrast to vertical transhumance, horizontal transhumance can be more easily disrupted by climate change or socioeconomic changes.

In Europe, vertical transhumant grazing between valley and high pastures is still practiced widely in Bavaria, Austria, and the Swiss Alps and other European highlands, although tourism and industry are presently playing a very important role in the local economy of these mountainous regions. In some of these mountainous areas, cattle are grazed by local farmers who still insist on the tradition of transhumant pastoralism, whereas in some other places in these mountainous areas, the cooperatives that own the pastures employ herdsmen to graze the livestock in the manner of seasonal migration. In some high valleys of the Pyrenees and the Cantabrian Mountains of the Iberian Peninsula, vertical transhumant pastoralism has been practiced as the sole support of the economy. In the Austrian highlands, unique social groups such as the *Pasiegos* in Cantabria, *Agotes* in Navarre, and Vaqueiros de alzada practice the lifestyle of transhumant pastoralism as a remnant of an older ethnic culture. In Scandinavia, transhumant pastoralism is still practiced as such: in summer, the herders move the livestock to a common mountain or forest

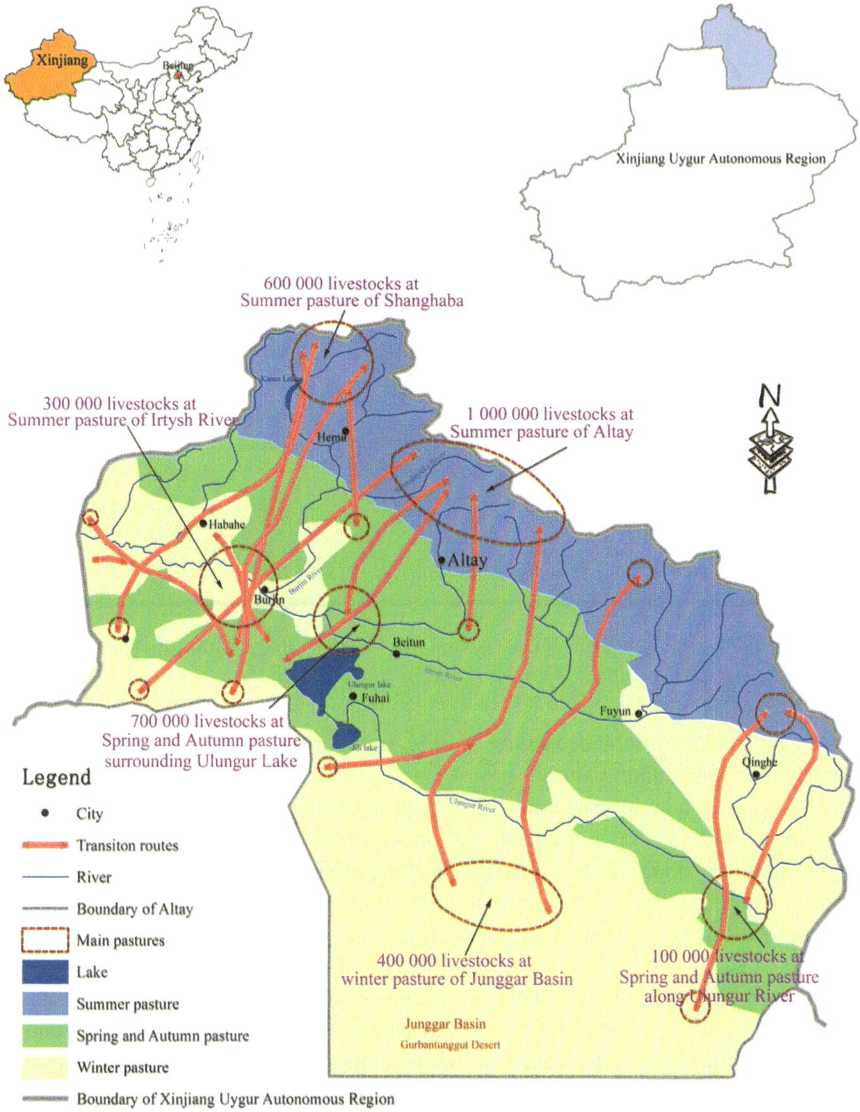

Fig. 1.7 Vertical transhumant grazing systems in the Altay region, Xinjiang Unger Autonomous Region

pasture, which is called *seter* (summer residence); in winter, they return the livestock to a home farm in valleys where the meadows are preserved for hay production. However, the arrival of motorized vehicles has been changing the character of traditional transhumance in this region. In the British highlands, livestock keepers used to spend summer on hillsides or in mountain areas and used to spend winter in valleys or low-lying meadows. Nowadays, most livestock keepers send their grazing

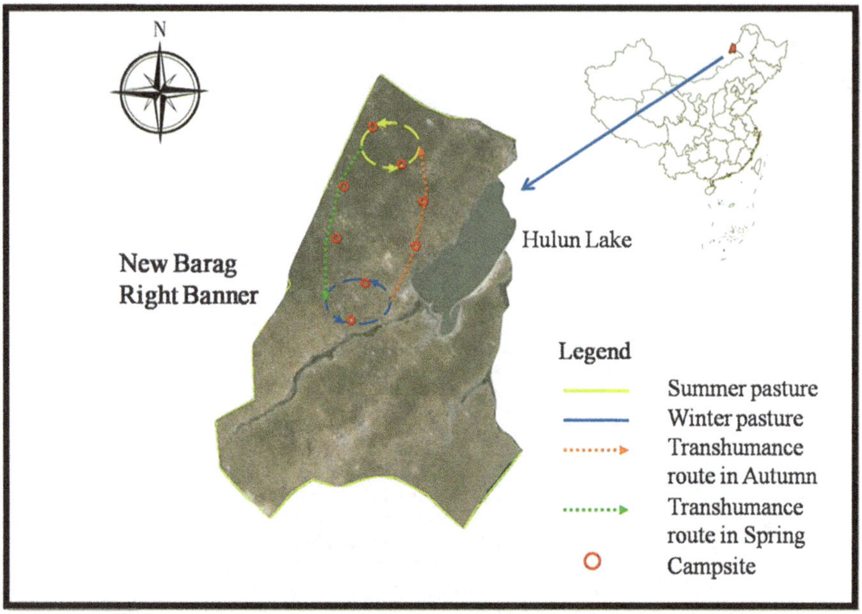

Fig. 1.8 Horizontal transhumant grazing systems for cattle and sheep on the Mongolian steppe in New Barag Right Banner, Inner Mongolia Autonomous Region, China

flocks by trucks to upland pastures during summer and lowland pastures during winter in a transhumant pattern.

In Asia, transhumant pastoralism has been maintained as the mainstay of subsistence economies in temperate and alpine zones more than over 2000 m above sea level on the southern slopes of the Himalayas and the alpine semiarid and arid zones over 3000 m above sea level on the northern slopes of the Himalayas, through the Qinghai–Tibetan Plateau in western China to the Eurasian steppe in Central Asia, including northern China and Mongolia. Along the Himalayan ranges, pastoral people of Tibetan origin such as the Zanskari in northwestern India, the Tamang in northern Nepal, the Brokpa in northern Bhutan, and Tibetans on the Qinghai–Tibetan Plateau of China still practice vertical transhumance, although in some cases nomadic pastoralism is also performed by these mountainous people. Across the Eurasian steppe in Central Asia, pastoral people such as Mongols in Mongolia and China, Kazaks in Kazakhstan and China, and Kyrgyzs in Kyrgyzstan predominantly practiced horizontal transhumant and nomadic pastoralism in some cases for centuries. In the mountain ranges of Central Asia and Southwest Asia, pastoralists move their herds seasonally back and forth between their homes in the valley and their temporary dwellings in the foothills every year. A typical example is the mobile life of Iran's Bakhtiari tribe, who practice vertical transhumant grazing from the Zagros mountain rangelands in Azerbaijan to the lowland pastures near the Arabian Sea (Rouhollah 1966).

In South America, transhumant pastoralism is practiced in the Andes of Argentina, Chile, Peru, and Bolivia (Andaluz-Westreicher et al. 2007), as well as in the Brazilian Pantanal (de Abreu et al. 2010). South America's transhumant pastoralism mainly

Fig. 1.9 Pastoral farming in Queensland in Australia: a grazier on a horse moving the herds of cattle (*left*); cattle grazing in different paddocks (*right*). (Photos by Shikui Dong, 2013)

involves movement of cattle in the Pantanal of Brazil and in parts of Argentina, whereas camelids are extensively raised on the altiplano. Goat raising is the major transhumant pastoralism in northern Neuquén and southern Mendoza, whereas sheep raising remains a major transhumant pastoralism on the Patagonian plains. In South America, the Criollos and other indigenous people are mostly pastoralists who are involved in transhumant grazing practices.

1.2.2.3 Pastoral Farming

Pastoral farming is a modern variation of nonnomadic pastoralism. It is often termed "farming/ranching aiming at producing livestock," rather than growing crops. Examples include dairy cattle farming, beef cattle farming, and wool sheep farming. Pastoral farming is practiced mostly in the ranches/pastures of developed countries such as Australia, Great Britain, Ireland, New Zealand, the USA, and Canada or developed regions of developing countries such as Argentina and Brazil. Pastoral farmers are also known as "graziers" and "ranchers" in most cases (Fig. 1.9). Some pastoral farmers grow crops purely as fodder for their livestock or purchase the fodder from crop farmers. This modern pastoralism is very different from the "traditional" pastoralism of the nomadic or transhumant system in terms of the level of investment in land and animals (Blench 2001).

1.3 Origin and History of Pastoralism

Pastoralism has been described as one of the great advances in human civilization, but the origin of pastoralism has been largely debated among scholars for centuries. According to Khazanov (1984), there are mainly three viewpoints. In the eighteenth and nineteenth centuries, a great number of scholars, including Montesquieu, Herder, Condorcet, Mortillet, Lubbock, Morgan, and Engels, were supporters of the "tripartite theory," which insists that pastoralism was derived from hunting and emerged earlier than agriculture. In their view, the hunters became nomads as they

Fig. 1.10 Stone sculpture of hunting in the Mongolian grasslands in prehistory. (Photo by Shikui Dong, 2010)

started to domesticate and herd some injured, weak, or baby animals on the grasslands. In contrast, Vico objected to this theory by saying that agriculture emerged earlier than pastoralism; however, his opinion did not receive recognition at that time. Until the second half of the twentieth century, some scholars (Bacon 1954; Lattimore 1967; Vainshtein 1980) stated that agriculturalists could have begun the breeding of animals and that hunters borrowed the domesticated animals from the neighboring agriculturalists. A possible explanation is that wandering hunters were unable to follow herds as they could not maintain the necessary speed of movement (Khazanov 1984). Moreover, the changeable composition of the herds made their domestication impossible (Khazanov 1984). Differently from these two viewpoints, some scholars believed that early humans had to diversify their livelihoods to cope with the pressures of climate change (Toynbee 1935; Zeuner 1956) and population growth (Lees and Bates 1974; Spooner 1975; Gilbert 1975); for example, some hunters became nomadic pastoralists to domesticate and herd animals. It would appear that the emergence of pastoral nomadism was complex and a multifaceted phenomenon that cannot possibly be explained by any one isolated factor (Khazanov 1984). However, it has been commonly accepted that pastoral nomadism mainly evolved from primitive hunting (Fig. 1.10).

The time at which pastoralism originated has not been fully agreed among scholars. Some scholars claim that the raising of domestic cattle in Northeast Africa occurred as early as 9000 B.P., although more solid dates are available for domestic llamas and alpacas in the Andes of South America from 6000 B.P. onward (MacDonald and MacDonald 2000). The earliest literature documented that the people who firstly appeared as pastoralists were the Amorites. In the first half of the

Fig. 1.11 One-center origination of pastoralism

second millennium B.C., those Amorites started to herd cattle, sheep, goats, and donkeys in the Near East (Cribb 1991). The pastoral culture that was recognizably described in sub-Saharan Africa can be dated back to Pliny, who firstly recorded blood and milk drinking in the Horn of Africa, whereas pastoralism in this region is believed to have originated much earlier than this record. Although the literature can give some clues about the first appearance of pastoralism, the exact time of occurrence of pastoralism can be concluded only from archaeology, particularly from careful osteometric work which can demonstrate the relationships between domesticated animals and their undomesticated wild relatives. According to the archaeological record, pastoral culture in both East Africa and West Africa appeared in 4500–4000 B.P. (Marshall 2000). However, the assumption for interpreting the osteometric evidence that the early herders were controlling breeding conflicts with the fact that the herders were involved in the management of wild animals at the earliest stages of pastoralism; for example, reindeer pastoralism is stil controlling beeding conflicts today through involvement in wild animal management.

The centers where pastoralism originated and the routes by which it spread have now been examined more specifically (Khazanov 1984), but there is still a lot of debate on these issues. Some scholars believe that nomadic pastoralism originated from one center, the mountainous Zagros region of Southwest Asia (northern Iraq and northwestern Iran), where the earliest herders domesticated goats and sheep about 9000 years ago (Miller 1998). Concomitant with cereal cultivation, which began somewhat earlier in the same region (Southwest Asia), animal husbandry quickly dispersed from this center of origin northward and eastward (Fig. 1.11).

Fig. 1.12 Multiple-center origination of pastoralism

Some scholars think that nomadic pastoralism originated from multiple centers in different parts of the world (Fig. 1.12). With the domestication of the horse about 6000 years ago, on the fertile steppes of southwestern Russia, nomadic pastoralism as a way of life really started to expand throughout Central Asia (including Mongolia). Some of these ancient nomads would undoubtedly also have penetrated into the western Himalayas, where alpine meadows would have provided good grazing for their livestock. Some of them would have reached the Tibetan grasslands from Central Asia to the west and north (Miller 1998). The Qiang, a nomadic tribe believed to be the ancestors of modern Tibetans, started herding animals on the rangelands of the Qinghai–Tibetan Plateau about 4000 years ago. These early nomads were known to the Chinese in the Hsia Dynasty (2205–1766 B.C.) as they sent rugs made from the "hair of animals" to "Hsia Emperor." The nomads originated from the Kurgan culture of southern Russia are believed to have expand into the Indian subcontinent about 3500 years ago, bringing with them not only the practice of nomadic pastoralism but also the Indo-European languages they spoke (Miller 1998). From the Nile Valley and North Africa, the pastoral culture spread to other places on the African continent, possibly through the agency of the ancestors of the present-day Berbers (Blench 2001). Although there is no common view on the exact routes by which and dates at which pastoralism reached southern Africa (Bousman 1998), it seems that pastoralists' herding of sheep firstly and cattle shortly thereafter occurred in pre-Iron Age transmission in nearly 2000 B.P. (Blench 2001). In the Andes of South America, the first domestication and herding of llamas and alpacas occurred in about 6000 B.P.

(MacDonald and MacDonald 2000). In terms of a great difference in sociocultural features (such as language, tradition, and herding practices) among different pastoralist groups across the world, we assume that a multiple-center origination is more probable for explaining the spread of pastoralism worldwide.

1.4 Distribution of Contemporary Pastoralism

According to the global pastroralism map developed by the World Initiative for Sustainable Pastoralism, pastoralism is presently predominant in sub-Saharan African, southern Africa, Central Asia (including the Himalayas), northern Europe (including the Russian Arctic), central South America, western North America, and Australia (Fig. 1.13).

In most of Europe, pastoralism usually occurs in high mountains, in arid zones, or on poor soil lands, where intensive cultivation is physically not possible (World Initiative for Sustainable Pastoralism 2007). These areas also very often have high biodiversity or very specialized plant and animal communities of high nature value. In the European Alps and highlands, pastoralism is one of the major components of the agricultural sector (Biber 2006). For example, Alpine pastures account for a quarter of farmland for 500,000 cattle raised by 70,000 farmers on 12,000 sites in Austria as transhumant pastoralism. Alpine pastures in Upper Bavaria host about half of 50,000 cattle from 1400 sites in Bavaria and alpine pastures in Allgäu host the other half under transhumant pastoralism. Alpine pastures on the Swiss highlands, amounting to 35 % of the nation's farmlands, are grazed in summer by about

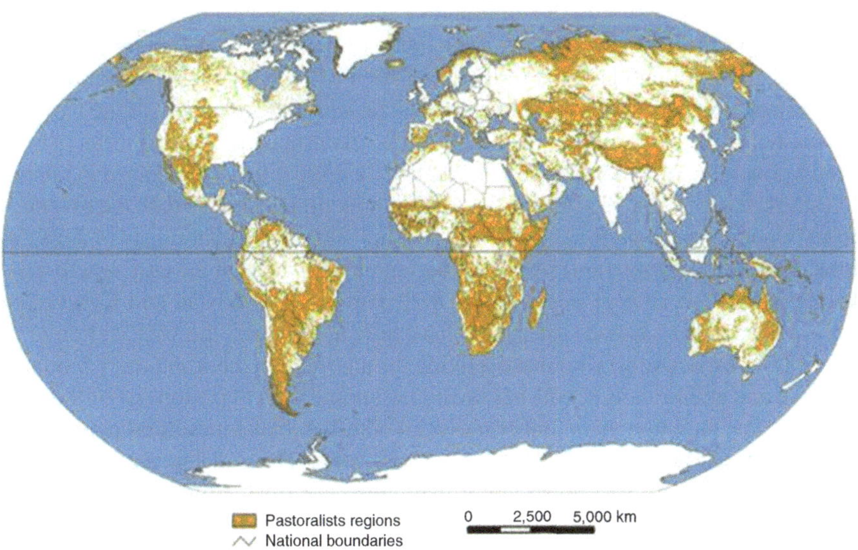

Fig. 1.13 Global distribution of pastoralism. (From http://www.iucn.org/wisp)

380,000 cattle (including 130,000 cows) and 200,000 sheep in transhumant pastoralism. The highlands of the Great British Isles (including Wales, Scotland, England, Ireland) maintain a great number of sheep and cattle by use of trucks as the transportation medium in upland–lowland migrating pastoralism. In Scandinavia, transhumant pastoralism is still largely practiced on sheep and cattle rearing in mountainous areas, such as Värmland, Dalarna, Härjedalen, Jämtland, Hälsingland, Medelpad, and Ångermanland in Sweden, in addition to reindeer rearing.

In Asia, grasslands/rangelands are composed of the largest contiguous landmass, stretching from the borders of eastern Europe to the Pacific Ocean. Asia's grasslands account for 25 % of the total grasslands/rangelands in the world (Kerven 2006). In Central Asia and the Far East, the rangelands/grasslands are climatically characterized by cold and snowy winters and warm summers. There, the low temperature (about −30 °C in Central Asia and inner Asia, and around −70 °C in Siberia) is the dominant environmental factor which limits the growth of grassland vegetation. In the grassland/rangeland areas of this region, pastoralism is practiced to raise common livestock such as sheep, goats, horses, and cattle, and some specific livestock, including yaks, Bactrian camels, and reindeer (Kerven 2006). The grazing pastures vary from permafrost tundra in the north to hot sandy deserts in the south, from the temperate tussock in low valleys of southern China to highland alpine meadows at altitudes of more than 4000 m on the Qinghai–Tibetan Plateau. The economic, social, and political position of pastoralism in this region varies considerably between the countries across the region. In China, more than 40 % of national landmass is covered by grasslands/rangelands, which occurs mostly in northern and western China. Inner Mongolia, Tibet, Xinjiang, Qinghai, Sichuan, Gansu, Yunnan, Liaoning, Jilin, and Heilongjiang are top ten provinces for pastoral production. The Chinese population involved in pastoralism is about, accounting for 15 % of the total national population of 1.3 billion. For pastoralism-specialized ethnic groups such as Tibetans, Mongols, and Kazaks, most of these populations are traditional pastoralists. In Mongolia, most of the population (85 % of the agricultural population) are pastoralists, and pastoral production provides the mainstay of the national economy, with a powerful political lobby. In Kazakhstan and Turkmenistan, a high proportion of the rural population (68 % and 43 % respectively) is engaged in raising livestock, but the relative value of livestock production is far less than the revenues from minerals in the national economy. In Tajikistan, most pastoralists (4 % of the total agricultural population) live in the high mountain areas in the east, where marketing is poor because of international trade barriers. Uzbekistan and Kyrgyzstan have considerable numbers of pastoralists (6 % and 7 % of the total agricultural population respectively) who make a living through selling meat and dairy products to urban and arable areas. China has the highest population of pastoralists (19.5 million) in Asia, although this number (only 2.4 % of the total agricultural population) is far lower than the population of Chinese cultivators. The grasslands/rangelands in Siberia in Russia are vast, even larger than those in China. However, these grasslands/rangelands are very in low productivity and support only a couple of million livestock and less than a million pastoralists (Kerven 2006).

In the Near East and South Asia, pastoralism is one of the major agricultural production systems in arid and semiarid areas (Gura 2006). In Iran and Jordan, up to 90 % of the country is dry land. The ratio of drylands to total land area is 45 % in Afghanistan, 60 % in Pakistan, 63 % in Iraq, and 55 % in Syria. Most of these drylands have, with exception of pastoralism, limited economic use (Gura 2006). In the Himalaya of South Asian countries such as Bhutan, Nepal, India, and Pakistan, pastoralism is still providing the mainstay for the regional near-subsistence economy, as agricultural production is impossible because of low temperatures at high elevation and steep terrains in the mountainous areas (Dong et al. 2009). For instance, regions such as Zanskar in northwestern India, Skardu in northern Pakistan, and Kham Magar in western Nepal still maintain nomadic pastoralism as the subsistence economies in the societies.

In Africa, pastoralism is distributed all over the continent. North Africa has vast areas of grasslands/rangelands, mainly steppe and arid Saharan land (Dutilly-Diane 2006). In terms of grassland/rangeland size, Morocco and Algeria are the top countries. They have about 20 million hectares of steppe, accounting for more than 40 % of the nation's territory. Tunisia ranks third in the size of grasslands/rangelands, which cover 25 % of the total territory. Two large desert counties, Egypt and Libya, the grasslands/rangelands covering about 1–2 % of the national territory (Dutilly-Diane 2006). The primary vocation on the grasslands/rangelands is livestock production. Therefore, the steppe in North Africa is called "the world of sheep." Pastoralism of small ruminant rearing is the traditional mode of valorization of the steppe. Although the reality may be overestimated to some extent, the document shows that 48 %, 62 %, and 75 % of the total small ruminant populations belong to pastoral production systems in Algeria, Tunisia (the data include animals in the center of the country as well) and Morocco respectively (Dutilly-Diane 2006).

In East Africa, pastoralists can be found in all countries, especially in the arid and semiarid dryland areas, where pastoralism is a major production system and livelihood strategy (Odhiambo 2006). In Kenya, there are about four millions pastoralists, accounting for more than 10 % of the nation's population. All of the arid and semiarid lands, constituting 80 % of the national landmass, are occupied by pastoral and agropastoral communities (Pastoralist Thematic Group 2001). In Uganda, there is a cattle corridor specialized for pastoral production. This corridor ranges from Mbarara in the southwest to Kaabong in the northeast of the country, covering 42 % of the country's landmass and 51 % of the national territory. Traditionally, most households (more than 60 %) along the cattle corridor are pastoralists (Odhiambo 2006). Outside the cattle corridor, such as in Kasese and Bundibugyo in the Western Rift Valley, there are also a large number of pastoralists. Totally, pastoralists constitute 22 % of the population of the whole nation (Odhiambo 2006). In Tanzania, the pastoral economy is the mainstay for supporting the livelihood of 10 % of the nation's population (Odhiambo 2006). The land use in most of the nation's arid and semiarid areas, such as Manyara, Arusha, Dodoma, Singida, Shinyanga, and Mwanza, is dominated by pastoralism. The pastoral groups own about 99 % of Tanzania's livestock population, and form the backbone of the livestock sector (Odhiambo 2006). In Sudan (including South Sudan), 20 % of the national population is involved in pastoralism, especially in arid and semiarid regions such as desert, and savanna, where pastoral production accounts for 80 % of the country's livestock wealth (Sin 1998) and contributes 25 % of foreign exchange earnings from livestock export (Odhiambo 2006).

In West Africa, pastoralism occurs mainly in Burkina Faso, Mali, Mauritania, Niger, Senegal. and Chad. located in vast Sahelian zone dominated by climatic hazards (Wane 2006). Although these countries are very similar in terms of pastoralism, mobile livestock production on the natural rangelands, there are some differences in the pastoral systems (i.e., nomadism, transhumance, or agropastoralism) which are determined by the mobility of the pastoralists' dwellings and the presence or absence of agricultural activities (Wane 2006). Within the vast Sahelian zone, pastoralism is principally the activity of the multivariety ethnic group of Fulani, and other ethnic groups such as the Touareg, Toubou, Wolof, and Serere have recently adopted pastoralism (Wane 2006). These pastoral people herd either monospecific or mixed groups of different livestock species, including bovines, ovines, goats, camelids, donkeys, and equines (Wane 2006). In the Sahelian zone, the livestock production characterized by pastoral mobility and full use of natural resources (the rangelands) contributes a great share to the national economy. According to the report prepared for World Initiative for Sustainable 1674 Pastoralism by (Wane 2006), the shares of agriculture in national GDP of Sahelian countries in 2003 were 31.0 % in Burkina Faso, 38.0 % in Mali, 20.0 % in Mauritania, 39.9 % in Niger, 17.6 % in Senegal, and 45.6 % in Chad, and the shares of livestock production in the national agricultural GDP were 24.7 % in Burkina Faso, 41.6 % in Mali, 70.0 % in Mauritania, 29.8 % in Niger, 37.3 % in Senegal, and 11.0 % in Chad (Wane 2006).

In southern Africa and the Horn of Africa, pastoralism is practiced widely from the Cape of Good Hope to Cairo. In this region, countries such as Somalia, Ethiopia, Botswana, Zimbabwe, South Africa, Namibia, Malawi, and Zambia have a tradition of pastoralism, although each of them is remarkably distinct in terms of livestock production and marketing systems because of variations in climate, natural resource endowment, colonial history, and current levels of national economic development (Behnke 2006). In Somalia, nearly 98 % of agricultural land is pastureland, most of which is rangeland, where three quarters of the total nation's population (9.7 million) make a living on livestock production. In Ethiopia, pastoral residents account for only 8 % of the total population, whereas the pastoral population occupies a large area (60 % of the country's land) and produces a higher share of national livestock outputs; that is, 73 % of the goats, 25 % of the sheep, 20 % of the cattle, and 100 % of the camels in the nation's livestock population (Aklilu 2002). In Botswana, rangelands account for 99 % of the nation's agricultural lands and support about 0.6 million people (47 % of the national population) who live on pastoral production. In Zimbabwe, pasturelands account for 84 % of agricultural lands, and 6 % of the nation's population lives on rangeland-based livestock production. Similarly, in South Africa, 84 % of agricultural lands are pasturelands ,and 16 % of the nation's population (more than 6.3 million) live in rangeland areas for livestock production. Despite South Africa being a relatively advanced industrial economy among all African countries, many pastoralists are unwilling to be involved in marketing production and insist on keeping the tradition of livestock production. Namibia ranks first among African counties in terms of ruminant meat production per capita. In this country, pasturelands account for 98 % of total agricultural lands, and 54 % of the total population lives on livestock production in semiarid rangeland areas. In contrast, Malawi ranks last among African countries in terms of meat and milk production. Although 42 % of Malawi's agricultural lands are pasturelands, only 2 % of the nation's population lives in rangeland areas for livestock production. In Zambia, pasturelands account for 85 % of the nation's agricultural lands, and 14 % of the national population lives on livestock production in semiarid rangeland areas (Behnke 2006).

In South America, indigenous pastoralism of herding camelids (llamas, alpacas, vicuña, and guanaco) has a long tradition as a form of livelihood and a production system in mountainous areas, particularly in the Andes (Westreicher et al. 2006). Nowadays, South American pastoralism exists mostly in the habitat known as "puna" or "altiplano" in the semiarid regions of the Andes between 3700 and 5000 m above sea level. Along the Andes, Argentina, Bolivia, Chile, and Peru are presently four major South American countries involved in pastoralist activities (Westreicher et al. 2007). In the central Andes, Bolivia and Peru are at the heart of South American pastoralism for historical, cultural, and geographical reasons, and pastoralist activities are importantly in their national economies (Westreicher et al. 2006). In Bolivia, rangeland-based alpaca production is the key pastoralism in the Andean highlands called "altiplano," particularly in the Cordillera Oriental, which is close to the Peruvian border. In Peru, the Andean highlands called "sierra" provide home for 41 % of the nation's population and serve as the production base for all of the nation's sheep, llamas, and alpacas and 70 % of the nation's cattle. In Argentina and Chile, pastoralists occupy marginal areas in the southern Andes and their economic relevance lies in their capacity to activate economic niches; that is, goat raising in northern Chile and southern Argentina, and camelid raising in northern Argentina.

In North America, pastoralism continues to be an important livestock production system practiced in mountainous areas and parts of the Great Plains (Huntsinger et al. 2010). Moreover, there is also reindeer pastoralism in arctic regions of North America. In the western USA, pastoralism relies in most cases on the use of public land resources, which are highland pastures under the jurisdiction of the US Forest Service and lowland steppes and deserts under the jurisdiction of the Bureau of Land Management (Sulak and Huntsinger 2007). The American ranchers traditionally move their herds up to highland pastures with the appearance of green grass in spring and summer, and graze their herds on the lowland steppe or desert in winter. In California and Texas, more ranches tend to have private land for pastoral production mainly because of the tradition of the Spanish land grant system. In these areas, livestock keepers, including ranch family members, hired shepherds, and hired cowboys, move to the mountains and stay in the line camps to herd their sheep and cattle on highland pastures during the summer, or visit the upland ranches regularly by using trailers to transport horses to the highland pastures (Huntsinger et al. 2010). In the southern Appalachians, livestock, especially sheep, are often grazed on grassy bald mountaintops where wild oats predominate. These balds might be the remnants of ancient bison grazing lands maintained by early Amerindians to some extent.

In Oceania, extensive pastoralism has been practiced on vast rangelands since European settlement (Earl and Jones 1996), allowing sheep and cattle to move as they choose the grazing lands over large areas. In Australia, this practice continues to survive in modern ranching systems throughout the country except in the central and coastal areas, where less pastureland can be found (Earl and Jones 1996). In New Zealand, this practice exists on one third of the country's land at high altitude (Lambert and Snow 2011).

1.5 Identity and Structure of Pastoralism

1.5.1 Characteristics of Pastoralism

Pastoralism generally has a mobile aspect, with herds being moved in search of fresh pasture and water (except for pastoral farming, in which pastoral farmers grow crops and improve pastures for their livestock). Pastoralism is often the optimal subsistence livelihood, which is generally independent of any particular local environment (O'Neil 2011). By nature, "pastoralists are flexible and opportunistic and can rapidly switch management systems as well as operating multiple systems in one overall productive enterprise" (Blench 2001). For example, pastoralists sell their herds or move them to new pastures when there is a drought. In contrast, agricultural cultivators rarely have these options. A pastoral subsistence pattern (especially nomadic pastoralism) is very often an adaptation to an irregular climate to reduce the risk in semiarid open country (O'Neil 2011). Mobile livestock grazing on the extensive grasslands/rangelands in arid and semiarid regions is a key feature of pastoralism (Blench 2001). The species of pastoral animals vary with different regions across the world, but they are all domesticated herbivores that are kept normally in herds and fed on rangeland forages or other abundant plant foods such as fodder trees (Blench 2001). The pastoral animals are herded as single species or mixed ones by pastoralists, depending on their traditional grazing practices. Sometimes, pastoralists keep nonpastoral species such as dogs and chickens, in addition to pastoral animals. Dogs are very important in protecting the livestock from wolves across a wide range of Eurasia from the Near East to Central Asia. In some parts of western Europe, pastoralists' still practice the tradition of using dogs to herd sheep (Fig. 1.14). This can help save a great amount of human labor inputs. In West Africa, Fulani nomads carry their chickens to feed on the rangeland worms when they graze their cattle and sheep, which adds significant value to livestock production for the pastoral households (Blench 2001).

Fig. 1.14 A dog herding the flocks of sheep on the Larzac Plateau, southern France. (Photo by Shikui Dong, 2011)

Different people who are involved in pastoralism across the world have the same identity of mobile livestock raisers, no matter in which region they live and to which ethnicity they belong. In addition to "pastoralists," these people are also named "nomads," "herders," "graziers" (mostly in Australia), and "ranchers" (USA, Canada, Brazil, Argentina, and South Africa) in the different literature. Historically, the culture of pastoralists has been interwoven with the culture of peripatetics, the other groups who move around the pastoralist communities to supply services to them. The most famous itinerants throughout much of Eurasia from Wales to India are the Gypsies, who are named "peripatetics" by Rao (1982, 1987). Although they do not herd any livestock on the rangelands/grasslands, peripatetics have played an important role in livestock trade in the long history of Eurasian pastoralism, particularly in Afghanistan and India, where both peripatetics and pastoralists are usually stereotyped as ethnically distinct in the same way and treated in the same category by national governments (Olesen 1994). Although both pastoral nomads and peripatetics adopt a similar mobile way of living, peripatetics are very different from pastoralists by identity. For example, in Afghanistan, pastoralists live in black goat-hair tents, but peripatetic live in white tents.

1.5.2 Social Structure of Pastoralism

The requirements of mobile grazing under extensive conditions have shaped the nature and structure of pastoral societies. Although there is an exception to any generalization, the basis of pastoral organization in every corner of the world is the clan, whose groupings can be very small and very shallow in time depth (Khazanov 1984). Control of the optimum territory by the clan is a function of the quality of pastures, the variability of the environment, and the livestock species herded. Livestock herds tend to be individually managed, whereas pastures are mostly collectively managed in the traditional pastoral areas of Africa, Asia, and South America. The livestock are normally herded by the individual households in the clan, but pastoralists can hire outsiders to graze their livestock if the herd sizes are too big to be managed by the household labor. For example, Fulɓe herders in modern Niger made slaves manage their great numbers of cattle herds. As such, "many pastoral societies in Africa and the Near East developed elaborate caste systems based on slaves and non-slaves in 19th century" as stated by Blench (2001). When the slaves received their freedom in the colonial era, they stayed with their original camps for some time, but the clans have been gradually broken up away to form independent households, particularly in the remote areas where traditionally authority cannot be brought to bear (Blench 2001).

Collective control of pasture resources is an ideal way of assuring pastoralists' mobility, as they must have access to a very large territory to reduce the risks of drought and inclement weather in arid regions. For Somali, as an example, to maintain mobility is so important that the territory is not strictly refined, and even the use of wells or pastures clearly defined to belong to an individual or other groups is possible if there is sufficient water or grass for all of the groups (Swift 1977). However, there are always some contradictions between the individual pastoralists who want to expand the family herds and the collective group that wants to equally share the pastures. The expansion of herds leads to unequal accumulation within pastoral communities, threatening group unity and pasture health. The individuals do need access to communal grazing lands and the aid of fellow pastoralists to help protect themselves from outsiders (Swift 1977).

These conflicts can be mitigated by a variety of institutions and beliefs. For example, the needs of the community are reinforced by a system of ideology, "livestock fetishism" (Bonte 1981), which reduces the inequity and promotes group unity. Moreover, these institutions and beliefs can sanctify the traditions of sacrifices, bridewealth giving, hospitability rules, and animal lending, which can not only reduce inequality of pastoral groups, but can also lessen the risk by permitting a wider dispersal of animals and by resolving labor bottlenecks. As an important aspect of pastoral life, hospitality plays a very important role in facilitating mobility and helping isolated herders obtain needed information. Segmentary lineages and similar forms of social organization are also well adapted to the needs of pastoral societies (Salzman 1978). Self-reorganized organizations (such as livestock associations), elected bodies (such as community committees), and norms, rules, and regulations derived from the traditions and practices have run the pastoral production systems in the pastoral realm worldwide for a long time.

1.5.3 Sex in Pastoral Society

The role of women in pastoral society has been widely debated, in part because pastoral societies are much more dominated by men than are most other subsistence systems (Blench 2001). There is an exception in the pastoral society of the Saharan Tuareg, which is mother (patrilineal) and women dominated. The use of labor within pastoral societies is very much sex specific. For pastoral production, women usually contribute more labor and play more important roles than men. However, pastoral women play less important roles than pastoral men or are totally ignored by the pastoral society in decision-making processes. In the male-dominated pastoral societies, elderly men usually make important decisions regarding herd mobility, turnover planning, conflict mitigation, and social relations among pastoral groups. In the family, men traditionally own the animals and control the money and the women have no rights to own animals and make budget plans.

In many pastoral societies, women primarily care for children and elderly people and perform domestic chores such as childcare, cooking, and weaving cloth (Fig. 1.15). In addition, women are customarily responsible for livestock rearing and herding (Fig. 1.15). Women are the key laborers in many pastoral societies for both processing and marketing milk and dairy products (Fig. 1.15), although women are not allowed to milk the animals or are allowed to milk only certain kinds of livestock in some societies (Blench 2001). When the animal herds are moved to new pastures, women may have to participate in dismantling and rebuilding their houses. In most cases, women herd the livestock and care for young and sick livestock kept near the homestead, whereas men herd the animals and sell meat animals in systems when a herd is split. Normally, the income from selling live animals and animal products goes to the men. Although women are responsible for most of the workload in pastoral societies in general, men usually acquire prestige and power in controlling the pastoral incomes and family consumption. Much evidence shows that empowering women remains a challenge in most of the pastoral regions across the world.

1.6 Importance (Values) of Pastoralism

Globally, the importance of pastoralism can be found in many dimensions, especially the socioeconomic, ecological, and cultural dimensions; for example, it supports huge human populations in rangeland areas, provides tremendous food and ecological services, makes significant contributions to the subsistence economy in some of the world's poorest regions, and maintains long-standing civilizations

Fig. 1.15 Female pastoralists' activities: (**a**) fuel (yak dung) collection in Qinghai, China; (**b**) livestock herding in Bolivia; (**c**) household chores such as water collection and carrying in Ethiopia; (**d**) milk processing in Tibet, China; (**e**) making of living materials (such as tents) in Tibet, China;

(Nori and Davies 2007). Davis and Hatfield (2007) stated that pastoralism can economically create existence values (intrinsic benefits for the global society), option values (retaining future opportunities), direct values (local benefits in social, economic, and environmental dimensions), and indirect values (associated with tourism and agriculture); see Fig. 1.16. In terms of economic values from the measured direct value point of view, Hatfield and Davies (2006) highlighted that pastoralism makes great contributions to agricultural and national GDP in countries such as Mongolia and Sudan, where pastoralism is as a predominant agricultural sector. According to official statistical data, pastoralism accounts for as much as 30 % of national GDP in Mongolia and 80 % of agricultural GDP in Sudan (Hatfield and Davies 2006).

Although few countries have official data on the contribution of pastoralist systems to national accounts, the available information indicates that the contributions of pastoralism to agricultural GDP are quite high in several African countries, such as Sudan, Senegal, Niger, and Kenya (Fig. 1.17). In East African counties, almost all Massai communities depend totally on pastoralism as the subsistence production system. It is believed that no system other than pastoralism can utilize the physical, climatic, and vegetative variations inherent in dry Africa as effectively, and the productivity of pastoral systems in Africa can be higher than that of other systems under the same conditions. African pastoralism has been shown to be between two and ten times more productive per hectare than ranching systems (Scoones 1995). These facts have changed the common belief that pastoralism adds little to national economic activities and is less productive than sedentary livestock raising to the new viewpoint that pastoralism is a viable economic system which can improve the livelihoods of millions of pastoralists and contribute to poverty reduction and environmental management in dry zones by promoting market access and enhancing mobility (Pastoralist Thematic Group 2001).

In addition to economic values, pastoralism has significant environmental value by providing all kinds of ecological services listed by Millennium Assessment 2003, including provisioning (such as food and fiber), supporting (such as soil formation and retention), regulation (such as climate regulation), and cultural (such as spiritual and religious) services. In terms of ecological services, a great amount of evidence shows that effective animal grazing can contribute to maintaining healthy rangeland vegetation, which generates rich biodiversity, promotes biomass production, captures carbon, reduces erosion, maintains soils, and facilitates water-holding capacity (Voisin 1959; Savory 1999; Frank et al. 1998). Large pastoral systems such as tropical savannas and temperate steppe represent a great (actual and potential) carbon sink, and pastoralism can effectively promote the potential of rangeland for capturing carbon. It was estimated that grasslands/rangelands store approximately 34 % of the global stock of CO_2, whereas only US$7 per hectare for the gas regula-

Fig. 1.15 (continued) (**f**) child caring in India; (**g**) milking animals in Afghanistan; (**h**) calf rearing in Qinghai, China; (**i**) handicraft making (such as embroidery) in Xinjiang, China. (Photos by (**a**) Xukun Su, 2014; (**b**) Tourrand, 2010; (**c**) Allan Degen, 2010; (**d**, **e**) Ruijun Long, 2008; (**f**) Shikui Dong, 2010; (**g**) Shaoliang Yi, 2010; (**h**) Xukun Su, 2014; (**i**) Xi Wang, 2011)

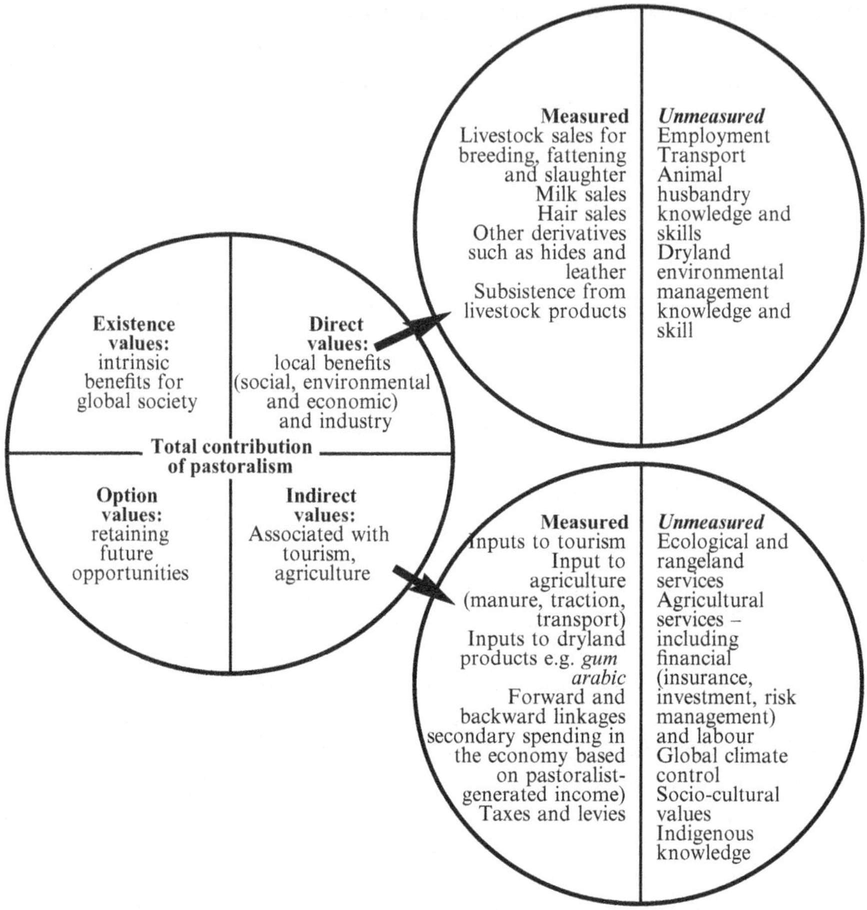

Fig. 1.16 Total economic values derived from pastoralism. (From Davis and Hatfield 2007)

tion function of this biome was given in a global valuation study (Costanza et al. 1997). Effective pastoral grazing management can be used as tool not only to improve grassland/rangeland biodiversity but also to prevent land degradation and desertification through maintaining rangeland ecosystem integrity (Niamir-Fuller 1999). The mean value of the maintenance of biodiversity in grasslands across different sites was estimated to be about US$7.5 per hectare per year (Yu et al. 2005), although the accurate estimation varies with many factors, such as the inclusion of all animal and plant species living in the grasslands/rangelands and the outsiders' willingness to pay for conserving grassland/rangeland biodiversity. Water-holding services are essential for pastoralism in the different grassland/rangeland areas, and effective pasture management, including grazing management, can improve

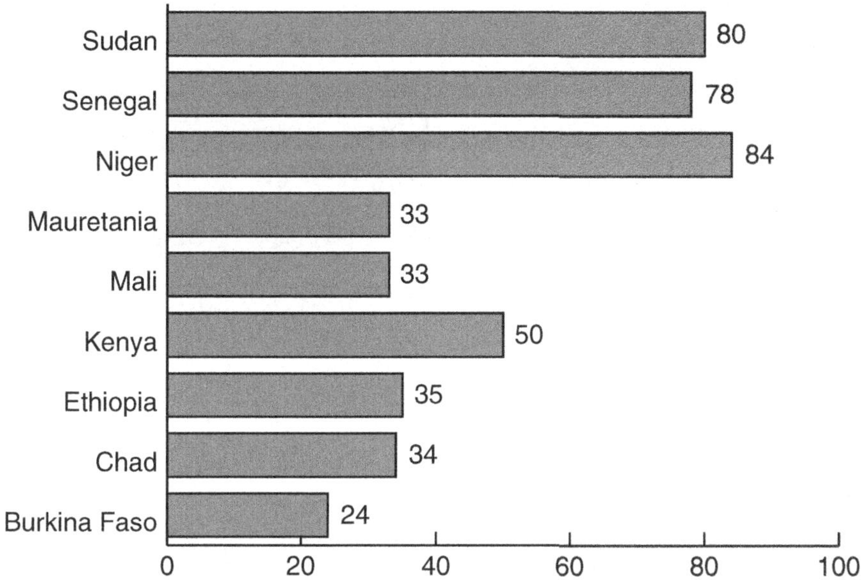

Fig. 1.17 Contribution of pastoralism to agricultural GDP in several African countries (Source: Davis and Hatfield 2007)

infiltration of water and reduce runoff, and thereby raise water tables. Although it is hard to quantify the water-holding services of pastoralism on a global scale because of data scarcity, the case study conducted by Yu et al. (2005) shows that water holding of the grasslands/rangelands (mostly grazing pastures) in the Qinghai–Tibetan Plateau of China was valued at US$1524 per hectare per year, which may provide an idea for gauging the value of the water-holding service provided by worldwide pastoralism. A great number of studies have shown that pastoralism plays important roles in maintaining ecosystem health and resilience and promoting water and mineral cycling in many grassland/rangeland ecosystems, but still no data are available for quantifying the value that pastoralism provides for maintaining water and mineral cycling. The case study conducted by Yu et al. (2005) in the grasslands/rangelands of China showed that the value of soil maintenance provided by pastoralism was US$3 per hectare per year.

Other key features of ecological services of pastoralism are sociocultural ones, traditional knowledge of the pastoralists (e.g., transhumant grazing by generations of traded knowledge on the carrying capacity of soils), pastoralist's coherent association with the landscape (usually mainly formed by pastoral activities), and the diversification of language and religions within pastoral societies in the world. A Mongolian saying states that "half of human history has been written in the grasslands." This does mean that pastoral civilization and agricultural civilization

Fig. 1.18 Religion associated with pastoralism: (**a**) Mongolian Obo, praying sites on grasslands; (**b**) Tibetan Buddhist temple, half-down sheep in the main gate; (**c**) Tibetan Maany stones, the praying rite on Tibetan rangelands; (**d**) holy mountain with scared yak status in Tibet. (Photos by (**a**) Wei Sha, 2012; (**b**) Shikui Dong, 2013; (**c, d**) Shikui Dong, 2012)

were equally important in human history. The diverse cultures created by different pastoral societies represent the rich resources of the arts (Fig. 1.18), sports (Fig. 1.19), religion, etc., across the different corners of developing world (Fig. 1.20). Although these values have often been underestimated or even overlooked by researchers and policymakers, there is great potential for adding extra values to pastoralism by converting these sociocultural resources into tourism and education.

Fig. 1.19 Sports associated with pastoralism: (**a**) horse racing among Tibetan pastoralists in Qinghai, China; (**b**) horse catching among Mongolian pastoralists in Inner Mongolia, China; (**c**) wrestling among Mongolian pastoralists in Inner Mongolia, China; (**d**) archery racing among Mongolian pastoralists in Inner Mongolia, China. (Photos by (**a**) Shikui Dong, 2012; (**b–d**) Wei Sha, 2013;)

1.7 Future of Pastoralism

In its long history, pastoralism has coevolved and coexisted with agriculture, and pastoralists have even successfully conquered agricultural societies. This is particularly true for the Mongol horse pastoralists, who conquered the agriculture-dominated

Fig. 1.20 Arts associated with pastoralism: (**a**) Kazak pastoralist's painting of donkey grazing in Xinjiang, China; (**b**) Kazak pastoralist's embroidering in Xinjiang, China; (**c**) Yugur pastoralist's dance in Gansu, China; (**d**) Kazak pastoralist's horse-feet-like violin in Xinjiang, China. (Photos by (**a**) Shikui Dong, 2013; (**b**, **d**) Xi Wang, 2013; (**c**) Chengzhang Zhao, 2014)

societies of China and Central Asia in the thirteenth century, and they also seized control of agriculture-dominated societies of Persia, Iraq, much of Russia, and the northern parts of South Asia in the fourteenth century (O'Neil 2011). Much evidence shows that the pastoralism as the mobile livestock production system has been adopted in recent centuries and will certainly remain in the future in the developing world (including developing regions in developed counties such as rangeland zones in Australia). Pastoralism will be kept as the backbone of the economy and as the mainstay of ecosystem protection in marginal and fragile areas, because it is generally regarded as an efficient, low-energy-requiring subsistence base for these areas.

However, the present shape evolved from very distinctive influences in the twentieth century (Table 1.2), and there will unlikely be a return to some prior imagined golden era. Similar situations can be found in the pastoral societies in various regions of the world; the drivers that have heavily changed African pastoralism are affecting pastoralism in Central Asia and other pastoral regions. Pastoralism has been declining because of agricultural expansion, industrial development, and sed-

Table 1.2 Key factors influencing pastoralism in the twentieth century (Blench 2001)

Factor	Impact
Modern veterinary medicine	Increases in productivity and greatly enlarged herds
Modern weapons	Major decline in predator threats, increasingly violent ethnic conflict, and high levels of insecurity
Enclaving	Collapse of traditional "safety nets" in terms of long-distance migration in periods of climatic extremes
International pressure for hygiene in slaughtering and dairying	Declining market for pastoralist products
Declining prestige of dairy products	Terms of trade running constantly against pastoral livelihoods
World market in livestock products	Governments import cheap meat, milk, etc., to satisfy urban demand at the expense of the pastoral sector
Ideological interference by the state	Inappropriate social and management strategies adopted and maintained by a combination of subsidized inputs and implied violence
Alternative calls on pastoral labor	Pressure for children to go to school and younger people to earn cash outside the pastoral economy
Modern transportation infrastructure	Replaces systems where transport is a major element of economic production (llamas, horses)
Introduction of high-input, high-output exotic breeds	Makes pastoralists dependent on effective infrastructure where input supply is irregular, creating periodic crises
Emergency relief, restocking, and rehabilitation programs	Keeps nonviable households in pastoral areas, thereby accelerating the cycle of deficits
Conservation lobby	Pressure to turn previously pastoral land over to reserved wildlife/biodiversity regions with corresponding hard currency income from tourism
Encroachment on rangeland	Rangeland is being eliminated through the use of politically attractive but often uneconomic irrigation systems

entary livestock farming in recent centuries. Most national governments in pastoral regions tried to settle pastoralists and reduce herd populations prevent overgrazing. Many pastoral communities are increasingly becoming sedentarized groups, often devoting themselves to small-scale cultivation, even though the quality and success rate of this type of cultivation are quite low. With the increasing sedentarization of pastoralists, the reduction in labor input in mobile livestock rearing may lead to a shift from multiple pastoralism toward solely pastoral farming or agropastoralism production, implying a terrible loss of diversity of pastoralism. As a result, the practices of pastoralism have been overwhelmed. If these situations continue, it is likely that pastoral societies across the world will have more unpleasant fates in the future.

Although pastoralism is changing to adapt to natural pressures and socioeconomic forces, most pastoral societies are marginalized by governmental policies and development strategies. In High Asia, the tragedy of responsibility associated with modernizing traditional pastoral practices and preserving modernist worldviews is currently challenging the sustainable development of pastoralism (Kreutzmann 2013). In Central Asian republics such as Kazakhstan, Turkmenistan, Uzbekistan, and Kyrgyzstan, decollectivization and the consequent loss of subsidies provided by the former Soviet regime may have the potential to bring about a return of more traditional systems of nomadic pastoralism. However, the collapsed public veterinary services, poor access to pastoral areas, and unstable market prices for livestock products reflect big problems for most places in the pastoral regions. In China, the increased pasture enclosure and the pastoralist settlement aiming at decreeing resource utilization strategies and implementing “modern” lifestyles through the interference of central authorities are nowadays altering the traditional land use practices of pastoralism in vast grassland areas. With the perception that modernity can be achieved only in urban settings, the central and regional governments in China are promoting the urbanization of township development in pastoral areas (Kreutzmann 2013). The associated consequences of overgrazing on some pastures are grassland degradation and desertification and those of undergrazing on some pastures are shrub encroachment and biodiversity loss in northern and western regions of the country. In Africa and the Near East, pastoralism is being much more marginalized by the gradual expansion of agricultural production with increasing rainfall. Increased grazing pressures on pastures are leading to the degradation of fragile grasslands/rangelands and may force some of pastoralists to quit their traditions of livelihood, mobile livestock rearing. In the Americas and the circum-Mediterranean region, infrastructure development and enclosed livestock production associated with regional development strategies are forcing out the remaining pastoralists. In some areas, the marginal lands that were previously used as pastures and homelands by pastoralists are increasingly being converted into reserves of biodiversity. These consequences have been accelerating impoverishment in many countries in the developing world, a situation intermittently remedied by mineral revenues but not through the development of pastoral systems.

Whatever the future of pastoralism is, the lessons learned in different pastoral areas across the world could be absorbed in the policy-making structures of sustainable pastoralism. Whether the importance of pastoralism is appreciated by global communities or not, the environmental services of pastoralism need to be widely recognized and the respective governments in the pastoral regions should act effectively to protect or restore such services. Irrespective of whether the emerging economies can benefit from the past mistakes made in the pastoralism sector, it is important to maintain the environmental benefits of pastoralism while it still exists. As stated by Davis and Hatfield (2007): "The key is to disseminate improved understanding of pastoral society as broadly as possible, making both policy and the effective management of pastoral systems as widespread as possible in the future." Moreover, McAllister et al. (2006) stressed that understanding past adaptation of pastoralism is important for planning and directing the future of pastoralism. Therefore, the lessons learned and experiences obtained in the past should be considered in the policy making for sustaining pastoralism in the future.

Acknowledgments The author acknowledges the funding organizations including Ministry of Science and Technology of the People's Republic of China, National Natural Science Foundation of China, and Asian Scholar Foundation, India China Institute of New School, Beijing Normal University for their financial supports. The author is grateful to numerous scholars who have provided a lot of reading materials, photos, paintings, graphs etc. to enrich the contents of this chapter. The author is most grateful to the authors whose articles, tables and figures are cited in this chapter.

References

Aklilu Y (2002) An audit of the livestock marketing status in Kenya, Ethiopia and Sudan, vol I and II. OAU/Interafrican Bureau for Animal Resources

Andaluz-Westreicher C, Mérega JL, y-Palmili G (2007) The economics of pastoralism: study on current practices in South America. Nomadic Peoples 11(2):87–105

Bacon E (1954) Types of pastoral nomadism in Central and Southwest Asia. SJA 10(1):44–68

Behnke R (2006) Review of the literature on Pastoral Economics and Marketing: the horn of Africa and southern Africa. Report prepared for the World Initiative for Sustainable Pastoralism, IUCN EARO

Biber JP (2006) Review of the literature on pastoral economics and marketing: Europe. Report prepared for the World Initiative for Sustainable Pastoralism, IUCN EARO

Blench R (2001) 'You can't go home again': pastoralism in the new millennium. FAO report

Bonte P (1981) Ecological and economic factors in the determination of pastoral specialization. J Asian Afr Stud 16(1–2):33–49

Bousman CB (1998) The chronological evidence for the introduction of domestic stock into southern Africa. Afr Archaeol Rev 15:133–150

Caballero R, Fernández-González F, Badia RP, Molle G, Pier Paolo Roggero PP, Bagella S, D'Ottavio P, Papanastasis VP, Fotiadis G, Sidiropoulou A, Ispikoudis I (2009) Grazing systems and biodiversity in Mediterranean areas: Spain, Italy and Greece. PASTOS 39(1):9–152

Costanza R, d'Arge R, de Groot R, Farber S, Grasso M, Hannon B, Naeem S, Limburg K, Paruelo J, O'Neill RV, Raskin R, Sutton P, van den Belt M (1997) The value of the world's ecosystem services and natural capital. Nature 387:253–260

Cribb R (1991) Nomads in archaeology. Cambridge University Press, Cambridge

Davis J, Hatfield J (2007) The economics of mobile pastoralism: a global summary. Nomadic Peoples 11(1):91–116

de Abreu UGP, McManus C, Santos SA (2010) Cattle ranching, conservation and transhumance in the Brazilian Pantanal. Pastoralism 1(1):99–114

Dong SK, Wen L, Zhu L, Lassoie JP, Yan ZL, Shrestha KK, Pariya D, Sharma E (2009) Indigenous yak and yak-cattle crossbreed management in high altitude areas of northern Nepal: a case study from Rasuwa district. Afr J Agric Res 4(10):957–967

Dutilly-Diane C (2006) Review of the literature on pastoral economics and marketing: North Africa. Report prepared for the World Initiative for Sustainable Pastoralism, IUCN EARO

Earl J, Jones C (1996) The need for a new approach to grazing management—is cell grazing the answer? Rangelands J 18:327–350

FAO (2002) Cattle and small ruminant systems in sub-Saharan Africa: a systematic review. Available at: ftp://ftp.fao.org/docrep/fao/005/y4176E/y4176E00.pdf

FAO (2001) Pastoralism in the new millennium. Animal production and health paper no 150. UN Food and Agriculture Organization, Rome

Frank DA, McNaughton SJ, Tracy BF (1998) The ecology of earth's grazing ecosystems. Bioscience 48(7):629–34

Gilbert AS (1975) Modern nomads and prehistoric pastoralists: the limits of analogy. JANES 7:54–71

Gura S (2006) Review of the literature on pastoral economics and marketing: Afghanistan, India, Iran, Iraq, Israel, Jordan, Pakistan, Palestine, Syria, and Turkey. Report prepared for the World Initiative for Sustainable Pastoralism, IUCN EARO

Hatfield R, Davis J (2006) Global Review of the economics of pastoralism. Report prepared for the World Initiative for Sustainable Pastoralism, IUCN, Nairobi

Huntsinger L, Forero L, Sulak A (2010) Transhumance and pastoralist resilience in the western United States. Pastoralism 1:10–37

IFAD (2008) Livestock and pastoralists. Available at: http://www.ifad.org/lrkm/factsheet/pastoralists.pdf

Kerven C (2006) Review of the literature on pastoral economics and marketing: Central Asian, China, Mongolia and Siberia. Report prepared for the World Initiative for Sustainable Pastoralism, IUCN EARO

Khazanov A (1984) Nomads and the outside world. Cambridge University Press, New York

Kreutzmann H (2013) The tragedy of responsibility in high Asia: modernizing traditional pastoral practices and preserving modernist worldviews. Pastoralism 3:7, http://www.pastoralismjournal.com/content/3/1/7

Lambert MG, Snow VO (2011) Management of grazing system in New Zealand and environmental impacts. In: Lemaire G, Hodgson J, Chabbi A (eds) Grassland productivity and ecosystem services. CAB International, London, pp 188–197

Lattimore O (1967) Inner Asian frontiers of China. Beacon, Boston (first published in 1940)

Lees SH, Bates DG (1974) The origins of specialized pastoralism: a systemic model. Am Antiq 39(2):187–193

MacDonald KC, MacDonald RH (2000) The origins and development of domesticated animals in arid West Africa. In: Blench RM, MacDonald KC (eds) The origin and development of African livestock. University College Press, London, pp 127–162

Marshall F (2000) The origins and spread of domestic animals in East Africa. In: Blench RM, MacDonald KC (eds) The origin and development of African livestock. University College Press, London, pp 191–221

McAllister RRJ, Abel N, Stokes CJ, Gordon IJ (2006) Australian pastoralists in time and space: the evolution of a complex adaptive system. Ecol Soc 11(2):41, http://www.ecologyandsociety.org/vol11/iss2/art41/

Miller DJ (1998) Fields of grass: portraits of the pastoral landscape and nomads of the Tibetan Plateau and Himalayas. International Centre for Integrated Mountain Development, Kathmandu

Millennium Ecosystem Assessment (2003) Ecosystems and Human Wellbeing:A Framework for Assessment [M]. Washington DC, Island Press

Niamir-Fuller M (1999) International aid for rangeland development: trends and challenges. In: Eldridge D, Freudenberger D (eds) Proceedings of the 6th international rangeland congress, Townsville, Australia, 17–23 July 1999, pp 147–152

Nori M, Davies J (2007) Change of wind or wind of change? Climate change, adaptation and pastoralism. The World Initiative for Sustainable Pastoralism, International Union for Conservation

of Nature, Nairobi, Kenya. http://cmsdata.iucn.org/downloads/c__documents_and_settings_hps_local_settings_application_data_mozilla_firefox_profile.pdf
Odhiambo M (2006) Review of the literature on Pastoral Economics and Marketing: Kenya, Tanzania, Uganda and the Sudan. Report prepared for the World Initiative for Sustainable Pastoralism, IUCN EARO
Olesen A (1994) Afghan craftsmen. Thames & Hudson, London
O'Neil D (2011) Pastoralism. http://anthro.palomar.edu/subsistence/sub_3.htm
Pastoralist Thematic Group (2001) Pastoralist poverty reduction strategy. Unpublished report, March 2001
Pratt DJ, Le Gall F, De Haan C (1997) Investing in pastoralism: sustainable natural resource use in arid Africa and the Middle East. World Bank Technical Paper 365, World Bank, Washington, DC, USA.
Rao A (1982) Non-food-producing nomads and the problem of their classification. East Anthropol 35(2):115–134
Rao A (1987) The other nomads. Bohlau, Cologne
Rouhollah R (1966) The northern tier: Afghanistan, Iran and Turkey. Van Nostrand , Princeton, p 85
Salzman PC (1978) Does complementary opposition exist? Am Anthropol 80:53–70
Savory A (1999) Holistic management: A new framework for decision making. Island, Washington
Scoones I (1995) 'New directions in pastoral development in Africa', in living with uncertainty. IT Publications, London
Sin MA (1998) 'SUDAN' in lane. In: Charles R (ed) Custodians of the commons: pastoral land tenure in East and West Africa. IIED, London, Chapter 6
Spooner B (1975) Nomadism in Baluchistan, in: Leshnik dans L. S. & Sontheimer G. D. (eds.), Pastoralists and Nomads in South Asia, Weisbaden, pp. 171–182
Sulak A, Huntsinger L (2007) Public lands grazing in California: untapped conservation potential for private lands? Rangelands 23(3):9–13
Swift JJ (1977) Pastoral development in Somalia: herding cooperatives as a strategy against desertification and famine. In: Glantz MH (ed) Desertification: environmental degradation in and around arid lands. Westview, Boulder, pp 275–305
Toynbee A (1935) A study of history, vol 1. Oxford University Press, London
USAID (2011) Comprehensive Africa Agriculture Development Programme (CAADP) climate-smart agriculture workshop: pastoral and agro-pastoral systems factsheet. Available at: http://www.caadp.net/pdf/1a.%20Pastoral%20and%20Agropastoral%20systems%20facthseet.pdf
Vainshtein S (1980) Nomads of South Siberia: the pastoral economies of Tuva. Cambridge University Press, Cambridge
Voisin A (1959) Grassland productivity. Island, Washington
Wane A (2006) Review of the literature on pastoral economics and marketing: West Africa. Report prepared for the World Initiative for Sustainable Pastoralism, IUCN EARO
Westreicher CA, Mérega JL, Palmili G (2006) Review of the literature on pastoral economics and marketing: South America. Report prepared for the World Initiative for Sustainable Pastoralism, IUCN EARO
World Initiative for Sustainable Pastoralism (2007) Pastoralists' species and ecosystems knowledge as the basis for land management. WISP Policy Brief 5:1–4
Yu G, Lu CX, Xie GD, Luo ZJ, Yang L (2005) Grassland ecosystem services and their economic evaluation in Qinghai-Tibetan plateau based on RS and GIS. IEEE Int 4:2961–2964
Zeuner FE (1956) The domestication of animals. Scientia vol xci. N-DXXV, series VI:23–28

Chapter 2
Vulnerability and Resilience of Human-Natural Systems of Pastoralism Worldwide

Shikui Dong, Shiliang Liu, and Lu Wen

Abstract This chapter describes the definitions of resilience, vulnerability, and human-natural systems, presents general views on vulnerability/resilience of pastoralism, provides the framework for assessing vulnerability/resilience of pastoralism, and identifies vulnerability/resilience of human-natural systems of pastoralism worldwide. Resilience is defined as the capacity of a system, community, or organization to withstand loss or damage and to recover from the impact of an emergency or disaster. Vulnerability is defined as the sensitivity of people, places, ecosystems, and species to contingencies and stress, and their capability to cope with them. A human-natural systems is defined as the integrated system in which people interact with natural components. Human–environment systems, social–ecological systems, ecological–economic systems, and population–environment systems are different forms of human-natural systems. The resilience/vulnerability of the human-natural systems concerns the resilience/vulnerability of interdependent systems of people and nature. The human-natral systems of pastoralism worldwide are reducing their resilience and enhancing their vulnerability to natural stress and human-induced shocks. An agroecosystem–livelihood–institution three-dimensional "vulnerability/resilience" framework and a pressure–state–response model can be used to examine the vulnerability/resilience of pastoralism worldwide. Ten case studies from seven major pastoral regions across six continents show that the vulnerability of pastoralism is very different across the world. Climate change and climate variability have driven fragile pastoral agroecosystems into more vulnerable conditions in the Great Plains of North America. Socioeconomic drivers such as land tenure change, agriculture policy reform, and human and livestock population growth have disrupted the pastoral institutions at local and national levels into marginalized ones in Central Asia, the South American Andes, the European Alps and highlands, Queensland in Australia, and the Arctic. Combined natural and human factors have driven pastoral agroecosystems and institutions into more vulnerable situations in the African Sahel and the Asian highlands. Social–ecological learning, technical and management

S. Dong (✉) • S. Liu
School of Environment, Beijing Normal University, Beijing 100875, China
e-mail: dongshikui@sina.com

L. Wen
Inner Mongolia University, Hohhot 010021, China

S. Dong et al. (eds.), *Building Resilience of Human-Natural Systems of Pastoralism in the Developing World: Interdisciplinary Perspectives*,
DOI 10.1007/978-3-319-30732-9_2

innovations, social–ecological system renewal, and reorganization of institutions are pathways to mitigate the negative causes and effects of the pastoralism's vulnerability.

2.1 Introduction

Sustainability of pastoralism in the developing world is presently confronted with a number of threats and pressures. Worldwide, population growth, economic development and urbanization, land use changes, government policy changes, and climate change are putting great pressures on pastoralism, the invaluable social, cultural, economic, and ecological asset in grassland/rangeland areas (Nori and Davies 2007). Key ecosystem services such as biodiversity conservation and food production provided by rangeland ecosystems through pastoralism may be vulnerable to these changes in the developing world (Schröter et al. 2005; Abildtrup et al. 2006). It is vital to identify the drivers and causes of pastoralism degradation and to develop a framework for assessing the vulnerability of pastoralism worldwide. Here, we highlight the driving forces of pastoralism transformation and evaluate the vulnerability/resilience of pastoralism across different continents by using sound assessment frameworks.

2.2 Defining Resilience and Vulnerability

Resilience and vulnerability are paired terms whose definitions vary greatly in different fields. Resilience originally appeared as a concept in the science of ecology to define the capacity of ecosystems with alternative attractors to remain in the original state subject to disturbances (Hollings 1973). In this concept, it is implied that a disturbance can drive the system over a stability domain of the original state, which does not mean returning qualitatively to the original state of the system. This concept has been repeatedly used in the field of ecology as the term of ecological resilience by many scholars (Hollings 1996; Gunderson 2000; Folke 2006; Scheffer 2009). Nowadays, the concept of resilience has been broadly used in other fields. In the field of engineering, the term "resilience" has been technically used as engineering resilience in a narrow concept of the return rate to equilibrium after disturbance (Holling 1996). In the field of psychology, resilience is referred to as the capability to protect individuals from developing serious problems as a result of exposure to stress or adversity, which are known as risk factors (Luthar 2006). In the field of social science, resilience is understood as the ability of human groups or communities to cope with external stresses and disturbances associated with social, political, and environmental changes (Adger 2000). In general, it is accepted that resilience is broadly the capacity of a system, community, or organization to withstand loss or damage and to recover from the impact of an emergency or disaster. Similarly, vulnerability can be defined differently in various fields. As a synthesized term,

Fig. 2.1 The relationship between resilience and vulnerability, linked but not opposite. (From Maguire and Cartwright 2008)

"vulnerability" was defined by Millennium Ecosystem Assessment (2005) as "the sensitivity of people, places, ecosystems and species to contingencies and stress, and their capability to cope with them".

It is generally accepted that these two terms are linked but not opposite. Resilience is a response to vulnerability (Kassam 2010). The higher the resilience, the less likely there will be damage and the faster and more effective recovery may be; the higher the vulnerability, the more exposure there may be to loss and damage (Fig. 2.1). As resilience and vulnerability can be defined in many ways as many complex systems have multiple attractors, the understanding of these two terms is often complicated by a lot of issues: what factors contribute to vulnerability and resilience, what levels exhibit vulnerability and resilience, what are the dynamics of vulnerability and resilience over time, and what are the changes of vulnerability and resilience from location to location. In recent perspectives, as summarized by Maguire and Cartwright (2008), there are mainly three views on resilience: resilience as stability (buffer capacity), resilience as recovery (bouncing back), and resilience as transformation (creativity). Similarly, there are three views on vulnerability: vulnerability to a hazard, vulnerability as a "state," and vulnerability as components of a community.

2.2.1 Views on Resilience

From the various studies across a range of disciplines, the views on resilience can be summarized into three major perspectives (Adger 2000; Folke 2006; Maguire and Hagan 2007). However, there is a common aspect in all perspectives that resilience is the ability to withstand and respond positively to stress or change (Maguire and Cartwright 2008).

2.2.1.1 Resilience as Stability

The stability view of resilience was firstly developed from ecological studies, in which resilience is defined as the ability of a community or ecosystem to return to a predisturbed state. In this view, resilience is measured as the amount of disturbance that a community or ecosystem can tolerate before it shifts into another state, which is often termed a "threshold." Beyond the threshold, a community or ecosystem is unlikely to return to its functional state (Folke 2006). A resilient community or ecosystem has a high threshold, implying it can absorb considerable stress before it reaches its threshold.

2.2.1.2 Resilience as Recovery

The recovery view defines resilience as the ability of a community to return to its original state from a change driven by the stressor. In this view, resilience is gauged as the time taken for a community to recover from a change or stressor (Maguire and Hagan 2007; Pimm 1984); that is, the shorter the time needed to return to the original state, the more resilient the community is; the longer the time needed to return to the original state, the less resilient the community is.

2.2.1.3 Resilience as Transformation

The transformation view considers resilience as the adaptive capacity of a community in response to a change, meaning that it shifts to a new stable state instead of returning to an original state in coping with disturbance. In this view, it is believed that a resilient community may creatively respond to a change by transforming fundamentally the basis of the community. For example, a grazing-based pastoral community may develop economic activities other than pastoral production (e.g., tourism) or explore innovative grazing practices to mitigate rangeland degradation. This view is particularly important to help understand how a resilient community can respond positively to change, as it acknowledges that the members of the community themselves can shape the "trajectory of change" and deal with the impacts caused by the change (Herreria et al. 2006).

2.2.2 Views on Vulnerability

Compared with resilience, vulnerability is usually more difficult to define (Schoon 2005). From various studies across a wide range of disciplines, recent views on vulnerability can be summarized into three perspectives.

2.2.2.1 Vulnerability to a Hazard

In the field of natural hazard studies, vulnerability is broadly defined as the frequency, magnitude, timing, and intensity of the hazard that a community faces (Fenton et al. 2007). From this perspective, a community's vulnerability is derived from the physical aspect of the stress itself, an outcome of a hazardous event. In this view, the definition of vulnerability overlooks the characteristics of the community, which shape the community's responses to a hazard or other shocks.

2.2.2.2 Vulnerability as a "State"

The view of vulnerability as a "state" is generally applied to assess whether a community is inherently vulnerable or not. When vulnerability is viewed as a "state", a community (or subsections of a community) is thought to be intrinsically vulnerable and less able to cope with stresses, shocks, and change (Brooks 2003). This view focuses mostly on the characteristics of the community that lead to the vulnerability; for example, poverty, inequality, low housing quality, and poor access to services. However, it acknowledges almost nothing about the importance of the resources and capacities of a community that help the community cope with stresses, changes, and shocks (Brooks 2003).

2.2.2.3 Vulnerability as Component of a Community

This view considers vulnerability as one component of a community. It incorporates the idea that vulnerabilities, resources, and adaptive capacities of a community are dynamic and multifaceted. In this view, resilience and vulnerability are not opposite aspects of a community, and they may exist in a community at the same time (Fenton et al. 2007). For example, a pastoral community in arid or semiarid areas might be vulnerable to forage shortage in a dry year. However, this community may have adaptive capacity to overcome forage shortage by buying foodstuffs from outside or selling more livestock.

2.3 Understanding Vulnerability and Resilience of Human-Natural Systems

2.3.1 Definition and Types of Human-Natural Systems

In 2007, the US National Science Foundation firstly established the Dynamics of Coupled Natural and Human Systems Program to recognize the need for enhancing public understanding of complex systems. This program was established to promote and financially support "quantitative, interdisciplinary analyses of relevant human and natural system processes and complex interactions among human and natural systems at diverse scales". As a result, there are major changes in progress concerning how the US scientific community develops approaches to address interdisciplinary and applied environmental problems. Since then, an increasing number of interdisciplinary programs have been integrating ecological and social sciences to study and better understand the dynamics of human-natural systems. Many scholars have stated that it is not effective to study human and natural systems separately when addressing social–ecological and human–environment interactions over the long term (Gunderson and Holling 2002; Redman 1999; Walker et al. 2004; Walker and Salt 2006). Nowadays, human-natural systems research is becoming an exciting and integrative field of cross-disciplinary scientific inquiry, with research projects covering a variety of coupled systems in locations spanning the globe. These projects studying human-natural systems are characterized as follows: they address complex interactions and feedback between human and natural systems; they are interdisciplinary, engaging biological, physical, and social scientists around common questions; they integrate various tools and techniques from the biological, physical, and social sciences; they are context specific and illustrate long-term temporal dynamics (Liu et al. 2007).

Human-natural systems, or coupled human and natural systems (CHANS) as defined by Liu et al. (2007), are integrated systems in which people interact with natural components (Fig. 2.2). Human-natural systems exist across multiple spatial,

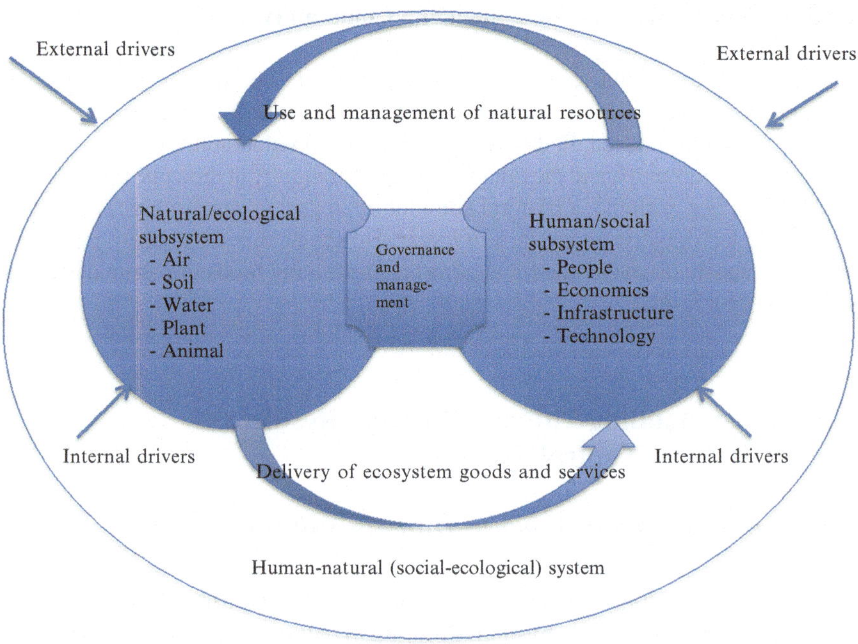

Fig. 2.2 Framework of a human–natural (social–ecological) system

temporal, and organizational scales, which may be hierarchically linked. They also exhibit nonlinear dynamics with thresholds, reciprocal feedback loops, time lags, resilience, heterogeneity, and surprises (Liu et al. 2007). There are different forms of human-natural systems, such as coupled human–environment systems, social–ecological systems, ecological–economic systems, and population–environment systems. Among them, the social–ecological system is the mostly documented one in the literature worldwide. The social–ecological system was defined by Anderies et al. (2004) as a coherent system in which an ecological system is intricately linked with and affected by one or more social systems, and the subsets of the ecological system and the social system contain units that interact interdependently and may be impacted by external and internal drivers (Fig. 2.2.). Up to now, the concept of a social–ecological system has been used by many scholars to emphasize the integrated concept of humans in nature and to stress that the delineation between social systems and ecological systems is artificial and arbitrary.

The social–ecological system, the most popular form of human-natural systems, is also defined as a set of critical resources whose flow (materials and energy) and use are regulated by a combination of ecological and social systems (Redman et al. 2004). In the social–ecological system, there are four components: the resource, the resource users, the public infrastructure providers, and the public infrastructures. These are intricately linked with the internal flows (including material and energy flow, and social and physical capital) and impacted by the external drivers of biophysical disruptions such as climate change and socioeconomic shocks such as economic depressions or inflation (Anderies et al. 2004). Pastoralism is a good example of the rangeland resources use regulated by the social–ecological system (Table 2.1). In the social–ecological system of pastoralism, the resource is rangeland that is used

Table 2.1 Entities involved in social–ecological systems of pastoralism

Entities	Examples	Potential problems
Resources	Rangeland	Complexity/uncertainty
Resource users	Herders using resources for grazing livestock Farmers using resources for growing crops	Overgrazing Overcropping
Public infrastructure providers	Executive and council of a local users' association Government bureau	Internal conflict or indecision about which policies to adopt Information loss
Public infrastructure	Engineering work	Wear out over time
Institutional rules	Government policies, customary laws, traditional norms	Memory loss over time, deliberate cheating
External environment	Weather, economy, political system	Sudden or slow changes that are not noticed

Modified from Anderies et al. (2004)

by various resource users such as herders and farmers. The public infrastructure providers could be a local users' association or a government bureau. Public infrastructure is the engineering work, which combines physical and social capital, two forms of human-made capital (Costanza et al. 2001). There are potential problems in each component derived either from the internal factors or from the external drivers (Table 2.1).

2.3.2 *Resilience of Human-Natural Systems*

A Human-natural systems is characterized by dynamic interactions between humans and nature, so the resilience of human-natural systems concerns the resilience of interdependent systems of people and nature. As the mostly documented form of human-natural systems, the social–ecological system has been widely studied in the dimension of resilience. The great acceleration of human activities on Earth is now making social–ecological resilience a global issue (Steffen et al. 2007). It is difficult to continuously separate ecological resilience from social resilience and it is irrational to try to explain them independently (Folke et al. 2010). There is a clear link between social resilience and ecological resilience in the realm of human-natural systems such as pastoralism, in which the social groups or communities depend greatly on ecological and environmental resources for their livelihoods (Fig. 2.3). The linkage of social resilience and ecological resilience may occur in the way of synergistic and coevolutionary relationships (Norgaard 1994; Adger 2000).

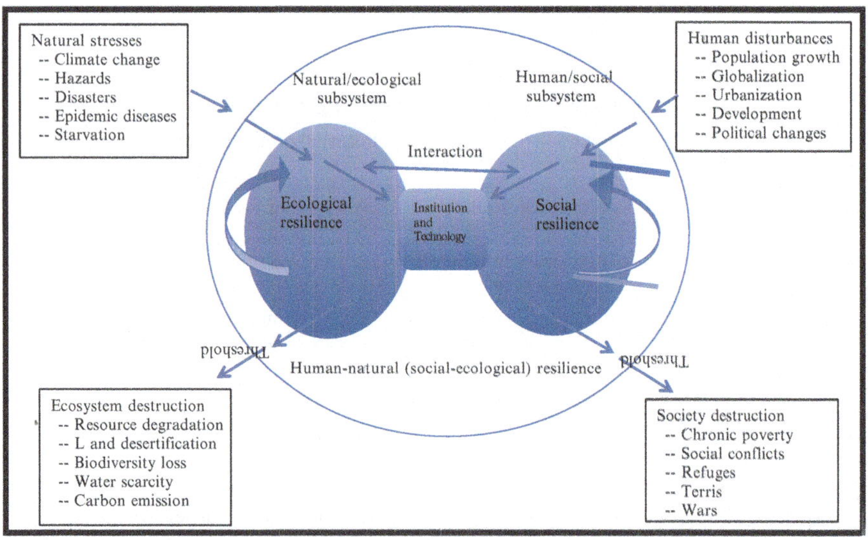

Fig. 2.3 Resilience of a human–natural (social–ecological) system

The strong link between ecological and social resilience may be explained by the dependence on ecosystems of communities, their institutional structures, and their economic activities. For example, sustainable or unsustainable use of rangeland resources can be highly related to the habitualized behavior, rules, and norms that govern society to use rangeland resources.

Social–ecological resilience, according to numerous scholars (Carpenter et al. 2001; Folke et al. 2002; Kassam 2010), has three specific elements: the amount of change a system can tolerate without reducing its function; the degree to which a system can self-organize in response to change for renewal; the degree to which a system can develop adaptation capacity through learning. Therefore, social–ecological resilience can be well understood through resilience thinking on the basis of three aspects: persistence, adaptability, and transformability (Walker et al. 2009; Folke et al. 2010). As defined by Folke et al. (2010), "persistence is the tendency of a social-ecological system subject to change to remain within a stability domain, continually changing and adapting yet remaining within critical thresholds. Adaptability is the capacity of a social-ecological system to adjust its responses to changing external drivers and internal processes and thereby allow for development within the current stability domain, along the current trajectory. Transformability is the capacity of a social-ecological system to create new stability domains for development, a new stability landscape, and cross thresholds into a new development trajectory." These three aspects interrelate and interact as multiscale resilience, which is, according to Folke et al. (2010), "fundamental for understanding the interplay between persistence and change, adaptability and transformability." Therefore, scholars often incorporate persistence, adaptability, and transformability as key ingredients of resilience thinking for social–ecological systems (Table 2.2).

In social–ecological systems, some questions such as "resilience of what, to what?" are usually asked for specified resilience (Carpenter et al. 2001). However, too much focus on the specified resilience may cause the system to lose resilience in other ways (Cifdaloz et al. 2010). For example, if the resilience of rangelands to overgrazing is emphasized for pastoral systems through grazing bans and eco-migration (e.g., in China), the resilience of pastoral societies to environmental changes (e.g., climate change and land degradation) may be reduced. In contrast, the resilience of social–ecological systems does not define either the part of the systems that might cross a threshold or the kind of shocks that have to be endured by the systems. It should be about coping with uncertainty in both social and ecological dimensions to create opportunities for reassessing the current situation, triggering social mobility, recombining sources of experience and knowledge for learning, and sparking novelty and innovation. As for pastoralism, herders with their unique cultural and ethnic identities can use the indigenous knowledge and skills to manage the rangeland resources in a sustained manner, may develop the adaptive ability to deal with environmental shocks and crises through social learning, and can enhance their capacity to cope with environmental uncertainties through social mobilization/transformation. Hence, persistence, adaptability, and transformability are also the key for addressing the multidimensional resilience of human-natural systems of pastoralism.

Table 2.2 Ingredients of resilience thinking for social–ecological systems

Term	Definition
Active transformation	The deliberate initiation of a phased introduction of 1 or more new state variables (a new way of making a living) at lower scales, while maintaining the resilience of the system at higher scales as transformational change proceeds
Adaptability (adaptive capacity)	The capacity of actors in a system to influence resilience
Adaptive cycle	A heuristic model that portrays an endogenously driven 4-phase cycle of social–ecological systems and other complex adaptive systems. The common trajectory is from a phase of rapid growth where resources are freely available and there is high resilience (*r* phase), through capital accumulation into a gradually rigidifying phase where most resources are locked up and there is little flexibility or novelty, and low resilience (*K* phase), thence via a sudden collapse into a release phase of chaotic dynamics in which relationships and structures are undone (Ω), into a phase of reorganization where novelty can prevail (α). The *r–k* dynamics reflect a more or less predictable, relatively slow "foreloop" and the Ω–α dynamics represent a chaotic, fast "backloop" that strongly influences the nature of the next foreloop. External or higher-scale influences can cause a move from any phase to any other phase
Forced transformation	An imposed transformation of a social–ecological system that is not introduced deliberately by the actors
General resilience	The resilience of any and all parts of a system to all kinds of shocks, including novel ones
Panarchy	The interactive dynamics of a nested set of adaptive cycles
Regime	The set of system states within a stability landscape
Regime shift	A change in a system state from one regime or stability domain to another
Resilience	The capacity of a system to absorb disturbance and reorganize while undergoing change so as to still retain essentially the same function, structure, and feedbacks, and therefore identity; that is, the capacity to change so as to maintain the same identity
Social–ecological system	Integrated system of ecosystems and human society with reciprocal feedback and interdependence. The concept emphasizes the humans-in-nature perspective
Specified resilience	The resilience "of what, to what"; resilience of some particular part of a system, related to a particular control variable, to 1 or more identified kinds of shocks
Stability domain	A basin of attraction of a system, in which the dimensions are defined by the set of controlling variables that have threshold levels (equivalent to system regime)
Stability landscape	The extent of the possible states of system space, defined by the set of control variables in which stability domains are embedded
Threshold (aka critical transition)	A level or amount of a controlling, often slowly changing variable in which a change occurs in a critical feedback causing the system to self-organize along a different trajectory; that is, toward a different attractor
Transformability	The capacity to transform the stability landscape itself so as to become a different kind of system, to create a fundamentally new system when ecological, economic, or social structures make the existing system untenable

From Folke et al. (2010)

2.3.3 *Vulnerability of Human-Natural Systems*

In parallel to the resilience of human-natural systems, the vulnerability of human-natural systems has been widely studied in the academic community. Similarly, the social–ecological system has been the mostly documented form of human-natural systems in the dimension of vulnerability. From the definition given by Millennium Ecosystem Assessment (2005) that "vulnerability is the sensitivity of people, places, ecosystems and species to contingencies and stress, and their capability to cope with them", the vulnerability of social–ecological systems can be understood as the sensitivity and adaptability of systems to shocks and stresses in both social and ecological dimensions.

2.4 Assessing Vulnerability/Resilience of Human-Natural Systems of Pastoralism

2.4.1 *General View on Vulnerability/Resilience of Pastoralism Worldwide*

Pastoralism illustrates significant features of human-natural systems or social–ecological systems where people interact with natural components, including plants, animals, and environment services (Vavra 1995; Nori 2007). Over centuries, rangeland residents have been continuously practicing pastoralism to accumulate a sophisticated ecological knowledgebase to facilitate the close tracking of environmental conditions and to harmonize the interactions between society and nature. Nowadays, natural stresses and social–political threats such as climate change, population growth, land use changes, and political system changes are decoupling human and natural systems, resulting in changes in subsistence patterns of pastoralist groups, marginalization of traditional territories, and decreased adaptive capacity of the pastoral ecosystem throughout much of the developing world (Dong et al. 2012). Failure to reconcile emergent issues at the interface between the ecological, economic, and social considerations has repeatedly resulted in management and policy actions that do not achieve the objectives of optimizing yield of rangeland products in a sustainable manner (Thurow 2008). Most pastoral management policies in place today in the developing world have not led to sustainable outcomes, and the success or failure of many policies and management practices is based on their ability to take into account the complexities of human-natural systems or social–ecological systems. Therefore, the tolerance to shocks, self-organizing degree for renewal, and adaptation capacity of pastoralism worldwide have been mostly degraded, leading to decreased persistence, adaptability, and transformability of pastoralism. From this perspective, the general view is that the human-natural systems of pastoralism worldwide are reducing their resilience and enhancing their vulnerability to natural stress and human-induced shocks.

2.4.2 Framework for Assessing Vulnerability/Resilience of Pastoralism

Since the stresses and threats for sustainable pastoralism vary across different regions, it quite difficult to draw a common conclusion on how vulnerable/resilient pastoralism is worldwide. It is particularly important to conduct a structured comparison study to see how the pastoral production systems in different regions across the world are changing their vulnerability and resilience to different threats and shocks. It is also extremely important to explore how development strategies and other socioeconomic changes in different regions can help pastoral production systems worldwide become more resilient and robust at a time of growing risk and uncertainty.

The vulnerability/resilience of pastoral production systems, similarly to that of agricultural production systems, can be assessed in the dimensions of agroecosystems, livelihoods, and institutions (Fraser 2007). By referring to Fraser's (2007) and Dougill et al.'s (2010) three-dimensional "vulnerability/resilience" coordination framework of agricultural production systems, we can conceptualize a three-dimensional vulnerability/resilience coordination framework of pastoral production systems to compare different geographical regions and examine trends over time by studying the paths through the octant as traced by changes in the dimensions of agroecosystems, livelihoods, and/or institutions (Fig. 2.4).

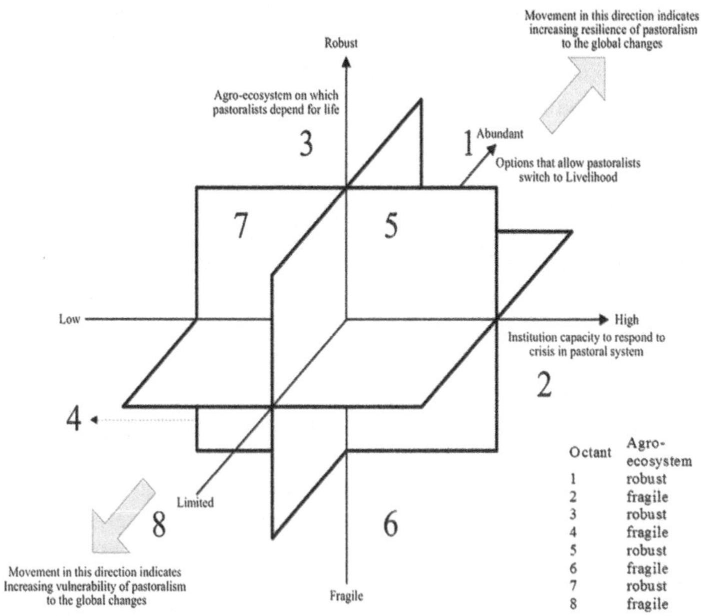

Octant	Agro-ecosystem	Livelihood option	Institution capacity
1	robust	abundant	high
2	fragile	abundant	high
3	robust	abundant	low
4	fragile	abundant	low
5	robust	limited	high
6	fragile	limited	high
7	robust	limited	low
8	fragile	limited	low

Fig. 2.4 Three-dimensional "vulnerability" coordination framework for assessing vulnerability/resilience of pastoralism

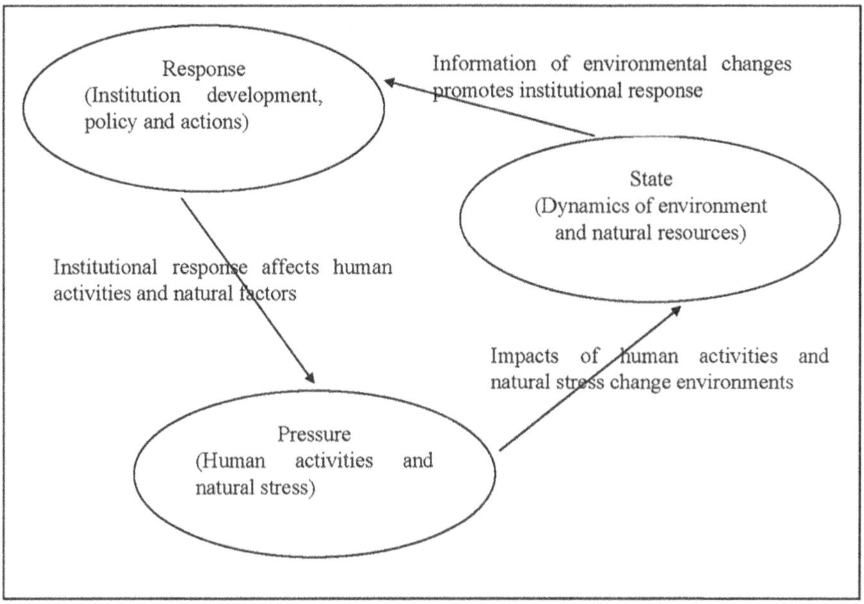

Fig. 2.5 Pressure–state–response framework modified from the OECD's analytic structure of policymaker information. (From Pearce and Freeman 1991)

In addition, supportive analysis is needed to fully understand cause–effect relationships within a three-dimensional vulnerability/resilience coordination framework of agricultural production systems. A widely used framework in natural resource use and environmental protection (Waheed et al. 2009), the pressure–state–response (PSR) framework developed by the Organisation for Economic Cooperation and Development is applicable to identify the cause and effect chains of the vulnerability/resilience of pastoralism. According to PSR theory, the human disturbances and natural stress will generate pressure on the natural environment, leading to changes in its state (Pearce and Freeman 1991). Information on environmental changes will promote the institutional responses, which in turn will affect human disturbance and natural stress; for example, reducing their effects and changing the driving forces or sources, so as to prevent or minimize the environmental responses that cause harm (Fig. 2.5).

By integrating the three-dimensional vulnerability/resilience coordination framework with the PSR framework, we will summarize and synthesize the worldwide case studies to provide a detailed analysis of how human–cultural systems of pastoralism in different regions (including Asia, Europe, Africa, South America, North America, Oceania, and the Arctic) are changing their vulnerability/resilience to natural and human-induced stresses, shocks, and changes.

2.5 Identifying Vulnerability/Resilience of Human-Natural Systems of Pastoralism in Major Pastoral Regions Across the World

2.5.1 *Cases from Asia*

Asia has the largest land cover of rangelands on Earth, stretching from the borders of eastern Europe to the Pacific Ocean, spanning 7000 km (Kerven 2004). Pastoralism ranges from reindeer keeping on permafrost tundra of the Arctic in the north to camel herding on hot sandy deserts in the south, and from the raising of sheep, goats, and horses on the plain steppe to yak and Tibetan sheep grazing in the highland meadows at altitudes of more than 4000 m above sea level. Here, we examine cases from the Central Asian steppes and the Asian highlands to address the vulnerability of pastoralism in Asia.

2.5.1.1 Central Asian Steppe: Marginalization of Pastoral Systems with Political Change and Command Economy Collapse

Central Asia's landmass is covered mostly by the steppe, the eastern sections of Eurasian steppe, one of the largest biomes on Earth, stretching from China, Mongolia, and southern Siberia across Xinjiang, Kazakhstan, southwestern Siberia, European Russia, Ukraine, and Hungary to Anatolia, Romania, Slovakia, and outlier steppes in Austria and Spain (Fig. 2.6). Central Asia has around an 250 million

Fig. 2.6 Map of the Eurasian steppe range

hectares of steppe pastures which are distributed in an even larger region of arid lands including central republics and parts of Russia, Mongolia, and China (FAO 1997). Since the last Ice Age, the steppe in this region has changed greatly in plant and animal populations, which firstly migrating into the steppe zone from refugia as the glaciers retreated and temperatures increased. After humans appeared on the steppe, they made a nomadic life by hunting animals on the steppe (e.g., antelopes, horses, and camels). Soon these hunters became nomads who exploited the steppe as grazing grounds for their earliest domesticated animals such as sheep and goats. Later, these nomads domesticated, bred, and raised horses and camels one after another on the steppe (Werger and van Staalduinen 2012). Until the early twentieth century, nomadic pastoralism was the core on which people built their livelihoods on the steppe in Central Asia. With the collapse of the communist command economy, the implementation of new policies on land and pasture in recent years (Baibagushev 2011; Doerre 2012; Kerven et al. 2012; Kraudzun 2012; Kreutzmann 2013; Robinson and Whitton 2010; Steimann 2012; Vanselow et al. 2012) has changed the livelihoods of the traditional steppe peoples and damaged or destroyed very large tracts of steppe land.

Among the five republics of Central Asia, Kazakhstan, Turkmenistan, Uzbekistan, and Kyrgyzstan have steppe pasturelands covering more than half or nearly half of the nation's lands (Table 2.3). Tajikistan is relatively small in terms of pastureland cover (about 23 % of the nation's lands, as shown in Table 2.3) but pastureland accounts for 75 % of the agricultural land area in the country (FAO 1997).

Table 2.3 Pastoral production in Kazakhstan, Kyrgyzstan, Uzbekistan, and Mongolia

Countries	Location	Permanent pasture (km^2)	Proportion of total land area (%)	Major pastoral ethnicity	Pastoralist population	Proportion of all agricultural population (%)
Kazakhstan	Central Asia	1,851,000	69	Kazakh	4,700,000	68
Kyrgyzstan	Central Asia	93,650	49	Kyrgyz	256,000	7
Uzbekistan	Central Asia	222,190	52	Uzbek	1,478,000	6
Turkmenistan	Central Asia	307,000	65	Turkmen	1,537,000	43
Tajikistan	Central Asia	31,980	23	Tajik	205,200	4
Mongolia	East and Central Asia	1,293,000	83	Mongol	2,051,000	84
China	East and Central Asia	4,000,000	41	Mongol, Tibetan Kazakh	19, 500,000	2.4

From FAO (1997) and International Livestock Research Institute (2002)

Historically, the main land use in this vast pastoral area was extensive migratory livestock production, without rigidly defined state-defined boundaries (Suleimenov and Oram 2000). Since the collapse of the Soviet Union in the early 1990s, with the promotion of a strategy focusing on the restructuring of agriculture to achieve food security and to adjust to market economy requirements, pasture management has been changed from the state-managed pastoral systems to de facto common property regimes and, more recently, to leasing or privatized systems based on leasing (Robinson et al. 2010). As a result of land privatization reform, land rights have been shifted from pastoral cooperatives to wealthy individuals and groups and the poorest population strata have been crowded out, leading to increased insecurity with regard to resource access and mobility options (Nori et al. 2005). Transformation of pasture use traditions associated mostly with population growth and policy reform has led to massive land degradation and increased carbon dioxide emissions in pastoral areas of Central Asian republics (Chuluun and Ojima 2002). "Drama of the commons" noted by Ostrom (2000) might be a more appropriate term than "tragedy of the commons" triggered by Hardin (1968) to capture the current situation of Central Asia's pastoralism in general (Kreutzmann 2013).

Although the effects of changing land rights on both rangeland and livestock management vary greatly among the five countries of Central Asia (Kazakhstan, Turkmenistan, Uzbekistan, Kyrgyzstan, and Tajikistan), many problems are commonly challenging the sustainable use of rangelands by traditional institutions and the pastoral livelihood in achieving food security during this economic transition period. In Kazakhstan, overgrazing of rangeland is still speeding up in spite of a dramatic decrease in livestock populations, because only 30–40 % of rangelands have been used for grazing and the remaining rangelands are constrained for grazing by a lack of drinking water and most remote rangeland is no longer used with the collapse of government special services (Suleimenov and Oram 2000). Independent Turkmenistan operates a leasehold pastoral production, which was derived from the Soviet Union in the late 1980s. However, shepherds tending state-owned animals have not been able to receive a salary from the state or their collective farm since the collapse of the Soviet Union (Kerven 2006), This may result in difficulties for the livelihoods of the shepherds. In Uzbekistan, the rangelands have been used more properly because a state-controlled system is still in place, but the widespread plantation of grain crops on marginal lands is accelerating soil erosion (Suleimenov and Oram 2000). In Kyrgyzstan, there have been the same problems as the most of rangelands have not been grazed since independence (Suleimenov and Oram 2000), and the common herding systems (the pastures have been continuously used by those communities that historically exploited them) following the collapse of state farms have promoted the partial use of marginal pastures (Farrington 2005). In Tajikistan, livestock has declined mainly because of civil conflict, the reduced capacity of households to access pastures and fodder, and the overall disruption of the state/collective sector, and access to veterinary services and protection against diseases are also problematic for the private sector (Kerven 2006).

Mongolia is one of the largest pastoral nations in the world, with 83 % of the land as permanent pasture, which is mostly the Eurasian steppe (Table 2.2). It probably has the highest pastoral population in the world, with 77 % of the national population living in rangeland areas (International Livestock Research Institute 2002). The pastoral economy has always been the mainstay of the national economy in Mongolia (Mearns 2004). During the socialist period, herders depended on the central government to provide them with regular salaries and means of transporting mobile livestock. The state collectives were responsible for allocating pastures, guiding seasonal movements, supplying veterinary services, fodder provision, and labor (Fernandez-Gimenez 2001). After the shift from a socialist economy to a market economy in Mongolia in the 1990s, all state collective farms collapsed and livestock were privatized; however, the public attitude of dependency was hard to discard (Muller and Bold 1996). With the demolishing of herding collectives in 1992, formal government institutions for pasture management such as financial, technical, extension, and marketing support were ended and customary institutions were too weak to fully replace them (Fernandez-Gimenez and Batbuyan 2004). As a consequence, a great amount of sown lands derived from the forced conversion of the most productive rangelands into croplands by Mongolia's socialist government in 1950s was abandoned, leading to a significant decline of crop production; for exmple, threefold reduction in 1995 in comparison with 1990 (Chuluun and Ojima 2002). In 1994, the postsocialist government passed the Land Law, which contained provisions for the regulation, management, and monitoring of pastureland, including the leasing of campsites, and possibly pasture. Leasing of winter and spring campsites began in 1998. However, the disappearance of vague and dynamic boundaries, which had been so adaptive in the past, is increasing herder vulnerability to climate change (Turner 1999; Fernandez-Gimenez 2002). The Land Law was revised and the new Law on Land came into effect in 2003. Unfortunately the new law preserved some of the ambiguities of the old law. In sum, both laws included provisions of certificates for possessing, essentially leasing, the winter and spring campsites, and potentially the winter and spring pastures as well. However, summer and autumn pastures are kept open for public use, which may have discouraged the pastoralists to make effective pasture management plans, thus negatively impacting their livelihoods (Fernandez-Gimenez and Batbuyan 2004; Upton 2008).

In China, steppe pasturelands cover about 59 % of the landmass in the northern territory, including most of the Xinjiang and Inner Mongolia autonomous regions and parts of Gansu, Qinghai, Shanxi, Shannxi, Hebei, Liaoning, Jilin, and Heilongjiang provinces (Li 1997). The total areas of steppe pasturelands in northern China is about 2,742,200 km^2, accounting for nearly 70 % of the total rangelands in China (Table 2.2). However, the steppe pasturelands in northern China were extensively degraded in past decades as a result of climatic variability and human activities such as overgrazing and overcultivation, which were driven by political changes. After 1949, the governments of the People's Republic of China the conversion of some of the most productive rangeland into cropland under collectivization programs as done in the USSR. These policies have not only reduced the amount of rangeland available for livestock production, but have also increased grazing intensity, often on less fertile grazing lands, leading to rangeland degradation and loss of soil fertility (Chuluun and Ojima 2002). After the ending of the collective systems in 1980s, the rangeland privatization facilitated by the land tenure reformation policy or the implementation of new Grassland Law, has enforced the enclosure of grazing pasturelands. A corresponding reduction in the spatial mobility of livestock herds has served to concentrate grazing pressures on some pasturelands, leading to the degradation of rangelands. For example, in the Xinjiang Uyghur Autonomous Region, degradation problems are worse on winter pastures where rangeland enclosure and individuation have been strongly enforced (Banks et al. 2003). Moreover, the process of allocation of pasturelands has led to increased social conflicts, inequity of access to water resources and good forages, a breakdown of traditional institutions, and a deepening division between rich and poor (Li and Huntsinger 2011). In addition to political changes, the pressures of population growth and economic development have modified the land use types, leading to significant changes in carbon dynamics and climate conditions, which may speed up the regional rangeland degradation (Chuluun and Ojima 2002). New conservation policies and management plans for pastureland use such as "Returning Cultivated Lands on the Slopes into Grasslands and Forest" (also known as "Grain for Green)" and "Retire Livestock, Return Grassland" (also known as "Grazing Ban) were formulated and enforced to drastically improve the situation, although the long-term effects of these polices on rangeland ecology and local livelihoods were questioned by both herders and professionals (Dong et al. 2007).

From the three-dimensional "vulnerability" space framework it can be seen that institutional changes in pastoralism (i.e., rangeland privatization and open access) in the Central Asian republics, Mongolia, and northern China resulted in a weakened institutional capacity to respond to crisis (e.g., disappearance of formal regulation of the socioeconomic dimension). These results would be reflected by a shift toward the eighth octant in Fig. 2.4, meaning increased vulnerability and decreased resilience of pastoralism in the dimensions of agroecosystems, institutions, and livelihoods in Central Asia.

2.5.1.2 Asian Highlands: Decline of Pastoral Systems with Social Transformation and Climate Change

The Asian highlands constitute an elevated and unique arena for shedding light on the spectrum of mountain pastoralism and rangeland management across a diverse spatial spectrum from the Pamirs, Tian Shan, Hindu Kush, and Karakorum to vast tracts of the Himalayas (including the Qinghai–Tibetan Plateau, QTP, of China). The Hindu Kush–Himalaya (HKH) forms the major body of the Asian highlands and is the world's highest mountain range. More than half of the HKH's land territory is covered by rangelands, with subtropical savannas at the Siwalik foothills, alpine meadows in the high-elevation Himalayan mountains, the extensive steppes on the QTP, and the cold, dry deserts in the Kunlun Mountains (Fig. 2.7). The rangelands of the HKH provide many important ecosystem services, such as food production, water supply, and biodiversity conservation, for millions of upstream and downstream people. With the specific features of geographical location, climate

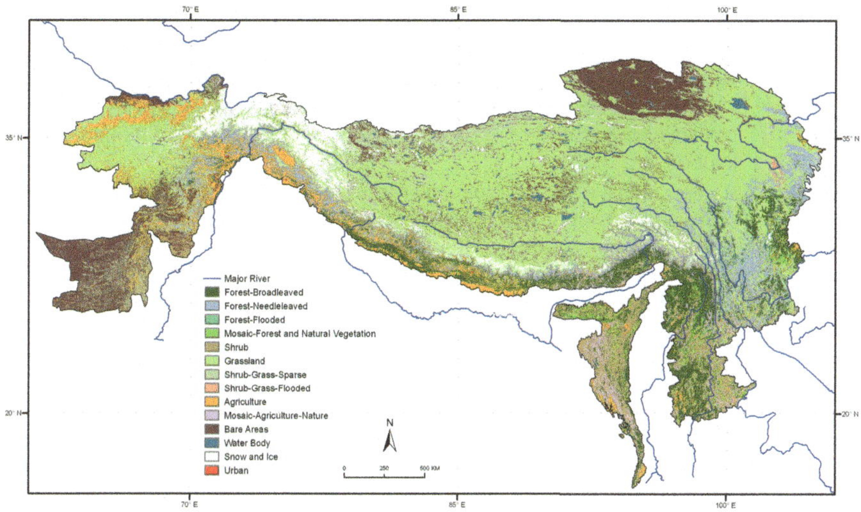

Fig. 2.7 Land use and land cover in the Hindu Kush–Himalaya region. (From Singh et al., 2011)

conditions, and vegetative cover (Vavra 1995), these rangelands serve as the grazing pastures for many distinct ethnic groups with high cultural diversity (Table 2.4). As a highly dynamic region of fragile and sensitive natural environments, of fundamental political changes, and of remarkable socioeconomic developments, the HKH is currently facing numerous challenges in sustainable use of rangeland resources and development of pastoral systems (Dong et al. 2010). Overgrazing is a serious issue in some pastoral areas where most of the primary vegetation in rangelands has disappeared because of heavy stresses from both pastoralists and their livestock (Dong et al. 2002). Overharvesting of medicinal plants in some high-altitude rangelands has threatened some important rangeland species with great economic and ecological value (Miller 1997a). Rapidly increased but not well-planned tourism has caused environmental problems in some rangeland areas (Miller 1997a). Rangeland degradation associated with overexploitation and overuse may have increased evapotranspiration rates, thus strengthening seriously negative effects of climate warming on pastoral production in this region (Du et al. 2004; Wang et al. 2006). Current protection and conservation policies and planning for sustainable rangeland development have overlooked the integration of ecology, production, and livelihood functions provided by pastoral systems (Miller 1997b; Dong et al. 2010). Similarly to the HKH, the whole Asian highlands are experiencing and will continue to experience social transformations, which will have dramatic impacts on all spheres of life in the fragile and sensitive natural environments in this region; for example, the pastoralists' traditional lifestyles, rangeland uses, and management practices are under rapidly increasing pressures from population growth and modernization processes (Kreutzmann 2012). From Afghanistan to Bhutan, as stated by Kreutzmann

Table 2.4 General information about pastoralism in the Hindu Kush–Himalaya (*HKH*) region

Location	It is located mostly in South Asia, extending across 8 Asian countries from Afghanistan in the west to Myanmar in the east, and from the Tibetan Plateau of China in the north to the Ganges Basin in the south. It is the world's highest mountain range, called "the roof of the world," with an area of 4.3 million square kilometers
Climate	It varies from a warm subtropical climate at the Siwalik foothills in the south to a cold alpine climate on the Qinghai–Tibetan Plateau in the north. Annual precipitation ranges from 100 to 1000 mm and falls mainly as snow and hail on the high mountains. The year-round temperature in most areas averages about 0 °C, dipping to −40 °C, in some areas in winter
Vegetation	More than 60 % of vegetation cover is rangeland, nearly 30 % of vegetation cover is forest, and around 10 % of vegetation cover is agriculture. Rangeland vegetation varies from subtropical savannas at low elevations to alpine meadows at high elevations
Land use	Pastoralism and agropastoralism are the dominate ways of utilizing the vast rangelands of the HKH. Agroforestry livestock grazing exists in some areas of the HKH
Animals	Grazing livestock such as yaks, sheep, goats, buffalo, zebu cattle, and horses as well as wild grazing mammals such as blue sheep, wild asses, and wild yaks
Population	It supports millions of pastoralists such as Tibetans in China, Gaddis and Gujjars in India, Tamangs in Nepal, and Brokpas in Bhutan
Social problems	Conflicts between the increased population and limited resources in pastoral areas. Change of pastoral livelihood driven by economic boom and social development

(2013), "the process of settlement continued, and in the true spirit of modernization theory, the convergence of lifestyles was envisage;… Modernization strategies have resulted in shrinking numbers of pastoralist."

Located in the northern HKH, the Qinghai–Tibetan Plateau (QTP) of China is a huge ecological area perfectly characterized by mountain pastoralism. However, the QTP's pastoralism is being threatened by rangeland degradation, a serious environmental problem associated with population growth, climate warming, and policy change. In the central part of the QTP, which is where the headwaters of three major Asian rivers (the Yangtze, the Yellow River, and the Mekong), are found, the rangelands have been overused by quick-growing human and livestock populations in recent decades (Ma et al. 1999; Wang and Chen 2001; Shang and Long 2005). It has been reported (Riley 2004; Fischer 2008) that the population growth of Tibetans on the QTP was about double that of the Han during 1982–2000, although it is difficult to accurately estimate the growth rate of the pastoralist population because of shortcomings in methods and time selection (Fischer 2008). Human population growth in pastoral areas may be highly associated with overstocking of the rangelands, because the number of livestock must be kept at a similar or even higher level to maintain an unchanged living standard of pastoralists (Harris 2010). The overexploitation of rangeland resources by local pastoralists and outside herb collectors without their caring about rangeland-carrying capacity has resulted in the massive degradation of the rangeland resources (Li and Huang 1995; Bai et al. 2002; Li et al. 2008). Nearly half of the QTP's alpine rangelands has been degraded in the past 40 years (Wang and Chen 2001) and about 26% of the QTP's alpine rangelands has been degraded severely to "black beach" or "black soil land," which is the bare land in the winter and land sparsely covered by annual weeds or poisonous plants in the summer (Li and Huang 1995; Ma et al. 1999, 2002; Shang and Long 2005).

In the eastern part of the QTP, rangeland health is currently being threatened by climate warming. For example, Klein et al.'s (2004) experimental study on alpine meadows and shrubland in the northeastern part of the QTP showed that the decline of plant species richness would be mostly associated with climate, and species losses can be accelerated by simulated grazing (i.e., clipping). Klein et al. (2007) also reported that 1.0–2.0 °C of warming in the growing season can lead to a drop of aboveground net productivity of alpine vegetation, particularly of palatable grass species. In contrast, Xu and Liu (2007) observed that climate warming led to a rise of the normalized difference vegetation index (a surrogate for plant biomass and productivity), which might be attributed to the enhanced woody plants in the vegetation composition on the QTP during 1982–2002. In addition, Baker and Moseley (2007) used a photo monitoring approach by comparing historical photographs and recent ones to show evidence that warming has resulted in glacier retreat and alpine tree-line advance on the high-altitude plateau of northwestern Yunnan in the southeastern part of the QTP. These changes may be associated with the decline of rangeland sizes and decrease of grazing pasture quality, thus threatening both rangeland health and pastoralists' livelihoods.

Besides climate change and population growth, changes of rangeland management policies in China have led to the alteration of land use and cover characteristics, thus promoting rangeland degradation and livelihood vulnerability in the pastoral areas. In recent decades, the Chinese government has launched a series of programs with multiple goals of reducing grazing pressure and improving pastoral livelihood, such as motivating pastoralists to adopt sedentary lifestyles; encouraging household responsibility for rangeland and livestock by clarifying the tenure of pasture land on a family basis; subsidizing construction of permanent winter homes, fences, and livestock shelters, and providing plots for growing supplemental winter fodder (Harris 2010). Although these programs have been promoted by the Chinese government with great ambitions, their long-term ecological and economic viability remains uncertain (Wu and Yan 2002; Yan et al. 2005; Davidson et al. 2008). The resettlement schemes have created numerous challenges for the local pastoralists regarding their pasture use rights (Ptackova 2011) or housing arrangements (Sulek 2012). More recent initiatives such as "Natural Forest Protection," "Grain for Green," and "Retire Livestock, Restore Pastures" may be effective in restoring degraded alpine rangelands, although they have encountered many obstacles; forexample., high monetary and labor costs, lack of skills and experience, and poor public services (Nyima 2003; Yeh 2003; Dong et al. 2007). Therefore, these programs may be disqualified from constituting a sustainable socioeconomic system in social and cultural terms (Walker and Salt 2006; Du et al. 2012). The environmental degradation processes have been triggered by social transformation and climate change, external development, and modernization strategies, which undermine the local and regional perceptions and participation in decision making (Kreutzmann 2012).

A similar situation is reflected in the impacts and scope of social transformation and climate change on the mountain pastoralism in the neighboring countries of the southern HKH, Afghanistan, Pakistan, India, Nepal, and Bhutan. A case study in Chitral, an area located in the eastern Hindu Kush of northern Pakistan, indicated that animal husbandry has lost importance because of socioeconomic and political changes such as population growth, land fragmentation, and division of herds as a result of the traditional law of inheritance (Nusser et al. 2012). Law and legal pluralism and uncertainty associated with loss of customary access regulations resulting from institutional and legal changes (Faizi 1999) have led to contested spatial territoriality and free grazing, resulting in pastureland degradation and forage deficiency (Nusser et al. 2012). The infrastructural improvements and agrarian developments have changed the socioeconomic conditions and local inhabitants' attitudes and behavior, land use patterns, and livelihood strategies, leading to increased transformation of the workforce from livestock rearing to crop cultivation on the valleys of plains and even off-farm jobs in the towns, low-land cities, or the Gulf countries (Nusser et al. 2012). In the Indian Himalaya such as Himachal Pradesh, pastureland-based animal husbandry is quite important, and pastoral production systems have been regarded as maladaptive and backward practices and largely overlooked by the local and national policymakers. State policies aiming at "modern, scientific" paradigms have increasingly restricted mobile forms of land use adopted by pastoralists (Saberval 1999; Bergmann et al. 2012). The explicit pastoral policy seems to be absent, as stated by Sharma (2003): "There are no official pastoral development policies; in fact both the Ministry of Agriculture and the Ministry of Environment and Forest are remarkable for their stance against pastoralists." The growing exclusion of valuable pasturelands through administrative acts of nature protection has challenged the sustainability of pastoralism as stated by Sharma (2003): "Today Himalayan pastoralism is perceived by decision-makers and politicians as an environmental threat to the Himalaya and the local pastoral groups are incessantly blamed for overgrazing and livestock increase." Besides, market orientation and globalization have created various problems for pastoralists, such as the privatization and commercialization of community-regulated resources (Bergmann et al. 2012), which may result in the further decline of pastoralism. Similar situations exit in the Nepalese Himalaya, where general perceptions of inefficient traditional management, nonadaptations of scientific knowledge, lack of investment, confusions over ownership, and conflicts have resulted in a low national priority and neglect of indigenous knowledge of, skills in, and techniques in pastoralism. The creation and expansion of protected areas contributed to the exclusion of herders from their inherited pasturelands, leading to a decline in pastoral production (Kreutzman 2012).

In addition to socioeconomic and political transformations, climate change has also deeply impacted pastoralism in the southern HKH. Numerous studies have documented that the HKH region has shown a trend of overall warming during the past 100 years (Yao et al. 2004; IPCC 2007). For example, warming in Nepal was

0.6 °C per decade between 1971 and 1994 (Shrestha et al. 1999). Increased variability is another feature of climate change; for example, the mean and maximum temperatures in winter increase constantly and the mean and minimum temperatures in summer decline consistently in the Karakoram and Hindu Kush mountains (Fowler and Archer 2006). The local pastoralists have to cope with climate change through adaptation and transformation. Case studies conducted by Yi et al. (2012) across Afghanistan, Pakistan, and Nepal along the Hindu Kush–Karakoram–Himalaya showed that the local pastoral people felt rising temperature, decreasing precipitation, and more unpredictable weather patterns, which have further caused the rapid movement of glaciers, reduction of fodder production, drying up of rivers, and water shortages, are threatening the very subsistence of pastoralism. Although the local pastoral communities have adapted to these changes passively or proactively by enhancing water resource management, changing the temporal and spatial pattern of seasonal migration, introduction of drought-resistant crops and animal varieties, or diversification of income-generation activities, their adaptability and transformability in response to climate change are severely limited by factors such as harsh physical conditions, poor economic capacity, and lack of adequate technology, skills, information, and social services (Yi et al. 2012). As a consequence, both the sustainability of rangeland ecosystems and that of pastoral communities have been threatened by ongoing climate change.

The abovementioned case studies in different sites across the HKH region indicate that rangeland degradation associated with overexploitation and climate change enhances the vulnerability of pastoral livelihoods on the Asian highlands and that institutional vulnerability associated with socioeconomic and political changes accelerated rangeland degradation and increased the vulnerability of pastoral societies and rangeland ecosystems. These results would be reflected by a move toward the eighth octant of the three-dimensional vulnerability space framework in Fig. 2.4, indicating increased vulnerability and decreased resilience of pastoralism in the Asian highlands for three key dimensions: agroecosystems, livelihoods, and institutions.

2.5.2 Cases from Africa

2.5.2.1 African Sahel: Degradation of Pastoral Ecosystems with Expansion of Agriculture and Modernization

The African Sahel covers much of Sudan, Chad, Niger, Mali, Mauritania, and a small part of Algeria from the east to the west and is a well-known semiarid transition zone between the Sahara and the subhumid savanna (Fig. 2.8). For centuries, the Sahel has been serving as one of the major pastoral production bases in Africa because of its specific geographical location, climate conditions, and associated vegetation composition (Table 2.5). Historically, pastoral production systems in the Sahel have been determined by pastoral communities through negotiated access to

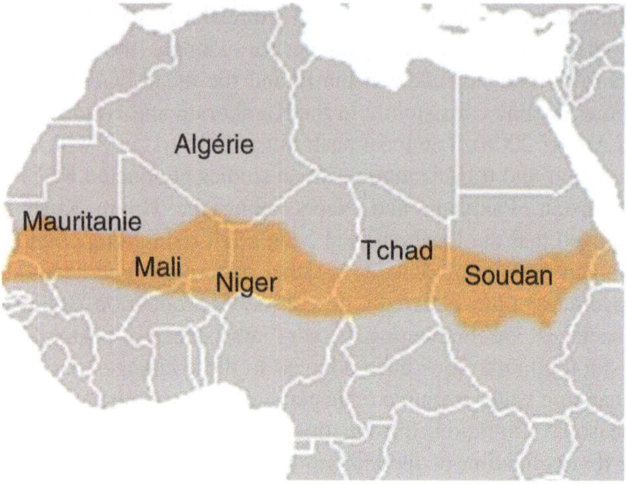

Fig. 2.8 Location of the African Sahelk

Table 2.5 General information about the African Sahel

Location	It represents the southern edge of the Sahara, extending from the Atlantic Ocean in the west to the Red Sea in the east, from Cape Verde to the south by the less arid Sudano-Sahelian belt, covering a surface area about 5.7 million square kilometers
Climate	It is a transitional zone between the arid Sahara in the north and the subhumid savanna zone in the south, with annual rainfall ranging from 200 to 600 mm
Vegetation	Vegetation cover of the Sahel is composed of bushes, grasses, and stunted trees that increase in density as one moves southward
Land use	Traditional way of utilizing the Sahel is mostly raising livestock in a system of seminomads; that is, farming and raising livestock in a system of transhumance
Animals	Grazing livestock of cattle, camels, sheep, and goats as well as wild grazing mammals such as the scimitar-horned oryx (*Oryx dammah*) and the dama gazelle (*Gazella dama*)
Population	It supports a population of about 58 million inhabitants, among them about 13 % are nomadic pastoralists; that is, Tuareg, Fulani, and other ethnic groups
Social problems	The expansion of agriculture and a shift to agropastoralism pushed nomadic pastoralists into more marginal regions

From Kandji et al. (2006)

water and pasture that did not have exclusive rights and by reciprocal arrangements between pastoralists and agriculturalists (Brooks 2006). These traditional pastoral production systems appear to have been well suited to the ecological and sociological conditions (Jarvis 1993), increasing flexibility through an enhanced ability to respond to a rapidly changing and increasingly unpredictable environment (Marshall and Hildebrand 2002). However, climate change, biophysical degradation of the environment, rapid population growth, and growing demands for agricultural production contribute negatively to the development of pastoralism, which largely depends on water availability and pasture productivity (Grouzis 1988; Watkinson and Ormerod 2001). Since the droughts of the 1970s, and particularly since the additional dry years of 1983–1985, the Sahelian pastoralists' adaptations to environmental conditions have been greatly weakened by inappropriate development practices (de Bruijn and van Dijk 1999; Warren 2005). The conditions experienced by pastoral communities, as stated by Thébaud and Batterbury (2001), are linked to "(a) the complexity of the activities they must use to ensure access to resources; (b) conflicts and cooperation between ethnic groups; (c) the inconsistent role of the state in assisting or constraining pastoral livelihoods; and (d) the negative discourse surrounding pastoralism that still circulates in some government and development policy circles."

According to data provided by some scholars (Ahmed et al. 2000; Thébaud and Batterbury 2001), the Sahel experienced several droughts in the early twentieth century, in 1913–1914, 1931–1933, and 1942, whereas there were unusually huge amounts of precipitation during the 1950s and 1960s (this was an exceptionally humid period relative to mean in the twentieth century), which provided pastoralists with enough water resources for abundant forage production, allowing the pastoralists to keep high stocking rates on the rangeland and farmers to spread northward into pastoral areas. This period was also a transition time for many African nations to become independent. In this period, as stated by Brooks (2006), "newly independent African nations focused on modern, technocratic solutions to development aimed at delivering economic growth and the traditional approaches to resource management and food security were increasingly marginalized." Both the political and the economic transitions in this period led to the expansion of agriculture northward into historically marginalized pastoral areas of the Sahel, which enhanced conflicts between agriculturalists and pastoralists in this region (Glantz 1996; Thébaud and Batterbury 2001). Increased agricultural sectors in agropastoral economies caused growing competition for both agricultural lands and pasturelands in the Sahel (Mortimore 1998; Bassett and Zueli 2000). The expansion of agricultural lands constrained transhumant herders' spatial movement for grazing management, impairing their rights for using the primitive pasturelands (Thébaud and Batterbury 2001). The livestock herds were no longer allowed to graze on harvested fields in some areas of the southern Sahel to improve soil conditions (by animal trampling and excreting urine and feces), as the farmers tended to keep their harvest residues in the harvested fields and fallows for themselves (Thébaud and Batterbury 2001).

The famines of the 1970s in the Sahel resulted from inappropriate development practices that were undoubtedly triggered by drought of the 1970s and 1980s (the Sahel desiccation), when there was a rainfall decrease of 29–49 % compared with the 1931–1960 baseline period according to the (Intergovernmental Panel on Climate Change's 2001) report. The severe droughts in 1973–1974 and 1983–1985 exerted heavy and long-lasting effects on pastoralists particularly (Thébaud and Batterbury 2001). As Brooks (2006) stated: "Over-extension of agriculture into historically marginal rangeland areas as a result of a failure to appreciate the nature of long-term (i.e., multi-decadal scale) climatic variability in the Sahel, resulted in massive losses of human life and livestock, the destruction of communities and livelihood systems, and massive societal disruption on a regional scale." Moreover, this agricultural expansion strategy for food production resulted in the degradation of the land resources in the Sahel; for example, overgrazing associated with the shrinking of pasturelands led to the devastation of the rangeland resources in many areas in the Sahel (Kandji et al. 2006). However, the policies and institutions for pastoral activities did not always deal well with itinerant herders, mobility, and common property rangeland management systems in Sahel (Lavigne Delville 2000; IIED 1999). The specific nature of pastoral land use was not fully acknowledged by modern laws in most Sahelian states, although various land reforms such as the Code Rural of 1993 in Niger and the Réorganisation Agraire et Foncèire in Burkina Faso touched on the issue of pastureland management. Recent legislation on natural resource management has proved to be not only inadequate but actually detrimental to pastoralists, as the policymakers often considered pastoral herding as nonprofitable in comparison with agricultural farming, exploitation of forest resources, and the creation of wildlife reserves (Thébaud and Batterbury 2001). The implementation of modern hydraulic projects (boreholes since the 1950s, and cement-lined wells since the 1970s) has weakened or even eroded property arrangements of water resources, as regulation of these public, open-access resources is far harder to achieve at sustainable levels (Thébaud and Batterbury 2001). As a consequence, Peul, Tuareg Tubu, and Arab communities in the pastoral Sahel often fight over access to water points (Thébaud and Batterbury 2001).

Although the future rainfall patterns in the Sahel remain uncertain and conflicting in different simulation studies (Kandji et al. 2006; Christensen et al. 2007), there will be large internal variability in precipitation, associated pastoral migration drifts, and population rearrangement according to some scholars' expectations (Bassett and Turner 2007; Galvin 2009). In the third assessment report of the Intergovernmental Panel on Climate Change, it was expected with the most rapid global climate change scenario that rainfall will increase in the Sahel (Carter et al. 2000; Hulme et al. 2001). If this expectation is the case, agricultural expansion into marginalized rangeland areas will be further encouraged by the development strategies such as the "model of agricultural production," one of the stimulants to change pastoralism into agricultural production (Warren 2005). Consequently, the social problems such as conflicts between agriculturalists and pastoralists, and environmental problems such as overgrazing of rangeland and degradation of land resources may be accelerated in the Sahel. Moreover, from past experiences, the growing incorporation of pastoral populations into modern societies in African countries may lead to the political, economic, and cultural marginalization of the pastoral society (Azarya 1996), and some pastoralists may live in a world of insecurity, war, famine, and drought (Baxter 1993). If the current and future public polices in the Sahel continue to stress development and "modernization" (Warren 2005), the vulnerability of pastoralists' livelihoods and pastoral institutions in this region will be enhanced.

This case can present three points in the three-dimensional vulnerability space framework for pastoral systems in the African Sahel: (1) change from the pastoral production system to an agricultural production system enhanced the fragile status of rangeland ecosystems in the agroecological dimension; (2) moving the pastoral populations into more marginalized regions because of the expansion of agriculture increased the risk of limited resource options in the livelihood dimension; and (3) modernization strategies and development models focusing on economic growth and agricultural production lowered the pastoral institution capacity to respond to crisis. As a result, this would be reflected in a shift toward the eighth octant in Fig. 2.4, indicating increased vulnerability and decreased resilience of pastoralism in the dimensions of livelihoods, agroecosystems, and institutions to the global changes in the African Sahel.

2.5.3 Cases from Europe

2.5.3.1 European Highlands: Erosion of Pastoral Systems with Depopulation and Land Abandonment

Pastoralism in Europe has a long history, in some areas for up to 10,000 years, and much of Europe's wildlife has developed alongside it (McCracken and Huband 2005). A wide variety of pastoral systems practiced in Europe, similar to the pastoralism in other regions of Earth, have been shaped by the climate, topographical conditions, and cultural traditions in this part of the world. It is considered as a low-production farming system, and pastoralism usually occurs in areas (high mountain habitats, arid zones, or otherwise poor soil areas) where higher production is physically not possible. These areas are also rich in biodiversity or endemic plant and animals with high nature value, they can be found throughout Europe, and they are a key feature of pastoralism. In addition, there are other features of pastoralism: traditional knowledge of the pastoralists, their close association with the landscape, and their need for diversification in pastoral activities (Biber 2006). However, the areas of high nature value pastoralism in Europe have declined over the past 30 years, and most high nature value pastoral systems are now confined to remote mountainous regions such as the Alps (McCracken and Huband 2005). European pastoralism, particularly that in the Alps and highlands of Europe, has been marginalized by depopulation of pastoralists and changes of land use practices.

The European Alps are one of the great mountain range systems, and stretch approximately 1200 km across eight countries, from Austria and Slovenia in the east, to Germany, Liechtenstein, and Switzerland in the north, to France and Monaco in the southwest, and to Italy in the south (Fig. 2.9). According to Lichtenberger (1994), pastoralism of the European Alps appeared firstly as long as 6000 years ago. Traditionally, the pastoralists in the Alps practiced the *alpeggio* system, a transhumant grazing system, to graze their livestock on the pastures near or above timberlines in summer and move their livestock back to valley bottoms at other times of the

Fig. 2.9 Location of the European Alps

year (Laiolo et al. 2004). In recent decades, the development of both industries in the valley bottoms (Laiolo et al. 2004) and ski-based tourism above or near the timberline (Brugger et al. 1984; Lasanta et al. 2007) has changed the traditional land management practices in the European Alps, making pastoralism in this area less necessary or economically unviable. Depopulation of pastoralists in the area due to their migration to the valley bottoms for better-quality life and reduction in stocking rates of grazing livestock associated with pastoralists' transformation to other livelihoods have resulted in changes in ecosystem functions and structures (Cernusca et al. 1999; Dirnböck et al. 2003); that is, shrub encroachment on the rangelands of the subalpine zone (Reyneri 2001; Laiolo et al. 2004). Shrub encroachment has led to significant changes in vegetation characteristics and animal populations (Beaufoy et al. 1994; Pain and Pienkowski 1997; Laiolo et al. 2004), as formerly open ground habitats have been lost and the diverse landscapes have been reduced. Although some scholars have stressed that mountain agriculture and livestock are quite important in tourism-based development models to mitigate the negative environmental effects (Wyder 2001; Laiolo et al. 2004), the facts show that plant diversity in ski runs of the Swiss Alps was greatly reduced in contrast to that of nearby rangelands (Urbanska et al. 1998), and that bird biodiversity in the ski runs of the western Italian Alps was decreased because of abandonment of the grazing pastures, whose edges can attract diverse avifauna (Laiolo and Rolando 2005). The movement of local pastoralists from the primary production system (agropastoralism) to the tertiary sector (industry) has also led to the death of the traditional agropastoral system,

Fig. 2.10 Old pastoral utilities displayed in a farmer's museum in the village of Peisey Nancroix in the French Alps. (Photo by Dong Shikui 2010)

especially in some unproductive and remote mountain valleys in the French Alps (Anthelme et al. 2001; Didier 2001). From our visit to Peisey Nancroix in 2010, a small village in the French Alps involved historically mainly in pastoralism, we found the local residents have abandoned the traditional pastoralism for ski-tourism development, and they have shifted their livelihoods from pastoralists to ski resort managers, hotel managers, migratory labors, etc. People have to see the tradition of pastoralism from a farmer's museum operated by an old villager who retired from being a traditional livestock grazer many years ago (Fig. 2.10).

Similarly to the Alps, the European highlands have served as the pastoral production base for thousands of years in human history. Since the highlands where pastoralism happens could very often not be used for more intensive forms of agriculture, the abandonment of land use for pastoral production resulted in the loss of pastoralism (Biber 2006). This is the general trend of pastoralism in the European highlands. In the Picos de Europa region of northern Spain, transhumant livestock farming is a traditional land use practice with a long history (Rescia et al. 2008), and increasing depopulation, a common phenomenon existing in mountain rural areas of Europe (Pereira et al. 2005), has led to the disturbance of this historical land use practice (MacDonald et al. 2000). As a consequence, the advancement of forest over rangelands and the spread of neighboring forests onto lands with herbaceous species have lead to shrinkage or disappearance of grazing pastures and interdigital fragmentation of rangelands in the Deva Valley of the Picos de Europa region (Forman 1995). In Rescia et al.'s (2008) survey it was shown that the vegetation cover in the 65 % of total sampling plots in the Picos de Europa region has been changed from rangeland to forest, and the gradual increase of unpalatable trees and woody shrubs over palatable grasses has resulted in rangeland degradation in this pastoral landscape. In recent decades, tourism has increased quickly to replace livestock grazing as an alternative to the pastoral economy, although some scholars have questioned the social and economic viability of tourism as a way of life for rural inhabitants (Izquierdo and Hanneman 2006; Rescia et al. 2008). Most members of the pastoralist population have become pensioners and only a small number of people are active in keeping traditional pastoral production systems at present. Rescia et al.'s (2008) survey indicated that there was a negative relationship between livestock farmers (even ex-livestock farmers) and most of the social agents, a common phenomenon among different components of the social–ecological systems in other rural areas (Kinzig et al. 2006). The conflicts between local pastoralists and other populations with different livelihoods may easily lead to the instability of social–ecological systems (Rescia et al. 2008).

Similarly, in Sistelo of the Peneda mountain range in northwestern Portugal, the depopulation associated with outmigration for better quality of life as wage laborers in foreign countries or other livelihood options has led to the abandonment of agricultural lands, especially pasturelands, on which pastoralists have spent their lives for centuries on the basis of *brandas*, a way of mobile grazing, in which they moved the livestock from the valleys in the winter to the higher mountains with better pastures in the summer (Pereira et al. 2005). However, depopulation and abandonment of grazing lands have continued as an onward trend from the 1950s to the present, with a 57 % decrease of the local population in Sistelo from 1960 to 2001 (Pereira et al. 2005). As a result, the populations of livestock declined and the tradition of seasonal mobile grazing was progressively abandoned despite the return of the *baldio*, traditional grazing pastures which were converted into forests by the appropriation of the state for afforestation in the 1940s, to the pastoral community in 1974 (Pereira et al. 2005). Rangeland ecosystem services such as provisioning of cattle as an income source have been decreasing, as other sources of income have replaced pastoralism as the provisioning services for local people (Pereira et al. 2005). Abandoned grazing fields were replaced by forests, and the species associated with pasturelands and farming lands decreased and those associated with forests increased (Parody et al. 2001), resulting in negative effects on biodiversity and the rural landscape. The related traditional knowledge, such as identifying the medicinal plants and the best forage plants for pasture, has been eroded (Pereira et al. 2005). It is apparent that the depopulation and pastureland abandonment were major drivers that caused the reduction of ecosystem services and livelihoods in Sistelo. These drivers are also affecting other mountainous areas of Portugal according to reports by other scholars (e.g., Ferreira et al. 1999).

These case studies in different sites in the European Alps and highlands show that land abandonment associated with depopulation led to diminishing ecosystem services of rangeland (loss of biodiversity and decline of livestock production) and displaced livelihood of pastoralists (changed into pensioners), which in turn weakened the indigenous institutions and regulations (traditional knowledge of biodiversity and identification of good pastures). These results represent a shift toward the fourth octant of the three-dimensional vulnerability space framework in Fig. 2.4: an increase in vulnerability and a decrease in resilience of pastoralism in the agroecosystem and institution dimensions, but an increase in resilience and a decrease in vulnerability of the pastoral system in the livelihood dimension.

2.5.4 Cases from South America

2.5.4.1 Bolivian and Peruvian Andeans: Decline of Pastoral Systems with "Modernizing" Agricultural Reform

South American pastoralism is confined to the semiarid regions of the Andes with the regular transhumance routes that sometimes are very ancient and mostly related to the herding of camelids including llamas, alpacas, vicufia, and guanacos (Westreicher et al. 2007). Andean pastoralism is known to be ancient, although information about the origins of pastoralism in the Andes remains sketchy (Westreicher et al. 2007). The evidence indicates that pastoralism existed in the Inca empire and in prehistory, when the domestication of llamas and alpacas by hunters who followed the movements of herds of wild animals between seasonally available pastures shifted to a pattern of transhumance (Westreicher et al. 2007). Nowadays, South American pastoralism exists primarily in the Andean regions of four South American countries: Argentina, Bolivia, Chile, and Peru (Fig. 2.11). Bolivia and Peru, especially their Andean highlands (sierra and altiplano) are the key areas for South American pastoralism (Kuznar 1991; Westreicher et al. 2007) in terms of pastureland size, animal production scale, and pastoralists' population size (Table 2.6). Although pastoral production contributes a significant share to the national economies of these two countries (Westreicher et al. 2007), agricultural reform policies in both nations aimed at "modernizing" pastoralists on the highlands have overlooked the importance of pastoralism, resulting in exacerbated degradation of environmental, economic, and social conditions for pastoral communities (Nori 2007).

Bolivia's Andean pastures were traditionally used corporately by large clusters of pastoral communities, known as *ayllus*, with strict rules of entry and resource management (Swift 2004). Under this system, the traditional pastoral productions of transhumance systems were kept as they had been historically to overcome the demographic constraints and the resource scarcity in most of the Andean highlands of South America (Swift 2004; Westreicher et al. 2007). Seeing the corporate tenure of pastures invariably as an irrational resistance to modernization or a stubborn attachment to "primitive" and "dysfunctional" ways of life, the Bolivian government initiated an agrarian/agricultural reform in 1953 (soon after the 1952 revolution in Bolivia) to provide the peasants with individual title to land (Swift 2004; Westreicher et al. 2007). The pastoralists had struggled to oppose this policy for decades, but the pastoralists and the state eventually compromised in the 1970s by subdividing the *ayllus* into smaller units (hamlets comprising a group of families), each of which received a land

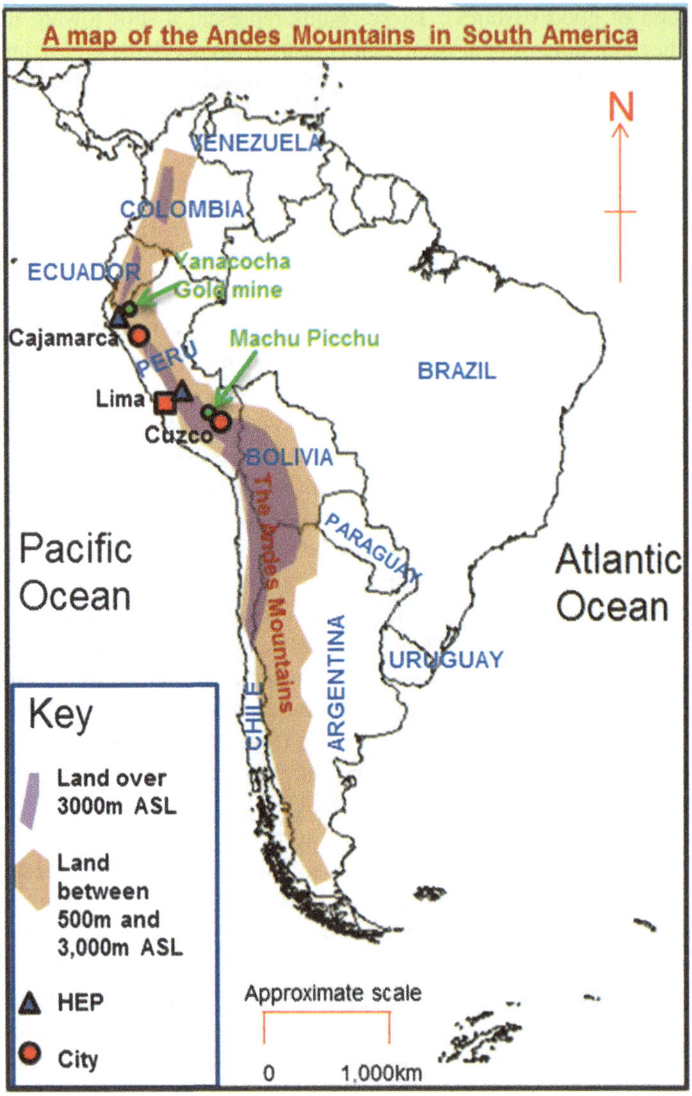

Fig. 2.11 Map of the Andean range

title (Swift 2004). As a consequence of this policy, customary tenure institutions have undergone considerable transformation and customary decision-making processes have been increasingly stressed by changing evolving political, economic, and social settings (Swift 1994). A combination of constraints of the natural environment, of historical burdens, and of current social, economic, and political problems have contributed to widespread poverty and underdevelopment in a period 1970s (Stadel 1995).

Table 2.6 General information about the Bolivian and Peruvian Andes of South America

Location	The Andean highlands are located about 3800 m above mean sea level in central Peru and Bolivia. The Peruvian highland is named *sierra*, and covers about 30 % of Peru's land area (1,285,220 km^2). The Bolivian highland is named as *altiplano*, and covers about 305,791 km^2 of the land, amounting to 28 % of the total territory of Bolivia
Climate	It is characterized by extreme cold and wind stress with mean annual temperatures ranging between 8 and 3 °C, and annual precipitation ranging from 800 mm in the north to 250 mm in the south
Vegetation	The vegetation is composed primarily of bunch grasses and low lying shrubs known as *tola*
Land use	In Peru, about 86 % of the land in the Andes is used exclusively as pasture. In Bolivia, pastoral management is dominantly present in the highlands
Animals	All of Peru's sheep, llamas, and alpacas are found here and 70 % of Peru's cattle are also produced in the region. The Bolivian wool marketing system was developed in this region, most especially since the middle of the 19th century
Population	About 41 % of the Peruvian population lives in the sierra and pastoralists account for more than 60 % of the sierra's rural population. In Bolivia, about 50 % of the population lives in the altiplano, and most of them are pastoralists
Social problems	Agrarian reform policy aimed to end exploitation by modernizing and mechanizing production and forced the transhumant pastoralists to settle in communities

From Pattie (1988), Kuznar (1991), and Westreicher et al. (2007)

In Peru, "modernization" and "mechanization" advocated by the 1969 agrarian reform have broken the customary pastoral institutions in the Andean highlands by forcing the pastoralists involved in transhumant grazing to settle in communities (Postigo et al. 2008), although these pastoralists have historically practiced effective transhumance systems with integrated herd mobility and seasonal pastureland use to recuse and share the risks in a harsh and dynamic environment (Brownman 1987; Postigo et al. 2008). Development projects led by these policies aimed to modernize livestock husbandry through improved pasture management, advanced alpaca breeding, and social capital improvement (Brownman 1983; Reineri et al. 2006), but the potential problems associated with these policies such as increased social gaps among different communities and no public access to some excluded pasture and water resources have become barriers for benefiting all of the rural population as a whole (Postigo et al. 2008). Since the early 1990s, Peru's agrarian reform has been redirected by neoliberal land policies focusing on a new concentration of land, capital, and knowledge in agribusiness (Postigo et al. 2008) by fostering decollection, which can allow individuals to own land title (Kay 2002). Although these polices have reinforced the community's identity as the true landowner, the struggles between new landowners and the community who had access to and control rights of the pastures historically have become increasing tensions (Postigo et al. 2008). These new policies have additionally weakened the governmental participation in agrarian development, resulting in the disappearance of agrarian extension, technical support, and credit from the government (Postigo et al. 2008). Further, the potential consequences of these policy reforms have led to both social problems such as increased inequity between the hired herders and property-owning pastoralists and environmental problems such as increasing pressure on pasture with the possible result of overgrazing (Brownman 1983; Lesorogol 2003).

These cases form the Bolivian and Peruvian Andeans of South America suggest that agrarian reform in these two countries has resulted in growing social gaps and poverty among the pastoral community, increasing pressures on the pastures from diminished governmental participation in pastoral development, and deterioration of the traditional land tenure system. This would be reflected by a shift toward the eighth octant of the three-dimensional vulnerability space framework in Fig. 2.4, indicating increased vulnerability and reduced resilience of pastoralism in the South American Andes in all three dimensions: agroecosystems, livelihoods, and institutions.

2.5.5 Cases from North America

2.5.5.1 American Great Plains: Deterioration of Pastoral Systems with Agricultural Expansion and Climate Warming

The Great Plains are the broad expanse of prairie and steppe which lie west of the Mississippi River and wast of the Rocky Mountains in the USA and Canada (Fig. 2.12). It has an area of approximately 1,300,000 km^2, spanning about 800 km from east to west and 3200 km from north to south. Much of the region was home to American bison (*Bison bison*) herds until they were hunted to near extinction by the European American population during the mid to late 1800s. As the bison population declined, much of the Great Plains became open range, hosting pastoralism/ranching operations where anyone was theoretically free to run cattle. Such pastoralism/ranching began in Texas and gradually moved northward, as cowboys drove Texan cattle north to railroad lines in Dodge City in Kansas and Ogallala in Nebraska, where they were shipped eastward. In the the Noirth America, the 1862 Homestead Act and the 1872 Dominion Lands Act were implemented to promote human settlement and agricultural development to secure the demands by increasing populations in the Great Plains. However, inappropriate cultivation associated with agricultural expansion together with extended drought and the financial crisis of the Great Depression in the late 1920s and early 1930s resulted in the environmental disaster known as the Dust Bowl in the region roughly centered on the Oklahoma Panhandle, including southeastern Colorado, southwestern Kansas, the Texas Panhandle, and extreme northeastern New Mexico, which forced many farmers to leave the land throughout the Great Plains. Since the 1950s, many rangelands in the Great Plains have been gradually converted into productive croplands by extensive irrigation on large landholdings, leading to the movement of pastoralism in the Great Plains to marginalized and degraded conditions (Wood 1998) (Table 2.7).

Fig. 2.12 Location of the Great Plains in North America (From an outline of American Geography)

Table 2.7 General information about the Great Plains of North America

Location	The Great Plains lie west of the Mississippi River and east of the Rocky Mountains in North America. The Great Plains makes up more than 15 % of the USA's land area through ten states
Climate	The Great Plains have a wide variety of weather throughout the year, with very cold winters and very hot summers. Wind speeds are often high
Vegetation	The native vegetation in the Great Plains is mainly composed of prairie and steppe
Land use	Much of the Great Plains became open range, hosting pastoralism/ranching operations to run cattle in the late 1800s. Humans have converted much of the prairies for agricultural purposes or ranches since the early 1900s
Animal	Livestock including both grazing and grain-fed-cattle operations dominate the Great Plains; for example, the Great Plains are home to more than 60 % of the US livestock. Some grazing mammals such as bison, elks, and mule deer also exist here
Population	There are about 10 million inhabitants, about 3 % of the US population
Social problems	Agricultural development and climate change are threatening ranching systems on the open ranges

From Wikipedia

In recent years, environmental issues of the rangelands in the Great Plains have been widely addressed with growing public concern about climate change in rangeland areas. In a report from the US Department of Agriculture, Parton et al. (2007) warned that the rangelands in the Great Plains risk increasing atmospheric CO_2. From combined modeling and experiments, Parton et al. (2007) expected that climate warming and CO_2 enhancement in the prairie of the Great Plains would continue for the following 5–10 years. They also estimated that vegetation production would increase with increasing atmospheric CO_2, although the quality of plants for livestock grazing would decrease; that is, decreasing nitrogen content in grasses with increasing CO_2 concentration (Parton et al. 2007). Moreover, Wagner (2007) warned that weakened/deteriorated rangelands would become more vulnerable to exotic plant invasion and spread of epidemic diseases. Some researchers concluded on the basis of a decade of measurements at the US National Science Foundation's Long-Term Ecological Research site in the short-grass steppe in northeastern Colorado that increased minimum temperatures in springtime were correlated with reduced abundance of buffalo grass (*Bouteloua gracilis*) and increased abundance of native and exotic forbs (Alward et al. 1999). Although it has been widely expected among the scientific community that climate warming will enhance the dominance of rangeland vegetation over woody vegetation, the evidence shows that forest vegetation along the northern edge of the North American Great Plains has expanded southward into the areas dominated by native rangeland over the past century (Peltzer and Wilson 2006). The changes in species composition of the rangeland associated with climate change, on the one hand, can promote the supporting services of rangeland ecosystems such as carbon and nitrogen storage/cycling (Liao et al. 2006; Hughes et al. 2006), but on the other hand, can lower the provisioning services of rangeland ecosystems such as the availability of a productive, palatable, drought-resistant grass such as buffalo grass that ranchers have to rely on for livestock production in the region (Parton et al. 2007). As the grazing animals need nitrogen-rich diets to meet their nitrogen requirement and facilitate forage digestion, ranchers have to supplement their ranching livestock's with hay or alfalfa to cope with a decease of the nitrogen contents in the native forages associated with enhanced atmospheric CO_2 concentration (Wagner 2007), which may result in an expansion of crop production across the Great Plains. The compound declines of rangeland quality and quantity (i.e., forage production) associated with climate warming and increased CO_2 concentration may weaken pastoralism in the Great Plains, and an experiment conducted by Liebig et al. (2005) shows that conversion of cropland or reclamation of mineland into pastureland can mitigate greenhouse gas emission through promoting the sequestration of carbon in soil.

From this case, it can be seen that agricultural expansion associated with regional development and the climate warming associated with CO_2 enrichment have driven the pastoral ecosystem (agroecosystem) in the North American Great Plains into a more fragile status. The ranchers' option of supplying the grazing livestock with hay or alfalfa enhanced the local institution's capacity of responding to avert a major crisis. Crop production in the Great Plains provided more livelihood options for local agropastoralists. These can be reflected by a trend toward the second octant of the three-dimensional vulnerability space framework in Fig. 2.4, implying decreased vulnerability and increased resilience of pastoralism in the livelihood and institution dimensions, but increased vulnerability and decreased resilience in the agroecosystem dimension.

2.5.6 Cases from Oceania

2.5.6.1 Australian Drylands: Degradation of Pastoral Systems with Production-Oriented Management

Rangelands are distributed in arid and semiarid areas of Australia and cover about 70 % of the national land territory (Fig. 2.13). Australian rangeland resources were used by indigenous hunter-gatherers about 40,000 years ago (Bowler et al. 2003). As a result of European settlement, the dominant rangeland use has been changed from hunting and gathering into extensive pastoralism, in which livestock of sheep

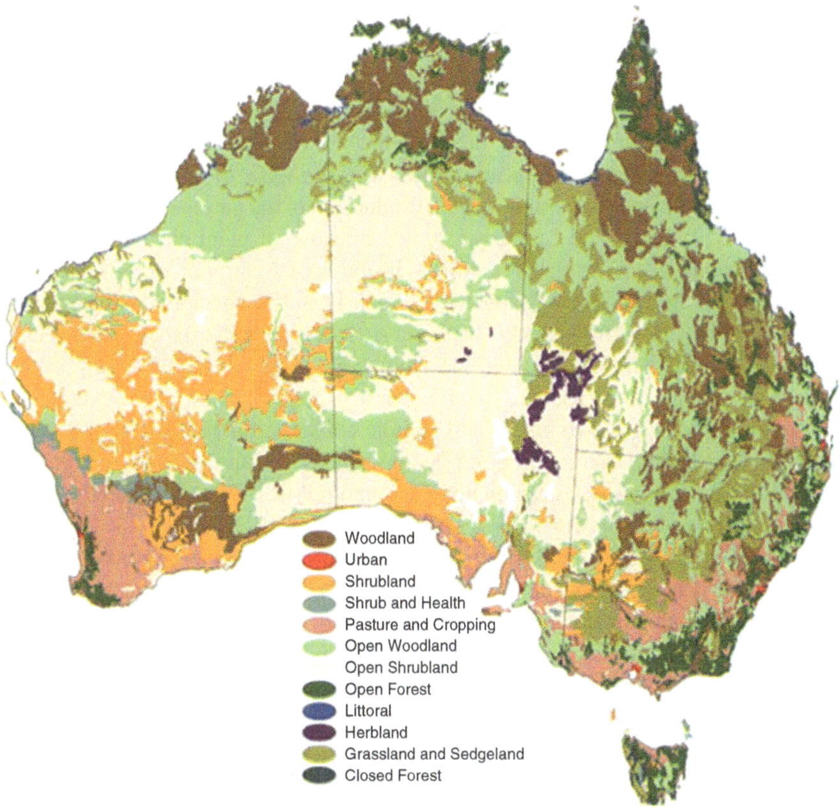

Fig. 2.13 Grassland types in Australia and Queensland

Table 2.8 General information about Queensland in Australia

Location	It is situated in the northeast of the mainland continent of Australia, bordered by the Northern Territory to the west, by New South Wales to the south, and by the Coral Sea and Pacific Ocean to the east. It has a total area of 1,852,642 km^2
Climate	The climate ranges from hot and dry desert in the southwest of the state to subtropical and tropical in the north, where the rainfall is summer dominant
Vegetation	Vegetation types range from semiarid tussock rangelands in the southwest to Mitchell grass downs and a range of woodlands from semiarid to tropical
Land use	The rangelands cover most of Queensland (>70 %). Pastoralism is the major land use, with the beef industry found throughout the rangelands and the sheep industry confined generally to the central western and southwestern areas of the rangelands
Animal	There are mainly grazing livestock such as sheep and cattle, as well as some soft-rooted native animals such as kangaroos and wallabies
Population	The total population is about 4.4 million (in 1999)
Social problems	Agricultural development and climate change are threatening ranching systems on the open ranges

and cattle move as they choose over the entire property (Earl and Jones 1996). In the past 200 years, the development of the pastoral system that the cattle and sheep owned by private ranchers graze native vegetation has displaced the indigenous grazing system in Australia (McAllister et al. 2006). Under the new grazing system developed by European migrators, the ground cover of vegetation have been totally removed by the livestock (sheep and cattle) with high grazing pressure, leading to severe soil erosion and vulnerable animal production in the nation (Allen Consulting Group 2001; Richards and Lawrence 2009). Because of plant clearance and poor livestock management, around 5.7 million hectares of Australian dryland has been affected by salinity, and this is expected to reach 17 million hectares within the next 50 years (Allen Consulting Group 2001). As such, a financially expensive and environmentally unsustainable situation has characterized much of Australia's pastoral industry for some time (Richards and Lawrence 2009). In the 1980s and 1990s, reassertion of Aboriginal rights to land and the conservation movement pushed the pastoral systems into marginaliztion (Heathcote 1994; Holmes 1994), resulting in conflicts between livestock grazing activities and public efforts to protect the environment (Buxton and Stafford 1996; Dale and Bellamy 1998).

To learn the current situation of pastoral industries in arid and semiarid regions of Australia, it is important to understand present land management practices in terms of the production paradigm. In Queensland (Table 2.8), similarly to other grazing areas in Australia, the productivism model dominates the pastoral systems (Richards and Lawrence 2009). However, this food production mode has been increasingly questioned because of public concerns about food quality and security, as well as environmental consequences (Lang and Heasman 2004). This productivism model, known as the "ideal typical" form, is normally characterized by production intensification and concentration along with product specialization (Argent 2002; Ilbery and Bowler 1998). This model has tried to stress food quantity over quality as a wider system of food production based on the assumption that consumers will be advantaged by the maximization of food production (Lang and Heasman 2004), resulting in unstable land use systems; for example, transforming perennial grass landscapes into the "breadbaskets of the world" (Gray and Lawrence 2001; Friedmann 2005). For Australia's ranchers, it is not easy to adopt a new production regime/productivism model for several reasons: poor investment in infrastructure and experimenting with "new" practices in difficult economic times (Lawrence et al. 2004); peer criticism of the ranchers who change their practices (Conacher and Conacher 1995; Guerin and Guerin 1994; Richards et al. 2005); and decreased ratio of prices paid for their livestock to the cost of inputs (Malcolm et al. 1996). Present trade systems, as stated by Richards and Lawrence (2009), have resulted in a decline in profitability at the property level, and consequently have restricted the ability of landholders to change to practices that are environmentally sustainable. This production mode is currently challenging the sustainable development of pastoralism in Queensland and even in the whole of Australia.

From this case it can be seen that the development of the productivism model will reduce the vulnerability of pastoralism and increase the resilience of the pastoralism in Queensland in Australia in the agroecosystem dimension by improving rangeland and livestock management, and in the livelihood dimension by promoting the outputs of the ranch. However, this will increase the vulnerability of pastoralism and reduce the resilience of pastoralism in the institution dimension as some difficulties have limited the ability of Australia's ranchers to adopt new production regimes. These can be reflected by a shift toward the third octant of the three-dimensional vulnerability space framework in Fig. 2.4.

2.5.7 Cases from the Arctic

2.5.7.1 Arctic Tundra: Decline of Pastoral Systems with Regional Development and Political Transformation

The Arctic is a polar region located at the northernmost part of Earth, covering parts of Canada, Russia, the USA (Alaska), Denmark (Greenland), Norway, Sweden, Finland, and Iceland in landmass. Arctic vegetation is composed of plants such as dwarf shrubs, graminoids, herbs, lichens, and mosses, which all grow relatively close to the ground, forming tundra, a typical biome in the Arctic (Fig. 2.14), where it is believed that arctic and subarctic people have domesticated and herded reindeer since the Bronze Age and the Iron Age. For centuries, reindeer pastoralism has played an important economic role for all the circumpolar peoples, including the Saami, Nenets, Khants, Evenks, Yukaghirs, Chukchi, and Koryaks, who raise reindeer for their meat, hides, and antlers and, to a lesser extent, for milk and transportation using a traditional nomadic herding system. However, the traditional reindeer pastoralism in the Arctic is being challenged by global changes such as regional development, political transformation, and climate change.

Fig. 2.14 Tundra in the Arctic

Globally, there are about five million domestic reindeer, of which about half are raised in the Russian far north, Siberia (Konstantinov 2005). With an area of 10,007,400 km^2, Siberia makes up about 58 % of Russia's territory, where 3,500,000 km^2 of tundra is used for domestic reindeer grazing (Vycius 1999). In Russian Siberia, some small ethnic groups live as reindeer pastoralists and sell the products from their reindeer herds. About one million people are traditionally involved in reindeer pastoralism, but the actual numbers of people who are currently engaged in reindeer pastoralism cannot be easily estimated (Kerven 2006). After the economic collapse in the region in the post-Soviet period, many of the reindeer pastoralists were arranged into ethnic associations and some of them have succeeded in marketing reindeer products of meat and antlers privately and they are currently active in developing new markets (Kerven 2006). Krupnik's (2000) observations show that in many areas across Siberia the pastoral economy of reindeer herding at the end of the first post-Soviet decade went into a sharp decline with the collapse of state support after the nation's transition to a market economy in the 1990s. By the early years of this century, some communities had experienced significant loss of their pastoral livelihoods. Reindeer pastoralism in Kamtchatka and Chukotka had crashed almost completely since 1990 (Stammler 2002). However, the collapse of the planned economy opened the door for private marketing development, which benefited reindeer pastoralism in some areas such as Yamal, Western Siberia, where the number of domestic reindeer was constantly increasing to the largest concentration in Russia according to Stammler's (2002) field study between 2000 and 2001. The reason was that reindeer pastoralism remained the stablest economy at the local scale in the first decade after the collapse of the Soviet Union, so more and more families in Yamal stayed for a life of reindeer grazing on the tundra rather than staying in the villages with material and social problems (Stammler 2002). Moreover, pastoral marketing for selling reindeer products (meat and fresh antlers) was accessible to the pastoralists as private commercial enterprises can compete well with state farms in collecting reindeer products (Stammler 2002). "Such a development is unique in the Post-Soviet Russian North, where reindeer herding in all other regions either collapsed or is experiencing a major crisis" as stated by Stammler (2002). Although the reindeer pastoralists in Yamal succeeded very well in selling their reindeer products to local market competitors soon after the breakdown of the planned economy, processing and selling the reindeer products for international markets at a good price are challenging the sustainability of reindeer pastoralism. This can be seen from Stammler's (2002) report on the Second Congress of Russian Reindeer Herders in 2002 in the summary of a speech made by one of the delegation from Yamal: "If enterprises would pay better prices for meat, they (herders) would slaughter. But now for some cents per kg (actually the highest price seems to be around 20 roubles, which currently equals 0.64 US cents), it is not worth doing it, so they (their herders) rather let their herds grow, although this is bad for the pastures (overgrazing)."

In the European Arctic, reindeer pastoralism is being threatened by regional development. For example, Marin (2006) described the issues of pastoral systems in the Finnmark highlands of northern Norway: "Pastoralism relies on semi-domesticated reindeer (*Ratigifer tatcttidus* a ruminant adapted to the arctic/sub-arctic environment, surviving the long winters by feeding mainly on mat-forming lichens. The semi-nomadic reindeer herders in this area are a part of the Saami minority who spreads over north-central Fennoscandia and part of the Kola Peninsula." This seminomadic reindeer pastoralism with features of flexible resource use patterns and land tenure regimes represents a good response to dynamic circumstances at both the temporal and the spatial scale, but it has been widely considered as an illegitimate or a kind of backward economic activity (Horowitz and Little 1987; Forrest 1998) and the central government launched a policy to confine, control, and settle the nomadic reindeer herders (Adams 2001). The traditional reindeer herding system of seminomadic pastoralism has been gradually replaced by a formalized grazing system reinforced by the government (Marin and Vedeld 2003). Meanwhile, state development strategies have aimed at controlling the population of grazing reindeer through destocking and commercial harvesting and reducing the grazing pressure of reindeer on the pastures through fencing and padlocking (Adams 2001), which resulted in conflicts between governments, who try to regulate the number of reindeer on the basis of an appraisal by biologists of the carrying capacity of pastureland, and the herders, who have hundreds of years of experience of how to regulate the number of reindeer (Bjorklund 1990). The state development strategies have also advocated changes toward sedentarization, formal land tenure, and capitalist production (Adams 2001), resulting in the privileging of "modern" forms of land use at the expense of traditional Saami's reindeer herding activities (Forrest 1998). The gaps between the policy prescriptions and the pastoral management strategies have resulted in disruption of local norms and rules of managing the resources and destitution of the communities in the Finnmark highlands of northern Norway (Marin 2006). This situation is, as reported by (Marin 2006), "not only threatening to the welfare of pastoral communities as a whole, but also to the environment where these processes must take place, making the sustainability goal seem both illusionary and hypocritical."

These cases indicate that the regional development and political transformation made pastoralism in the Arctic more challenging by disrupting local norms and rules of managing the resources and facilitating the degradation of pasture resources by either overgrazing or destocking, but they promoted the incomes of the pastoral communities through accessible marketing or subsidies. Therefore, the pastoral systems in the Arctic will shift toward the fourth octant of the three-dimensional vulnerability space framework in Fig. 2.4, indicating an increase of pastoralism's vulnerability and a decrease of pastoralism's resilience in both the agroecosystem dimension and the institution dimension, and a decrease of pastoralism's vulnerability and an increase of pastoralism's resilience in the livelihood dimension.

2.6 Causes and Effects of Vulnerability/Resilience of Pastoralism Worldwide

Although it has been reflected from the three-dimensional vulnerability space model that pastoralism's vulnerability and resilience are very different in the different corners of world, the results from all the cases reviewed here promote the public understanding that global changes have posed great stresses and will put growing pressures on pastoralism worldwide. Climate change and variability in pastoral areas are increasing, croplands and forests are increasingly encroaching onto rangelands, seasonal movements of grazing herds are declining because of settlement, and pastoral economies are embedded into nation states that are undergoing tremendous changes. These stresses do not necessarily individually and separately affect pastoralism, and the responses do not do so in isolation, but rather they are the results of multiple people acting in response to multiple stresses (Nelson et al. 2007). By comparing all of these case studies, we can find three types of causes and effects of vulnerability/resilience of pastoralism worldwide in the agroecosystem and institution dimensions: (1) climate change and climate variability have driven fragile pastoral agroecosystems into more vulnerable conditions—this can be found mainly from the case of the Great Plains of North America; (2) socioeconomic factors, such as land tenure change, agriculture policy reform, and human and livestock population growth, have disrupted the pastoral institutions at local and national levels into marginalized ones—this can be found from the cases in Central Asia, the South American Andes, the European Alps and highlands, Queensland in Australia, and the Arctic; (3) combined natural and human factors have driven pastoral agroecosystems and institutions into more vulnerable situations—this can be seen from the cases in the African Sahel and the Asian highlands.

In addition to degradation of pastoral agroecosystems and the destruction of pastoral institutions, pastoral livelihood on a global scale has been negatively influenced by global changes, such as the destruction of pastoral communities and great famine resulting from coupled climate invariability and agricultural expansion in African Sahel drylands, increasing conflicts over access to and use of rangelands among pastoralists associated with political changes and social transformations in the Central Asian steppe, decline or loss of grazing pastures for pastoral communities because of population growth and climate change in the Asian highlands, disruption of local norms and rules for managing rangeland resources and the destitution of pastoral communities associated with regional development in the European highlands, increased inequity and conflicts among pastoral communities resulting from land tenure changes in the South American Andes, loss of palatable forages and cultivation of fodder crops for grazing livestock production associated with climate change in the Great Plains of North America, and poor adaptation of pastoralists to anew production model and reduced benefits from pastoral production because of policy change in central Queensland in Australia. These are different vulnerability or resilience states of pastoral livelihood worldwide to cope with the pressures, stresses and changes.

The cause–effect analysis for pastoral systems at the global scale indicates that the vulnerability/resilience of pastoralism in any one of the institution, livelihood, and agroecosystem dimensions can influence that in other dimensions, and thus may easily result in instability of human–natural (social–ecological) systems of pastoralism. It is important to respond and address the causes and impacts of pastoralism's vulnerability/resilience through either alteration of the driving forces or sources and prevention/minimization of the environmental flows causing harm, or the lowering of the effects after they occur. In other words, it is critical to employ response in the PSR framework to mitigate the negative causes and effects of pastoralism's vulnerability and enhance the positive causes and effects of pastoralism's resilience. In such a way, the stability of human-natural systems (social–ecological systems) of pastoralism worldwide can be maintained.

2.7 Enhancing Resilience of Human-Natural Systems of Pastoralism Worldwide

Resilience, as described in the early sections in this chapter, can be addressed from three perspectives: persistence, adaptability, and transformability. Therefore, the resilience of a system can be enhanced through the promotion of persistence, adaptability, and transformability. Presently, adaptation and transformation are being widely highlighted for building resilience of social–ecological systems. As stated by Folke (2006), "social-ecological resilience involves transformation, encompassing the capacity for learning, innovation, renewal, re-organization and attainment of a state that is sustainable in the current (social, political, biophysical) environment." It is thus clear that social–ecological learning, technical and management innovations, social–ecological system renewal, and reorganization of institutions are pathways to mitigate the negative causes and effects of pastoralism's vulnerability and enhance the positive causes and effects of pastoralism's resilience. Adaptation is another important strategy to enhance the resilience of a system as stressed by Brooks (2003): "We may view reductions in social vulnerability as arising from the

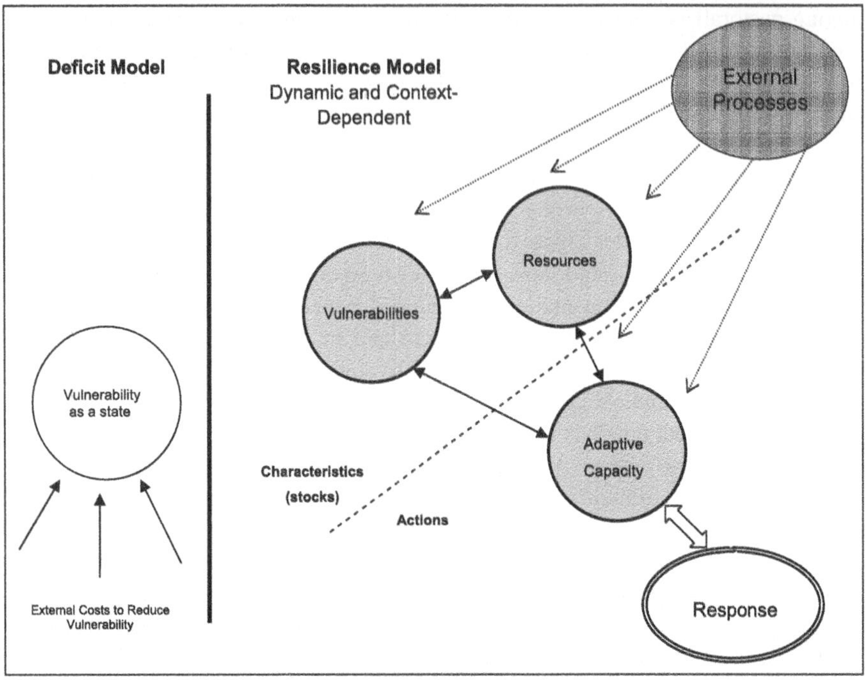

Fig. 2.15 Conceptual framework of resilience, vulnerability, and adaptive capacity. (From Folke et al. 2010)

realization of adaptive capacity as adaptation." Adaptive capacity is a concept closely associated with both resilience and vulnerability and can be converted into the concept of adaptation (Folke et al. 2010). According to Brooks (2003), adaptation can be understood as "the adjustments in a system's behavior and characteristics that enhance its ability to cope with external stresses." Adaptation includes actions taken to reduce vulnerabilities and to increase resilience (Smit and Wandel 2006), and adaptive capacity is the ability to take those actions (Folke et al. 2010). Both adaptation and adaptive capacity are related to reducing vulnerability and enhancing resilience (Fig. 2.15), and the relevant actions need to explore the factors and processes which allow adaptive capacity to be translated into adaptation (Brooks 2003). As suggested by Folke et al. (2010), "a crucial component of the ability to translate adaptive capacity into actual adaptation is the presence of redundancy in the system…; it is through this flexibility and redundancy that a community can translate its resources and adaptive capacity into adaptation and thereby demonstrate resilience." Therefore, it is important to find a good way to translate adaptive capacity of pastoral systems into actual adaptation and thus to reduce their vulnerability and enhance their resilience, as pastoralism is most often an adaptation to semiarid open country or high-altitude dryland throughout the world.

Finding a way to reduce the vulnerability of pastoral systems and to enhance the resilience of pastoralism in the era of global changes requires thorough research to understand the responses of pastoralism to all drivers, and to develop an integrated transdisciplinary framework for sustaining pastoralism. To achieve this goal, different actors, including professionals, practitioners, and policymakers, need to work together to find feasible options to enhance the resilience of pastoralism; that is, the capacity of pastoral systems to absorb shocks, stresses, and disturbance and to maintain the fundamental functions and basic structures (Walker and Salt 2006). The CHANS approach forwarded by Liu et al. (2007) can be applied to facilitate effective collaborations among social scientists, natural scientists, practitioners, managers, users, and policymakers to protect and sustain pastoralism worldwide. As stated by Dong et al. (2010), "untangling the complexities of CHANS, such as reciprocal effects, the influence of differing scales of biological and social organization, and emergent properties, could lead to novel scientific discoveries that are essential for the development of effective policies for ecological and socioeconomic sustainability of pastoralism." Interdisciplinary approaches advocated by CHANS are applicable to offer important insights into complex systems of pastoralism that are hard to understood well or effectively manage within a single dimension. The integrated CHANS approaches are also applicable to address the existing problems in pastoral ecosystems and pastoralist's livelihoods that cannot be addressed solely through technical innovations, political reformations, or economic development (Yang and Dong 2010). The resilience enhancement strategies, adaptation and transformation from resilience theory, with a "ball-and-cup" analogy and a "bouncing back" metaphor can be implemented through integrated CHANS approaches as a way of "fostering communication across disciplines and between science and practice" (Vetter 2009, p. 32). This can be proved by the profound case studies on enhancing sustainable pastoralism across the world in the following chapters.

Acknowledgments The authors express their great thanks to the funding organizations including Ministry of Science and Technology of the People's Republic of China, National Natural Science Foundation of China, and Asian Scholar Foundation, India China Institute of New School, Beijing Normal University for their financial supports. The authors are most grateful to the scholars whose articles, tables and figures are cited in this chapter.

References

Abildtrup J, Audsley E, Fekete-Farkas M, Guipponi C, Gylling M, Rosato P, Rounsevell M (2006) Socioeconomic scenario development for the assessment of climate change impacts on agricultural land use: a pairwise comparison approach. Environ Sci Policy 9:101–115

Adams W (2001) Green development environmental and sustainability in the third world. Routledge, London

Adger WN (2000) Social and ecological resilience: are they related? Prog Hum Geogr 24(3):347–364

Ahmed MM, Sanders JH, Nell WT (2000) New sorghum and millet cultivar introduction in sub-Saharan Africa: impacts and research agenda. Agric Syst 64:55–65

Allen Consulting Group (2001) Repairing the country: leveraging private investment. Business Leaders Roundtable, Sydney

Alward RD, Detling JK, Milchunas DG (1999) Grassland vegetation changes and nocturnal global warming. Science 283(5399):229–231

Anderies MA, Janssen MA, Ostrom E (2004) A framework to analyze the robustness of social-ecological systems from an institutional perspective. Ecol Soc 9(1):18, http://www.ecologyandsociety.org/vol9/iss1/art18

Anthelme F, Grossi JL, Brun JJ, Didier L (2001) Consequences of green alder expansion on vegetation changes and arthropod communities removal in the northern French Alps. For Ecol Manage 145:57–65

Argent N (2002) From pillar to post? In search of the post-productivist countryside in Australia. Aust Geogr 33:97–114

Azarya V (1996) Nomads and the state in Africa: the political roots of marginality. Avebury, Aldershot

Bai WQ, Zhang YL, Xie GD, Shi ZX (2002) Analysis of formation causes of grassland degradation in Maduo county, in the source of the Yellow River. Chin J Appl Ecol 13:823–826

Baker BB, Moseley RK (2007) Advancing treeline and retreating glaciers: implications for conservation in Yunnan, P.R. China. Arct Antarct Alp Res 39(2):200–209

Banks T, Richards C, Li P, Yan ZL (2003) Community-based grassland management in western China: rationale, pilot project experience and policy implications. Mt Res Dev 28:132–140

Bassett T, Turner M (2007) Sudden shift or migratory drift? FulBe herd movements in the Sudano-Guinean savannas of West Africa. Hum Ecol 35(1):33–49

Bassett TJ, Zueli KB (2000) Environmental discourses and the Ivorian Savanna. Annals of the Association of American Geographers 90: 67–95

Baxter PTW (1993) The "new" East African pastoralist: an overview. In: Markahis J (ed) Conflict and the decline of pastoralism in the horn of Africa. Macmillan, London, pp 143–163

Beaufoy G, Baldock D, Clark J (1994) The nature of farming. Joint Nature Conservation Committee, Peterborough

Bergmann C, Gerwin M, Nusser M, Sax WS (2012) State policy and local performance: pasture use and pastoral practices in the Kumaon Himalaya. In: Kreutzmann H (ed) Pastoral practices in High Asia: agency of 'development' effected by modernisation, resettlement and transformation. Springer, New York, pp 175–194

Biber JP (2006) Review of the literature on pastoral economics and marketing: Europe. Report prepared for the World Initiative for Sustainable Pastoralism, IUCN

Bjorklund I (1990) Sami Reindeer Pastoralism as an Indigenous Resource Management System in Northern Norway http://www.sristi.org/cpr/cpr_detail.php3?Mode=Institutions&page=44&limit=1

Bowler JM, Johnston H, Olley JM, Prescott JR, Roberts RG, Shawcross W, Spooner NA (2003) New ages for human occupation and climatic change at Lake Mungo, Australia. Nature 421(6925):837–840

Brooks N (2003) Vulnerability, risk and adaptation: a conceptual framework. Tyndall Centre working paper no 38. University of East Anglia

Brooks N (2006) Climate change, drought and pastoralism in the Sahel. Discussion note for the World Initiative on Sustainable Pastoralism. http://community.eldis.org/.5994ce60/WISP_climate_change_en.doc

Brownman DL (1983) Andean arid land pastoralism and development. Mt Res Dev 3:3241–3252

Brownman DL (1987) Pastoralism in highland Peru and Bolivia. In: Browman DL (ed) Arid land use strategies and risk management in the Andes. A regional anthropological perspective. Westview, Boulder, pp 121–149

Brugger E, Furrer G, Messerli P (1984) The transformation of Swiss mountain regions. Haupt, Bern

Buxton R, Stafford SM (1996) Managing drought in Australia's rangelands: four weddings and a funeral. Rangelands J 18:292–308

Carpenter SR, Walker BH, Anderies JM, Abel N (2001) From metaphor to measurement: resilience of what to what? Ecosystems 4:765–781

Carter TR, Hulme M, Crossley JF, Malyshev S, New MG, Schlesinger ME, Tuomenvirta H (2000) Climate change in the 21st century—interim characterizations based on the new IPCC emissions scenarios. The Finnish environment 433. Finnish Environment Institute, Helsinki, p 148

Cernusca A, Tappeiner U, Bayfield N (1999) Land-use changes in European mountain ecosystems. Blackwell, Berlin

Christensen JH, Hewitson B, Busuioc A, Chen A, Gao X, Held I, Jones R, Kolli RK, Kwon WT, Laprise R, Magana Rueda V, Mearns L, Menendez CG, Raisanen J, Rinke A, Sarr A, Whetton P (2007) Regional climate projections, in climate change 2007: the physical science basis. Contribution of working group I to the fourth assessment report of the Intergovernmental Panel on Climate Change. Cambridge University Press, Cambridge

Chuluun T, Ojima D (2002) Land use change and carbon cycle in arid and semi-arid lands of East and Central Asia. Sci Sin Vitae 45:48–54

Cifdaloz O, Regmi A, Anderies JM, Rodriguez AA (2010) Robustness, vulnerability, and adaptive capacity in small-scale social–ecological systems: the Pumpa irrigation system in Nepal. Ecol Soc 15(3):39, http://www.ecologyandsociety.org/vol15/iss3/art39/

Conacher A, Conacher J (1995) Rural land degradation in Australia. Oxford University Press, Melbourne

Costanza R, Low BS, Ostrom E, Wilson J (2001) Institutions, ecosystems, and sustainability. Lewis, New York

Dale A, Bellamy JA (1998) Regional resource use planning in the Rangelands: an Australian review. Land and Water Resources Research and Development Corporation, Canberra

Davidson G, Behnke RH, Kerven C (2008) Implications of rangeland enclosure policy on the Tibetan plateau. UPDATE magazine no 2, October 2008. International Human Dimensions Programme on Global Environmental Change, pp 59–62

de Bruijn ME, van Dijk HJWM (1999) Insecurity and pastoral development in the Sahel. Dev Change 30:115–139

Didier L (2001) Invasion patterns of European larch and Swiss stone pine in subalpine pastures in the French Alps. For Ecol Manage 145:67–77

Dirnböck T, Dullinger S, Grabherr G (2003) A regional impact assessment of climate and land-use change on alpine vegetation. J Biogeogr 30:401–407

Dong SK, Gao HW, Xu GC, Hou XY, Long RJ, Kang MY, Lassoie PJ (2007) Farmer and professional attitudes to the large-scale ban on livestock grazing of grasslands in China. Environmental Conservation 34 (3): 246–254

Dong SK, Wen L, Zhu L, Li XY (2010) Implication of coupled natural and human systems in sustainable rangeland ecosystem management in HKH region. Front Earth Sci China 4:42–50

Dong SK, Lassoie J, Wen L, Zhu L, Li XY, Li JP, Li YY (2012) Degradation of rangeland ecosystems in the developing world: tragedy of breaking coupled human-natural systems. International Journal of Sustainable Society 4(4): 357–371

Dougill AJ, Fraser EDG, Reed MS (2010) Anticipating vulnerability to climate change in dryland pastoral systems: using dynamic systems models for the Kalahari. Ecology and Society 15(2): 17. [online] URL: http://www.ecologyandsociety.org/vol15/iss2/art17/

Du MY, Kawashima S, Yonemura S, Zhang XZ, Chen SB (2004) Mutual influence between human activities and climate change in the Tibetan Plateau during recent years. Global Planet Change 41:241–249

Du FC (2012) Ecological Resettlement of Tibetan herders in the Sanjiangyuan: A case study in Madoi County of Qinghai. Nomandics 16(1): 116–133

Earl J, Jones C (1996) The need for a new approach to grazing management—is cell grazing the answer? Rangelands J 18:327–350

Millennium Ecosystem Assessment (2005) Ecosystems and human well being, vol I: current state and trends. Findings of the Condition and Trends Working Group. p 899

Faizi (1999) Indigenous resource management in Chitral. IUCN-CCS study. IUCN Pakistan, Chitral

Forrest S (1998) Territoriality and Sami-State Relations. Arctic Circle. Ed. Norman Chance. http://arcticcircle.uconn.edu/ArcticCircle/HistoryCulture/Sami/samisf.html

FAO (1997) FAO production yearbook and Agrostat, 1996, vol 50. UN Food and Agriculture Organization, Rome

Farrington J (2005) De-development in eastern Kyrgyzstan and persistence of semi-nomadic livestock herding. Nomadic Peoples 9(1–2):171–197

Fenton M, Kelly G, Vella K, Innes J (2007) Climate change and the Great Barrier Reef: industries and communities. In: Johnson JE, Marshall PA (eds) Climate change and the Great Barrier Reef: a vulnerability assessment. Great Barrier Reef Marine Park Authority and Australian Greenhouse Office, Australia

Fernandez-Gimenez ME (2001) The effects of livestock privatization on pastoral land use and land tenure in post-socialist Mongolia. Nomadic Peoples 5(2):49–66

Fernandez-Gimenez ME, Batbuyan B (2004) Law and disorder: local implementation of Mongolia's Land Law. Dev Change 35:141–161

Ferreira J, Roca M, Ventura E (1999) Prospects for sustainable development in mountain areas in Portugal: conceptual and policy-related issues. In: Bowler I, Bryant C, Firmino A (eds) Progress in research on sustainable rural systems. International Geographical Commision on "Sustainability of Rural System" Série Estudos Number 2. Centro de Estudos de Geografia e Planeamento Regional, Universidade de Lisboa, Lisbon, pp 163–173

Fischer AM (2008) "Population invasion" versus urban exclusion in the Tibetan areas of western China. Popul Dev Rev 34:631–662

Folke C (2006) Resilience: the emergence of a perspective for social–ecological systems analyses. Glob Environ Chang 16:253–267

Folke C, Carpenter S, Elmqvist T, Gunderson L, Holling CS, Walker B (2002) Resilience and sustainable development: building adaptive capacity in a world of transformations. Ambio 31(5):437–440

Folke C, Carpenter SR, Walker B, Scheffer M, Chapin T, Rockström J (2010) Resilience thinking: integrating resilience, adaptability and transformability. Ecol Soc 15(4):20, http://www.ecologyandsociety.org/vol15/iss4/art20/

Forman RTT (1995) Land mosaics: the ecology of landscapes and regions. Cambridge University Press, Cambridge

Fowler HJ, Archer DR (2006) Conflicting signals of climatic change in the Upper Indus Basin. J Climate 19:4276–4293

Fraser EDG (2007) Travelling in antique lands: studying past famines to understand present vulnerabilities to climate change. Clim Change 83:495–514

Friedmann H (2005) From colonialism to green capitalism: social movements and emergence of food regimes. In: Buttel F, McMichael P (eds) New directions in the sociology of global development. Elsevier, Amsterdam, pp 227–264

Galvin KA (2009) Transitions: pastoralists living with change. Ann Rev Anthropol 38:185–198

Glantz M (1996) Drought follows the plough: cultivating marginal areas. In: Ribot JC, Magalhaes AR, Frontmatter SSP (eds) Climate variability, climate change and social vulnerability in the semi-arid tropics. Cambridge University Press, Cambridge, pp 125–128

Gray I, Lawrence G (2001) Neoliberalism, individualism and prospects for regional renewal. Rural Soc 11:283–298

Grouzis M (1988) Structure, productivity et dynamique des systemes 'ecologiques sah'eliens (mare d'Oursi, Burkina Faso).' ORSTOM, Paris, p 336

Guerin L, Guerin T (1994) Constraints to the adoptions of innovations in agricultural research and environmental management. Aust J Exp Agric 34:549–571

Gunderson LH (2000) Ecological resilience: in theory and application. Annu Rev Ecol Syst 31:425–439

Gunderson LH, Holling CS (eds) (2002) Understanding transformations in human and natural systems. Island, Washington

Harris RB (2010) Rangeland degradation on the Qinghai-Tibetan plateau: a review of the evidence of its magnitude and causes. J Arid Environ 74:1–12

Hardin G (1968) The tragedy of the commons. Science 162: 1243–1248

Heathcote RLH (1994) Manifest destiny, mirage and Mabo: contemporary images of the rangelands. Rangeland J 16(2):155–166

Herreria E, Byron I, Kancans R, Stenekes N (2006) Assessing dependence on water for agriculture and social resilience. Bureau of Rural Sciences, Canberra

Holling CS (1973) Resilience and stability of ecological systems. Annu Rev Ecol Syst 4:1–23

Holling CS (1996) Engineering resilience versus ecological resilience. In: Schulze P (ed) Engineering within ecological constraints. National Academy Press, Washington, pp 31–44

Holmes JH (1994) Changing rangelands resource values: implications for land, tenure and rural settlement. In: Outlook of natural resources, Australian Bureau of Agricultural and Resource Economics, Canberra, vol 94, no 2, p 160–176

Horowitz MM, Little PD (1987) African pastoralism and poverty: some implications for drought and famine. In: Glantz M (ed) Draught and hunger in Africa: denying famine a future. Cambridge University Press, Cambridge, pp 59–82

Hughes RF, Archer SR, Asner GP, Wessman CA, McMurtry C, Melson J, Ansley RJ (2006) Changes in aboveground primary production and carbon and nitrogen pools accompanying woody plant encroachment in a temperate savanna. Glob Chang Biol 12:1733–1747

Hulme M, Doherty R, Ngara T, New M, Lister D (2001) African climate change: 1900–2100. Climate Res 17:145–168

Ilbery B, Bowler I (1998) From agricultural productivism to post-productivism. In: Ilbery B (ed) The geography of rural change. Longman, Harlow, pp 57–84

IPCC (2007) Climate change 2007: the physical sciences basis. In: Summary for policy makers, IPCC: 21. IPCC, Geneva

Intergovernmental Panel on Climate Change (2001) Climate Change 2001: Synthesis Report - Third Assessment Report of the Intergovernmental Panel on Climate Change

IIED (1999) Land tenure and resource access in West Africa: issues and opportunities for the next twenty "ve years. International Institute for Environment and Development, London, pp. 44

Izquierdo LR, Hanneman RA (2006) Introduction to the formal analysis of social networks using Mathematica. http://faculty.ucr.edu/hanneman/mathematica_networks.pdf. Accessed 9 Jun 2008

Jarvis LS (1993) Overgrazing and range degradation in Africa: is there need and scope for government control of livestock numbers. East Afr Econ Rev 7(1):95–116

Kandji ST, Verchot L, Mackensen J (2006) Climate change and variability in the Sahel region: impacts and adaptation strategies in the agricultural sector. Working paper by UNEP and ICRAF

Kassam KS (2010) Pluralism, resilience, and the ecology of survival: case studies from the Pamir Mountains of Afghanistan. Ecol Soc 15(2):8, http://www.ecologyandsociety.org/vol15/iss2/art8/

Kay C (2002) Agrarian reform and the neoliberal counter-reform in Latin America. In: Chase J (ed) The spaces of neoliberalism: land, place and family in Latin America. Kumarian, Bloomfield, pp 25–52

Kerven C (2006) Review of the literature on pastoral economics and marketing: Central Asian, China, Mongolia and Siberia. Report prepared for the World Initiative for Sustainable Pastoralism, IUCN

Kinzig AP, Ryan P, Etienne M, Allison H, Elmqvist T, Walker B (2006) Resilience and regime shifts: assessing cascading effects. Ecol Soc 11(1):20, http://www.ecologyandsociety.org/vol11/iss1/art20/. Accessed 9 Jun 2008

Klein JA, Harte J, Zhao XQ (2004) Experimental warming causes large and rapid species loss, dampened by simulated grazing, on the Tibetan Plateau. Ecol Lett 7:1170–1179

Klein JA, Harte J, Zhao XQ (2007) Experimental warming, not grazing, decreases rangeland quality on the Tibetan Plateau. Ecol Appl 17:541–557

Konstantinov Y (2005) Reindeer-herders: field-notes from the Kola Peninsular (1994–1995). Department of Cultural Anthropology and Ethnology, Uppsala University

Kreutzmann H (2012) Pastoralism: a way forward or back? In: Kreutzmann H (ed) Pastoral practices in High Asia: agency of 'development' effected by modernisation, resettlement and transformation. Springer, New York, pp 323–336

Kreutzmann H (2013) The tragedy of responsibility in High Asia: modernizing traditional pastoral practices and preserving modernist worldviews. Pastoralism: Research, Policy and Practice 3:7 http://www.pastoralismjournal.com/content/3/1/7

Krupnik I (2000) Reindeer pastoralism in modern Siberia: research and survival during the time of crash. Polar Res 19(1):49–56

Kuznar LA (1991) Mathematical models of pastoral production and herd composition in traditional Andean herds. J Quant Anthropol 3:1–17

Laiolo P, Rolando A (2005) Forest bird diversity and ski-runs: a case of negative edge effect. Anim Conserv 7:9–16

Laiolo P, Dondero F, Ciliento E, Rolando A (2004) Consequences of pastoral abandonment for the structure and diversity of the alpine avifauna. J Appl Ecol 41:294–304

Lang T, Heasman M (2004) Food wars: the global battle for mouths minds and markets. Earthscan, London

Lasanta TM, Laguna M, Laguna SM (2007) Do tourism-based ski resorts contribute to the homogeneous development of the Mediterranean mountains? A case study in the central Spanish Pyrenees. Tour Manage 28:1326–1339

Lawrence G, Richards CA, Herbert-Cheshire L (2004) The environmental enigma: why do producers professing stewardship continue to practice poor natural resource management? J Environ Pol Plann 6:251–270

Lavigne Delville P (2000) Harmonising Formal Law and Customary Land Rights in French-Speaking West Africa. In: Toulmin C, Quan JF (Eds.) Evolving Land Rights, Policy and Tenure in Africa. DFID/IIED/NRI, London, pp. 97–122

Lesorogol CK (2003) Transforming institutions among pastoralists: inequality and land privatization. Am Anthropol 105:3531–3542

Li B (1997) The rangeland degradation in northern China and its preventive strategies. Sci Agric Sin 30(6):1–9

Li XL, Huang BN (1995) The cause of "black soil patch" grassland in Qinghai province and management. Grassl China 51:64–67

Li XL, Yuan QH, Wan LQ, He F (2008) Perspectives on livestock production systems in China. Rangeland J 30:211–220

Li WJ, Huntsinger L (2011) China's grassland contract policy and its impacts on herder ability to benefit in Inner Mongolia: tragic feedbacks. Ecology and Society 16(2): 1. [online] URL: http://www. ecologyandsociety.org/vol16/iss2/art1/

Liao JD, Boutton TW, Jastrow JD (2006) Storage and dynamics of carbon and nitrogen in soil physical fractions following woody plant invasion of grassland. Soil Biol Biochem 38:3184–3196

Lichtenberger E (1994) Die Alpen in Europa. Österreochiske Academie der Wissenshaften. Veröffentlichungen Komm Humanökologie 5:53–86

Liebig MA, Morgan JA, Reeder JD, Ellert BH, Gollany HT, Schuman GE (2005) Greenhouse gas contributions and mitigation potential of agricultural practices in northwestern USA and western Canada. Soil Till Res 83:25–52

Liu JG, Dietz T, Carpenter SR, Alberti M, Folke C, Moran E, Pell AN, Deadman P, Kratz T, Lubchenco J, Ostrom E, Ouyang ZY, Provencher W, Redman CL, Schneider SH, Taylor WW (2007) Complexity of coupled human and natural systems. Science 317(5844):1513–1516

International Livestock Research Institute (2002) Mapping poverty and livestock in the developing world. In: Thornton PK, Kruska RL, Henninger N, Kristjanson PM, Reid RS, Atieno F, Odero AN, Ndegwa T (eds) A report commissioned by the UK Department for International Development, on behalf of the Inter-Agency Group of Donors Supporting Research on Livestock Production and Health in the Developing World, Nairobi, Kenya. http://www.ilri.cgiar.org/INFOSERV/WEBPUB/FULLDOCS/MAPPINGPLDW/media/index.html

Luthar SS, John Wiley & Sons, Inc. (2006) Resilience in development: a synthesis of research across five decades. In: Cicchetti D, Cohen DJ (eds) Developmental psychology, vol 3. Risk, disorder, and adaptation, 2nd edn.

Ma YS, Lang BN, Li QY, Li YF, Li FJ (1999) The present status of the grassland eco-environment at the headwater areas of Qinghai-Tibetan Plateau and resume strategies of degraded grassland. Grassl China 55:59–61

Ma YS, Lang BN, Li QY, Shi JJ, Dong QM (2002) Study on rehabilitating and rebuilding technologies for degenerated alpine meadow in the Yangtze and Yellow River source region. Pratacult Sci 19:1–5

MacDonald D, Crabtree JR, Wiesinger G, Dax T, Stamou N, Fleury P, Gutierrez Lazpita J, Gibon A (2000) Agricultural abandonment in mountain areas of Europe: environmental consequences and policy response. J Environ Manage 59:47–69

Maguire B, Cartwright S (2008) Assessing a community's capacity to manage change: a resilience approach to social assessment. Social Sciences Program from Bureau of Rural Sciences, Commonwealth of Australia

Maguire B, Hagan P (2007) Disasters and communities: understanding social resilience. Aust J Emerg Manage 22:16–20

Malcolm B, Sale O, Egan A (1996) Agriculture in Australia: an introduction. Oxford University Press, Melbourne

Marin AF (2006) Confined and sustainable? A critique of recent pastoral policy for reindeer herding in Finnmark, northern Norway. Nomadic Peoples 10(2):209–232

Marin A, Vedeld P (2003) The political ecology of managing the common reindeer ranges in Finnmark, sub-arctic Norway. Presented at Joining the Northern Commons: lessons for the world, 17–21 August 2003, Anchorage. http://dlc.dlib.indiana.edu/dlc/bitstream/handle/10535/345/Marin%2c_Andrei.pdf?sequence=1. http://dlc.dlib.indiana.edu/archive/00001184/00/Marin,_Andrei.pdf

Marshall F, Hildebrand E (2002) Cattle before crops: the beginnings of food production in Africa. J World Prehist 16:99–143

McAllister RRJ, Abel N, Stokes CJ, Gordon IJ (2006) Australian pastoralists in time and space: the evolution of a complex adaptive system. Ecol Soc 11(2):41–56

McCracken DI, Huband S (2005) Nature conservation value of European mountain farming systems. Glob Change Mt Reg 23(V):573–582

Miller D (1997a) Conservation biodiversity in the HKH rangelands. ICIMOD News- letter 27 (Spring), 8–11

Miller D (1997b) Range management and pastoralism: new perspectives and their implications. ICIMOD Newsletter 27 (Spring), 4–7

Mearns R (2004) Sustaining livelihoods on Mongolia's pastoral commons: insights from a participatory poverty assessment. Dev Change 35(1):107–139

Mortimore MJ (1998) Roots in the African Dust. Sustaining the Sub-Saharan Drylands. Cambridge University Press, Cambridge, UK

Muller F, Bold B (1996) On the necessity of new regulations for pastoral land use in Mongolia. Appl Geogr Dev 48:29–51

Nelson DR, Adger WN, Brown K (2007) Adaptation to environmental change: contributions of a resilience framework. Annu Rev Environ Resour 32:395–419

Norgaard RB (1994) Development betrayed: the end of progress and a coevolutionary revisioning of the future. Routledge, London

Nori M (2007) Mobile livelihoods, patchy resources and shifting rights: approaching pastoral territories. Issues paper of International Land Coalition (ILC), World Initiative for Sustainable Pastoralism (WISP) and Canadian International Development Agency (CIDA), April 2007. ILC, Rome,

Nori M, Davies J (2007) Change of wind or wind of change? E-conference climate change, adaptation and pastoralism. The World Initiative for Sustainable Pastoralism,. http://www.iucn.org/wisp/documents_english/WISP_CCAP_final_en.pdf

Nori M, Switzer J, Crawford A (2005) Herding on the brink: towards a global survey of pastoral communities and conflict. An occasional working paper from the IUCN Commission on Environmental, Economic and Social Policy. www.iisd.org/pdf/2005/security_herding_on_brink.pdf

Nusser M, Holdschlag A, Rahman F (2012) Herding on high grounds: diversity and typology of pastoral systems in the eastern Hindukush (Chitral, northwest Pakistan). In: Kreutzmann H (ed) Pastoral practices in High Asia: agency of 'development' effected by modernisation, resettlement and transformation. Springer, New York, pp 31–52

Nyima T (2003) China case study 3: pastoral systems, change, and the future of grazing lands in Tibet. In: Suttie JM, Reynolds SB (eds) Transhumant grazing systems in temperate Asia. Plant production and protection series 31. Food and Agriculture Organization of the United Nations, Rome, pp 151–187

Ostrom E (2000) Private and common property rights. http://allserv.rug.ac.be/~gdegeest/2000book.pdf

Pain DJ, Pienkowski MW (1997) Farming and birds in Europe. Academic, London

Parody JM, Cuthbert FJ, Decker E (2001) The effect of 50 years of landscape change on species richness and community composition. Glob Ecol Biogeogr 10(3):305–313

Parton WJ, Morgan JA, Wang GM, Gross SD (2007) Projected ecosystem impact of the prairie heating and CO_2 enrichment experiment. New Phytol 174:823–834

Pattie PS (1988) Agriculture sector assessment for Bolivia. Chemomis International Consulting Division, Washington

Pearce DW, Freeman S (1991) Information requirements of policy decision-makers. In: Proceedings of the environmental information forum, Montreal, Canada, 21–24 May, 1991 (State of Environment Reporting, Environment Canada. Minister of Supply and Services Canada, Ottawa), pp 56–101

Peltzer DA, Wilson SD (2006) Hailstorm damage promotes aspen invasion into grassland. Can J Bot 84:1142–1147

Pereira E, Queiroz C, Pereira H, Vicente L (2005) Ecosystem services and human well-being: a participatory study in a mountain community in Portugal. Ecol Soc 10(2):14, http://www.ecologyandsociety.org/vol10/iss2/art14/

Pimm SL (1984) The complexity and stability of ecosystems. Nature 307:321–326

Postigo JC, Young KR, Crews KA (2008) Change and continuity in a pastoralist community in the high Peruvian Andes. Hum Ecol 36:535–551

Ptackova J (2011) Sedentarisation of Tibetan nomads in China: Implementation of the nomadic settlement project in the Tibetan Amdo area; Qinghai and Sichuan Provinces. Pastoralism: Research, Policy and Practice 1: 4

Redman CL (1999) Human impact on ancient environments. University of Arizona Press, Tucson AZ, pp 288

Redman C, Grove MJ, Kuby L (2004) Integrating social science into the Long Term Ecological Research (LTER) network: social dimensions of ecological change and ecological dimensions of social change. Ecosystems 7(2):161–171

Reineri C, Frank E, Toro O (2006) Camélidos sudamericanos domésticos: investigaciones Recientes. Centro de Estudios y Promoción del Desarrollo (DESCO), Lima,

Rescia AJ, Pons A, Lomba I, Esteban C, Dover JW (2008) Reformulating the social–ecological system in a cultural rural mountain landscape in the Picos de Europa region (northern Spain). Landsc Urban Plann 88:23–33

Reyneri A (2001) Integrazione tra attività agricola e ricreativa nelle vallate alpine. Le aree a verde per i centri turistici dell'ambiente alpino. Assessorato Agricoltura e Risorse Naturali, Regione Autonoma Valle d'Aosta, Aosta, Italy. Relazione Eecnica

Richards C, Lawrence G (2009) Adaptation and change in Queensland's rangelands: cell grazing as an emerging ideology of pastoral-ecology. Land Use Policy 26:630–639

Richards C, Lawrence G, Kelly N (2005) Beef production and the environment: is it really 'hard to be green when you are in the red'? Rural Soc 15:192–209

Riley NE (2004) China's population: new trends and challenges. Popul Bull 59:1–36
Robinson RM, Whitton S, Biber-Klemm S, Muzofirshoev N (2010) The impact of land-reform legislation on pasture tenure in Gorno-Badakhshan: from common resource to private property? Mt Res Dev 30(1):4–13
Saberval VK (1999) Pastoral policies: shepherds, bureaucrats and conservation in the western Himalaya. Oxford University Press, Delhi
Scheffer M (2009) Critical transitions in nature and society. Princeton University Press, Princeton
Schoon M (2005) A short historical overview of the concepts of resilience, vulnerability and adaptation. Working paper W05-4
Schröter D, Cramer W, Leemans R, Prentice IC, Araujo MB, Arnell NW, Bondeau A, Bugmann H, Carter TR, Gracia CA, De La Vega-Leinert AC, Erhard M, Ewert F, Glendining M, House JI, Kankaanpaa S, Klein RJT, Lavorel S, Lindner M, Metzger MJ, Meyer J, Mitchell TD, Reginster I, Rounsevell M, Sabate S, Sitch S, Smith B, Smith J, Smith P, Skykes MT, Thonicke K, Thuiller W, Tuck G, Zaehle S, Zierl B (2005) Ecosystem service supply and vulnerability to global change in Europe. Science 310:1333–1337
Shang ZH, Long RJ (2005) Formation reason and recovering problem of the "black soil type" degraded alpine grassland in Qinghai-Tibet Plateau. Chin J Ecol 24:652–656
Shrestha AB, Wake CP, Mayewski PA, Dibb JE (1999) Maximum temperature trends in the Himalaya and its vicinity: an analysis based on temperature records from Nepal for the period 1971–94. J Climate 12:2775–2787
Sharma VP, Kohler- Rollefson I, Morton J (2003) Pastoralism in India: a scoping study. A reoprt from Livestock Production Programme of the United Kingdom Department for International Development
Singh SP, Bassignana-Khadka I, Karky BS, Sharma E (2011) Climate Change in the Hindu Kush-Himalayas: The State of Current Knowledge. Kathmandu: ICIMOD
Smit B, Wandel J (2006) Adaptation, adaptive capacity, and vulnerability. Global Environment Change, 16: 282–292
Stadel C (1995) Development needs and the mobilization of rural resources in highland Bolivia. Yearb Conf Latin Am Geogr 21:37–48
Stammler F (2002) Success at the edge of the land: Present and past challenges for reindeer herders of the West-Siberian Yamal-Nenets autonomous Okrug, Nomadic Peoples, 6:51–71. http://www.eth.mpg.de/dynamic-index.html?http://www.eth.mpg.de/people/stammler/c-project.html
Steffen E, Crutzen PJ, McNeill JR (2007) The Anthropocene: are humans now overwhelming the great forces of nature? Ambio 36:614–621
Suleimenov M, Oram P (2000) Trends in feed, livestock production, and rangelands during the transition period in three Central Asian countries. Food Policy 25(6):681–700
Sulek ER (2012) 'Everybody Likes Houses. Even Birds Are Coming!" In: Kreutzmann H (ed) Pastoral practices in High Asia: agency of 'development' effected by modernisation, resettlement and transformation. Springer, New York, pp 235–257
Swift J (1994) Dynamic ecological systems and the administration of pastoral development. In: Scoones I (ed) Living with uncertainty: new directions in pastoral development in Africa. Intermediate Technology Publications, London, pp 153–173
Swift JJ (2004) The global drylands imperative: pastoralism and mobility in the drylands. UNDP Drylands Development Centre, Nairobi, Kenya
Thébaud B, Batterbury S (2001) Sahel pastoralists: opportunism, struggle, conflict and negotiation. A case study from eastern Niger. Glob Environ Chang 11:69–78
Thurow TL (2008) Sustainability and optimization: issues of perspectives and scale. In: Organization Committee (eds.) Multifunctional grasslands in a changing world, vol I. Proceeding of XXI international grassland congress-VIII international rangeland congress, Guangzhou, China. Guangdong People's Publishing Houses
Turner MD (1999) The role of social networks, indefinite boundaries and political bargaining in maintaining the ecological and economic resilience of the transhumance systems of Sudan-

Sahelian West Africa. In: Niamir-Fuller M (ed) Managing mobility in African rangelands. Intermediate Technology Publications, London, pp 97–123

Upton C (2008) Social capital, collective action and group formation: development trajectories in post-socialist Mongolia. Hum Ecol 36:175–188

Urbanska KM, Erdt S, Fattorini M (1998) Seed rain in natural grassland and adjacent ski run in the Swiss Alps: a preliminary report. Restoration Ecol 6:159–165

Vetter S (2009) Drought, change and resilience in South Africa's arid and semi-arid rangelands. S Afr J Sci 105:29–33

Vycius J (1999) Rangeland resources of the former Soviet Union territory. In: Eldridge D, Freudenberger D (eds) People and rangelands. Proceedings of the VI international rangeland congress, Townsville, Australia, July 19–23, pp 796–797.

Wagner CG (2007) Great Plains grasslands at risk. Futurist 41(5):13

Waheed B, Khan F, Veitch B (2009) Linkage-based frameworks for sustainability assessment: making a case for driving force-pressure-state-exposure-effect-action (DPSEEA) frameworks. Sustainability 1:441–463

Walker B, Salt D (2006) Resilience thinking: sustaining ecosystem and people in a changing world. Island , Washington

Walker BH, Holling CS, Carpenter SR, Kinzig A (2004) Resilience, adaptability and transformability in social–ecological systems. Ecol Soc 9(2):5, http://www.ecologyandsociety.org/vol9/iss2/art5

Walker BH, Abel N, Anderies JM, Ryan P (2009) Resilience, adaptability, and transformability in the Goulburn-Broken Catchment, Australia. Ecol Soc 14(1):12, http://www.ecologyandsociety.org/vol14/iss1/art12/

Wang GX, Chen GD (2001) Characteristics of grassland and ecological changes of vegetations in the Source Regions of Yangtze and Yellow Rivers. J Desert Res 21:101–107

Wang GX, Wang YB, Qian J, Wu QB (2006) Land cover change and its impacts on soil C and N in two watersheds in the center of the Qinghai-Tibetan Plateau. Mt Res Dev 26:153–162

Warren A (2005) The policy implications of Sahelian change. J Arid Environ 63:660–670

Watkinson AR, Ormerod SJ (2001) Grasslands, grazing and biodiversity: editors' introduction. J Appl Ecol 38:233–237

Werger MJA, van Staalduinen MA (2012) Eurasian steppes: ecological problems and livelihoods in a changing world. Springer, New York

Westreicher CA, Mérega JL, Palmili G (2007) The economics of pastoralism: study on current practices in South America. Nomadic Peoples 11(2):87–105

Wood WR (1998) Archaeology on the Great Plains. University Press of Kansas, Lawrence

Wu N, Yan ZL (2002) Climate variability and social vulnerability on the Tibetan Plateau: dilemmas on the road to pastoral reform. Erdkunde 56:2–14

Wyder J (2001) Multifunctionality in the Alps. Mt Res Dev 21:327–330

Xu WX, Liu XD (2007) Response of vegetation in the Qinghai-Tibet Plateau to global warming. Chin Geogr Sci 17:151–159

Yan ZL, Wu N, Yeshi D, Ru J (2005) A review of rangeland privatization and its implications in the Tibetan Plateau, China. Nomadic Peoples 9(1):31–51

Yang ZF, Dong SK (2010) Understanding coupled human and natural systems in a changing world. Front Earth Sci China 4(1):1–2

Yao TD, Wang YQ, Liu SY, Pu JC, Shen YP (2004) Recent glacial retreat in high Asia in China and its impact on water resources in northwest China. Sci China D Earth Sci 47(12):1065–1075

Yeh ET (2003) Tibetan range wars: spatial politics and authority on the grasslands of Amdo. Dev Change 34:499–523

Yi SL, Ismail M, Yan ZL (2012) Pastoral communities' perspectives on climate change and their adaptation strategies in the Hidukush-Karakoram-Himalaya. In: Kreutzmann H (ed) Pastoral practices in High Asia: agency of 'development' effected by modernisation, resettlement and transformation. Springer, New York, pp 307–322

Chapter 3
Maintaining the Human–Natural Systems of Pastoralism in the Himalayas of South Asia and China

Shikui Dong, Shaoliang L. Yi, and Zhaoli L. Yan

Abstract This chapter presents an overview of pastoralism in the Himalayas and summarizes the current situation and trends of human–natural systems of pastoralism in the Nepalese Himalaya, in the Indian Himalaya, and on the Qinghai–Tibetan Plateau of China. The human–natural system of pastoralism has lasted in a relatively stable manner for centuries in the Himalayas, especially through flexible responses to the variability of climate conditions in the short term. However, a great number of external and internal driving forces are currently threatening the sustainability of the long-term nature of pastoralism. They complicate interactions and feedbacks between human and natural components of pastoralism in coping with the stresses, and the integration of various tools and strategies from the ecological and social sciences as well as other disciplines in sustainable pastoral development. In the Himalayan region of northern Nepal, local institutions of collective action and indigenous property right systems for pastoral resource management are the key adaptive strategies to overcome the difficulties in pastoral management associated with poor cooperation and collaboration between the government and the pastoral society. In the Indian Himalaya, well-organized local institutions and commonly agreed norms and rules among the pastoral societies have promoted the sustainable use of pastoral resources in coping with pressures and threats of climatic, socioeconomic, and political changes. On the Qinghai–Tibetan Plateau of China, local pastoralists have developed adaptive actions of mobility, specificity, preparedness, diversification, exchange, collaboration, and partnership based on their knowledge and wisdom to cope with rangeland degradation driven by climate change and human overexploitation.

S. Dong (✉)
School of Environment, Beijing Normal University, Beijing 100875, China
e-mail: dongshikui@sina.com

S.L. Yi
Environmental Change and Ecosystem Service, ICIMOD, Kathmandu, Nepal

Z.L. Yan
Chengdu Institute of Biology, Chinese Academy of Sciences, Chengdu 610041, China

S. Dong et al. (eds.), *Building Resilience of Human-Natural Systems of Pastoralism in the Developing World: Interdisciplinary Perspectives*,
DOI 10.1007/978-3-319-30732-9_3

3.1 An Overview of Pastoralism in the Himalayas

The Himalayas range across eight Asian countries, from Afghanistan in the west to Myanmar in the east, and from the Qinghai-Tibetan Plateau of China in the north to the Ganges Basin in the south. The latitude range is some 11°, from about 38°N in Pakistan and Afghanistan to 27°N in Bhutan. This vast distance obviously involves considerable changes in climate and vegetation. More than 60 % of the total area of 4.3 million square kilometers in the Himalaya is covered by rangelands which are primarily or directly used for pastoral livestock grazing. Mobile grazing guided by customary rules and institutions has been practiced in the Himalaya since early human civilization and is considered to be one of the important livelihood activities, especially for people living in high-altitude areas. For centuries, the alpine meadows, the major type of rangelands in the high Himalayas, have been used as grazing grounds by livestock of local communities and the animals from adjacent lower valleys during summer (Farooquee 1994). The grazing zones go further northwest to the Karakoram foothills and the Hindu Kush. Animal husbandry has been an integral part of the subsistence economy in the Himalaya, and livelihood's dependence on livestock increases with increasing elevation (Sundriyal 1995). Although the amount of animal husbandry in the Himalaya is not exactly documented, livestock grazing at the upper levels of mountains is dominated by mobile pastoral communities; for example, in Bhutan, fewer than 14,000 yak herder households with fewer than 50,000 yaks contribute about 3 % to the national products (Derville and Bonnemaire 2010). Pastoralism sustains about 150 million people, who are mostly ethnic minorities with a unique culture and tradition in the region, and impacts three times as many people living in downstream regions of the Himalaya. With rich diversities from both the cultural perspective and the biological perspective, pastoralism in the Himalayas is characterized as a typical human–natural system (social–ecological system). Similarly to the other human–natural systems of pastoralism in the world, the Himalaya's pastoralism is critically important in this region for the human populations it supports, the food and ecological services it provides, the economic contributions it makes to some of the world's poorest regions, and the long-standing civilizations it helps to maintain (Dong et al. 2011).

In the western Himalaya, which includes the Himalayan ranges in northwest India to the west of the Nepalese border, the Himalayan ranges in northern Pakistan, the Hindu Kush in northern Afghanistan, and other mountain ranges where they run down to the plain, transhumant or nomadic grazing systems are widely practiced to locate the best herbage resources from pastures and rangelands (predominantly *Artemisia* steppe). These grazing systems are particularly prevalent in the ethnic herder groups such as the Gujjars, Bakarwals, Gaddis, and Changpas in India, the Gujjars and Bakarwals in Pakistan, and the Kuchis (who are part of the Pushtun majority) in Afghanistan. Throughout the region, these herders adopt almost the same grazing pattern of vertical migration, in which they graze their livestock during winter in warmer zones such as the plains, foothills, and the desert fringe, and move upward as when the weather becomes warm until they reach mountain or alpine pastures in the summer. Nomadic pastoralism is practiced by the unsettled herders, who follow a migratory cycle between high pastures and lowlands throughout the year, and transhumant pastoralism is practiced by the lowland-settled herders, who send their livestock in summer to within reach of high pastures. Lowland overwintering can give herders access to both pastoral markets and opportunities for seasonal employment. The dates of migration have traditionally been fixed on the basis of herders' perceptions of seasonal changes. Although the migratory herders have the grazing rights for most of the rangelands as the traditional way of use and can access the lowland pastures on oral agreement with local residents, they do not normally own the grazing lands. Small livestock such as sheep and goats are normally the basis of the herding systems in this region, although buffalo and cattle are grazed by the migratory Gujjars in Pakistan and India, camels are herded by the Balochistan in Pakistan and Afghanistan, and yaks are grazed by ethnic minority herding groups in some high-elevation areas of Afghanistan, Pakistan, and India (Fig. 3.1). In most cases, selling wool and live animals for meat (a small portion is usually reserved for home consumption on special occasions) is the only source of income for the herders to trade for the daily necessities and food (mainly cereals).

In the eastern Himalaya, pastoralism is very important in rangeland areas of northern Nepal, northern Bhutan, and Sikkim in India, whereas the extreme east is forest rather than rangeland. This region is generally wetter than the western

Fig. 3.1 Mobile grazing goats in Himachal Pradesh, India. (Photo by Shikui Dong, 2012)

Himalaya and there large grazing livestock such as yaks and chauri (cattle and yak hybrids) are much more important than small grazing animals such as goats and sheep. Migratory herding has been well adopted by the ethnic groups, such as the Bhotias and Sherpas in middle Nepal, the Tamangs and Kirats in eastern Nepal, the Bhutias in Lachen and Lachung of Sikkim, India, and the Brokpas in northern Bhutan. Herding systems in the eastern Himalaya are more stratified altitudinally, and herders keep the animal species according to their preferences for the altitude. The migratory herding of yaks and chauri dominates in the alpine–cool–temperate areas at relatively higher altitude, and the mobile grazing of cattle and buffalo is very important in the temperate–subtropical areas at relatively lower elevation (Fig. 3.2). The same grazing pasture may be shared by different livestock species and different herding groups in different seasons of the year; that is, subalpine meadow dominated by sedges may be the summer pastures of the chauri and the winter grazing lands of the yaks, and the temperate rangelands associated with oak or mixed forest of oak and ble pine may be the winter pastures of the chauri and the summer grazing lands of buffalo and cattle. In these mixed herding systems, some herder families remain relatively stationary to engage in agricultural farming as agropastoralists and they entrust their grazing livestock to others for part of the year. These pastoral systems take advantage of the variations in climate, vegetation, and labor. In most cases, selling diary products and live animals for family

Fig. 3.2 Herders moving yaks from subalpine shrublands to alpine meadows in Langtang of Rasuwa District, northern Nepal, in summer. (Photo by Shikui Dong 2007)

income is the major purpose of herding livestock, in addition to home consumption.

In the northern Himalaya, the Qinghai–Tibetan Plateau (QTP) appears as a huge geographical unit where most land territories are covered by rangelands, which have been utilized by ethnic Tibetans in China as grazing pastures for centuries. Because of the diverse climate and landscapes across the vast region of the QTP, the herding systems differ greatly between the east and the west. In the western part of the QTP, where the climate is arid or semiarid and the landscapes are dominated by alpine steppe or alpine desert, the nomadic or transhumant herders generally graze their yaks, sheep, and sometimes goats in a migration cycle between the high-elevation pastures in summer and the low-elevation pastures in winter, which is similar to what the herders do in the western Himalaya. However, the herders have to move their livestock on the plateau throughout year and have no access to low plains or valleys because of geographical barriers. In the central and northern parts of the QTP, especially the headwater areas of three rivers (the Yellow, the Yangtze, and the Mekong), where the climate is semiarid and semihumid and the landscapes are dominated by alpine steppe or alpine meadow and alpine shrub meadow, the herders similarly graze their yaks and sheep in a nomadic or transhumant migration cycle between the summer pastures at high elevation and the winter pastures at low elevation (Fig. 3.3). In the far eastern region of the QTP (e.g., the Hengduan Mountains), where climate is much wetter and

Fig. 3.3 Tibetan nomads transporting goods and living materials with yaks to a camp on the higher summer pasture (above 4000 m) on the Qinghai–Tibetan Plateau in the warm season (Photos by James Lassoie, 2012)

the landscapes are more diverse, the herders use the migratory herding system which is similar to that adopted by the herders in the eastern Himalaya; that is, they graze their yaks and sheep on the pasture of alpine meadow at the highest elevation in summer and on the pasture of subalpine meadow at high elevation in winter; they graze the chauri on the pasture of subalpine meadow at high elevation in summer and on the temperate rangelands associated with oak or mixed forest of oak and ble pine at low elevation in winter. In these mixed herding systems, some herders also practice agricultural farming as agropastoralists in low-altitude valleys near their residential areas for part of the year.

It is clear that mobile pastoralism characterized by the vertical movement of livestock in a cyclic manner is common throughout the Himalayas, where the pastoral communities make full use of rangeland resources in different ways, ranging from mountain nomadism through transhumance to combined mountain agriculture (agropastoralism). Over centuries, the stability of pastoralism across the Himalayas has been maintained through the balance among pastures, livestock, and pastoralists/herders, but the balance is not simple. The pastoral groups have applied diverse strategies such as livestock and human mobility and dispersal to overcome the problem of overgrazing rangelands. Herders' movement in the Himalayas is heavily focused toward obtaining specific production or other functions with featured socioeconomic organization and property rights. However, Himalayan pastoralism is not solely bonded with a certain kind of economic system (i.e., consumption-oriented or market-oriented production). Moreover, it is not tied to a specific type of land tenure: some pastoral communities have control over the land territory within which they migrate for mobile livestock grazing, whereas others have to graze their livestock with a formally or informally contracted migration route on public or private lands, of which they do not have political or legal ownership. Pastoral groups are also very diverse in their political structure, ranging from state-controlled peasants, to community-based collectives, to individualized households. In most cases, the permanent and essential resources such as grazing pastures and drinking water are shared by different pastoral groups, whereas the livestock herds are owned privately by individual pastoralists.

Mobile pastoralism in the Himalayas is not only a subsistence pattern, but also an effective means of exploiting marginal environments. Different types of pastoralism in the Himalayas can be understood as different adaptation strategies, which are closely associated with the geography, ecology, and socioeconomic development levels in the locality. It is only through pastoral mobility that the local people can efficiently use all potential resources in the large ecological zones, especially to convert the low values of plant resources in remote areas into high values of animal products through migratory grazing. Low energy and transportation availability is generally associated with low population density and high mobility of a pastoral society. Pastoralists can increase the reproduction and survival rates of livestock through good investments in animal breeding and health care and can make more profit by investing more human labor in milk and wool production than in meat production. The pastoral production systems rarely focus on a single product, whereas they make full uses of both "continuing" (calves, lambs, and kids; milk, butter, and cheese; transport and traction; manure; hair and wool) and "final" (meat, wool, hides, and skins) products (Bhasin 2011). This is the way that pastoralists can make the most profit from use of resources in marginal areas of the Himalayas where farming is not feasible.

Within a pastoral ecosystem, there is ecological diversity represented by various ecological habitats, flora, and fauna, and cultural diversity reflected by different ethnicities, religions, and customs. These two components are interrelated and interact in the various pastoral production systems across the Himalayas. Culturally, pastoral populations living in the harsh environments of the Himalayas have developed many adaptive resource use strategies to overcome the problems of water and land limitations depending on the sociocultural characteristics of the local population. As a key form of an adaptive natural resource use strategy, pastoralism is a long-lasting livelihood option for many indigenous people living in the Himalayas to balance the relationships among pastures, livestock, and people. The pastoral technologies require that the herders' daily life practices be in line with the requisites of the grazing livestock (i.e., pasture, water, salt, and protection from predators). Pastoralists remain mobile all year round to sustainably use the permanent rangelands, which are essential sources for grazing livestock, so that they cannot make large investments in other assets such as personal goods, houses, and land. The social structures, functional groups, and administrative institutions of pastoral society in the mobile way of life have been developed on the basis of the needs demanded in the migratory production mode.

As a coupled human–natural system, traditional pastoralism has lasted in a relatively stable matter for centuries in the Himalayas, especially through flexible responses to the variability of climate conditions in the short term. However, a great number of external and internal driving forces are currently threatening the sustainability of the long-term nature of pastoralism, and are triggering adaption strategies possibly to transform this system. With the increasing trend of globalization of the marketing system, the pasturelands have been increasingly commercialized and/or turned into national parks, resulting in many problems for the pastoral groups. With the expansion of agriculture and forestry into rangeland areas, the herders have been forced by political marginalization to abandon their traditional and customary rights to these grazing lands. With the advent of socioeconomic reforms and economic development, the pastoral economy and marketing systems have been significantly changed, leading to breakage of traditional trade relationships between highland pastoral communities and lowland agricultural communities. With more avenues for earning, opportunities for waged labor, and the attraction of a better life in urban areas, outmigration is a growing trend in the pastoral areas of the Himalaya. Rangeland degradation associated with climate change and overexploitation are resulting in a decline in pastoral production in the Himalayan ranges. However, current policies and strategies related to pastoral production in the Himalaya have overlooked the integration of social, economic, and environmental factors, which will likely intensify social–ecological problems. On the basis of case studies from the QTP of China, the Nepalese Himalaya, and the Indian Himalaya, we summarize the current situations and trends of human–natural systems of pastoralism in the

Himalayas and new approaches that would promote better management, enhance the security of local environments, and mitigate the regional environmental problems.

3.2 Case Study from the Nepalese Himalaya: Importance of Indigenous Knowledge and Institutions in Sustainable Pastoral Management

3.2.1 Background

Nepal is situated in the southern Himalaya, bordering China in the north, India in the west and south, and Bhutan in the east. Around 12 % of the nation's territory is defined as either rangelands, or pasturelands or "grazing lands" (Land Resource Mapping Project 1986; Rajbhandary and Pradhan 1990; Rai and Thapa 1993; Shrestha 2001), which are mostly located in the hilly and mountainous areas of the Himalaya in northern Nepal (Table 3.1). These rangelands play critically important roles in economic development and human well-being in the Nepalese Himalaya. Pastoralism of buffalo, zebus, yaks, chauri, sheep, and goats based on rangeland grazing is a relatively small but important part of the farming practices among ethnic populations living in the Himalayan areas of northern Nepal (Rai and Thapa 1993). Pastoralists are involved in milk, wool, hair, hide, abd blood production or keeping live animals as transportation tools or work energy sources.

For a long time, rangeland livestock grazing has been the dominant land use by indigenous communities in remote mountains and valleys of northern Nepal (Alirol 1979). Some scholars (Chand et al. 1991; Dong et al. 2007) have stressed that livestock grazing and pastureland management in this region have continuously suc-

Table 3.1 General information about case study sites

Information	Dhunche	Gatlang	Langtang
Location (elevation)	Lowland (1900 m)	Middle land (2200 m)	High mountain (3300 m)
Climatic zone	Subtropical-Temperate zone transition zone	Temperate zone	Subalpine zone
Farming systems	Multiple farming of livestock, crops, fodder, and vegetables	Crop–livestock mixture farming	Livestock farming (tourism)
Total households	164	223	61
Livestock composition in individual households	1–2 cattle, 2–3 buffalo, 4–5 sheep and goats, 10–15 yaks and chauri (only 10 % of households engage in yak farming)	1–2 cattle, 10–20 sheep and goats, 10–15 yaks and chauri (half of households engage in yak farming)	20–30 sheep, 2–3 horses, 10–15 yaks and chauri (80 % of households engage in yak farming)

ceeded in exploring the indigenous knowledge of local people, which is rooted in the geographical, physical, climatic, and ecological characteristic of the Nepalese Himalaya. With all of this indigenous knowledge, local people have converted many constraints into opportunities (Tamang 1993; Dong et al. 2007). Moreover, sustainable utilization of natural resources and improvement of local livelihood are generally highly related to the pastoral institutions that govern the natural resource utilization, particularly property rights and collective action, which shape the local people's natural resource use patterns, which in turn impact the outcomes of the pastoral production systems in the region, as stressed by Dong et al. (2007), who stated that "together, mechanisms of collective action and property rights define the incentives people face for undertaking sustainable and productive management strategies, and they affect the level and distribution of benefits from the use of the natural resources", and Meinzen-Dick and Gregorio (2004), who state that "the close linkages between property rights, collective action, and natural resource management are critically important for technology adoption, economic growth, food security, poverty reduction, and environmental sustainability."

Although indigenous knowledge and local institutions have played very important roles in sustaining the rangeland management and livestock production in pastoral systems of the Nepalese Himalaya over centuries (Dong et al. 2009), traditional resource management practices have been ignored or overlooked by centralized governments in the past, and there is a huge gap between local people's traditional practices and the knowledge of professionals and practitioners (researchers, planners, and policymakers; Tamang 1993). These problems and gaps have resulted in significant resource degradation and social conflicts over natural resource use (Shrestha 2001). Hence, it is essential to understand indigenous practices of natural resource management, local institutions for property rights and policy formulation, and the collective decision-making processes and to build on this knowledge and information to achieve sustainable development of pastoral systems in the Nepalese Himalaya. We conducted and updated this case study to investigate and document the indigenous knowledge, local institutions, and their roles in sustaining pastoral resource management and enhancing the resilience of human–natural systems of pastoralism in the Himalayan region of northern Nepal.

3.2.2 Methods

This case study was conducted in Rasuwa District (Fig. 3.4), a high Himalayan and mountainous district of Nepal, whose name means "grazing land for sheep and cattle." This district represents the pastoral areas of the Nepalese Himalaya very well in the indigenous pastoral production systems, the historical traditions of pastoralism, and the socioeconomic importance of pastoral systems to local livelihoods. This district is located in the northwest of Nepal's Central Development Region (one of four regions in Nepal), bordering the Tibet Autonomous Region of China in north and the Sindhupalchowk, Nuwakot, and Dhading districts of Nepal in the southeast, south, and

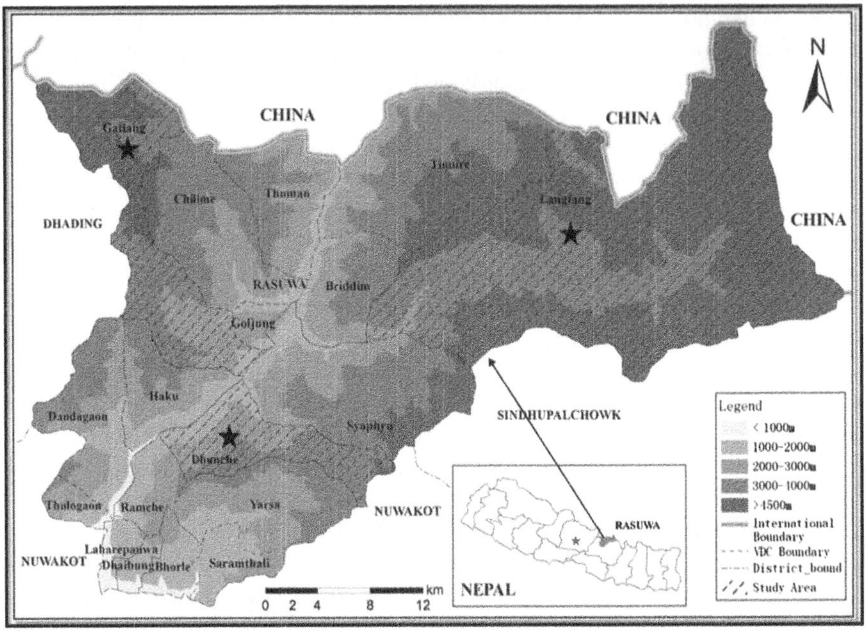

Fig. 3.4 Location of case study sites in Rasuwa District, Nepal

west respectively. The total population of the 18 village development communities (VDCs) in this district is about 44,000 from around 8700 households with a mean size of 5.05 people (Tourism for Rural Poverty Alleviation Program 2005). Most of the population (65 %) in this district belongs to the Tamang ethnic group, with evident Tibetan origin, whose major production system is extensive livestock grazing on native rangelands. Dhunche, Gatlang, and Langtang were selected as the representative VDCs in this district for the field investigation after consideration of the spatial variations of the geographical locations, climate conditions, and farming systems (Fig. 3.4, Table 3.1).

All the data in this case study were collected and updated with use of various data sources, including research publications, reports, newsletters, and a field investigation between 2006 and 2007. Integrated approaches including participatory rural appraisal, open-ended questions and pretested questionnaires (10, 14, and 11 households in Dhunche, Gatlang and Langtang respectively), key-person interviews (6, 6, and 8 individuals in Dhunche, Gatlang and Langtang respectively), and group discussion (21, 14, and 12 participants in Dhunche, Gatlang and Langtang respectively) were used in the investigation. Information and knowledge about indigenous pastoral practices, traditional herding management strategies, pastoral land tenure and resource property systems, and pastoral institution and governance arrangement were gathered in the investigation. Supplementary information about problems, constraints, challenges, opportunities, and changes in pastoral management systems, external public support, and partnerships were collected and recorded from both primary sources through group discussion and personal communications and secondary sources through desk study and literature review. The data quality was controlled by careful investigation and cross-checks with different sources. Systematic qualitative techniques recommended by Patton (1990) and Miles and Huberman (1994) were used to analyze all the data.

3.2.3 Results

3.2.3.1 Indigenous Practices

The information collected from three case study sites shows that local pastoralists have been continuously applying a vertical transhumant gazing system, a recurrent feature of indigenous grazing management systems across the Nepalese Himalaya. This grazing system is characterized by the moving of livestock toward high alpine pastures in the monsoon season and to lower pastures or forests during the winter (Fig. 3.5) so as to make good use of climate conditions and feed availability between different ecoclimatic zones along altitudinal gradients. Through the seasonal movement in a yearly

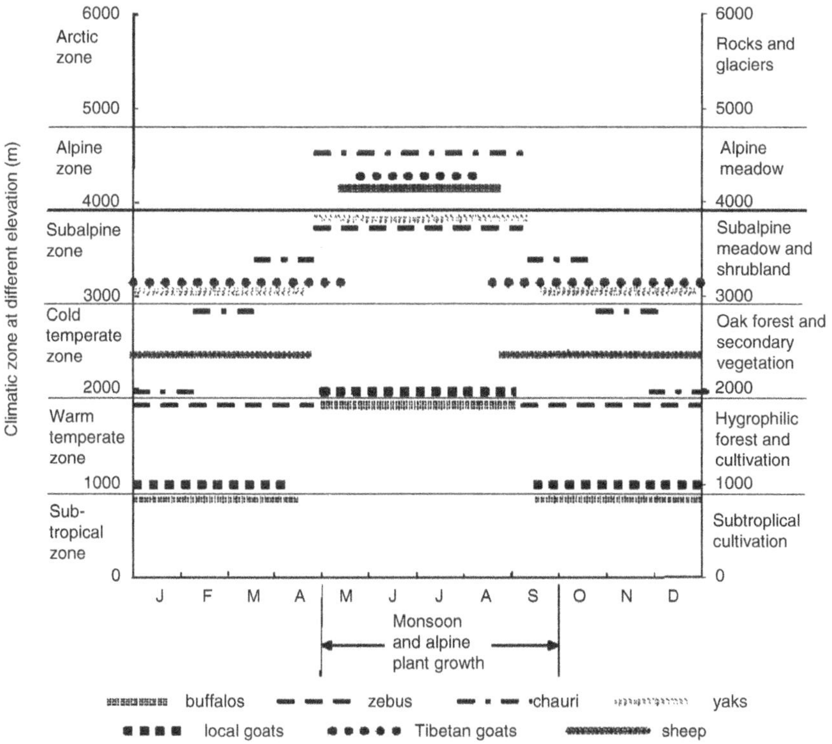

Fig. 3.5 Herding systems along altitudinal gradients characterized by different climate and vegetation in the eastern Himalaya (e.g., Rasuwa District of northern Nepal). (From Dong et al. 2007)

cycle, the pastoralists can secure consistent feed requirements by grazing livestock for maintenance, movement, growth, production, and reproduction.

Herding different livestock species according to their ecological and niche requirements for climate, vegetation, and altitude is a key practice in this indigenous transhumant grazing system; for example, the local pastoralists move their chauri from subalpine meadows at an altitude of about 3000–4000 m in summer gradually down to oak forests at an altitude of about 2000 m in winter, and they move their yaks from the alpine meadows at an altitude of about 4000–5000 in summer gradually down to subalpine meadows or shrublands at an altitude of about 3000–4000 m. To overcome the problems of feed deficit in the winter season, most herders in the case study sites maintained only the number of livestock that could be fed adequately using rangelands forages with a small amount of fodder supplements. In such a way, the local pastoralists can efficiently use the rangeland resources at the different altitudes and well maintain the production of different types of livestock with different habitat preference in different seasons.

Rotational grazing of the livestock between different plots of the same pasturelands (summer, winter, or transitional pastures) on the basis of feed availability is another key indigenous practice adopted by local pastoralists over centuries. The movement of livestock from one plot to another normally occurs every 10–15 days depending on the herders' judgments of grass cover and height. The carrying capacity of pastureland is estimated annually on the basis of climatic variability through a well-defined method among the herder groups to ensure the stability of each plot for a fixed number of animals. The same plot can be repeatedly used in the same grazing season if the grass cover and height have recovered very well. The campsites are protected with stone or reseeded with the native grasses to reduce the risk of soil erosion when the livestock are moved to another grazing plot. In such a way, as stated by local pastoralists, the relationship between the grazing pressure of livestock and the carrying capacity of pasturelands can been balanced and the pasturelands can be protected from overgrazing. Moreover, the local pastoralists stressed that this rotational grazing practice is helpful to reduce the potential spread of external and internal parasites.

3.2.3.2 Collective Actions

Local people have developed their own institutional arrangements for shared use of pastoral resources in a collective way over a long time. There are basically two sets of local organizations: an elected community committee composed of 11–12 people, which acts as the leader and decision maker at a community level; and a couple of civil associations, which are self-identified groups of households with common interests or with the same resource pools (e.g., livestock, vegetables, crops, and forest) at a group level. In some cases, five to seven people are elected from same type of associations to form a subcommittee, which acts as a representative for each type of association to deal with other associations (Fig. 3.6). The community committee is mostly responsible for controlling and regulating the access to pasturelands and fodder resources through enforcement of

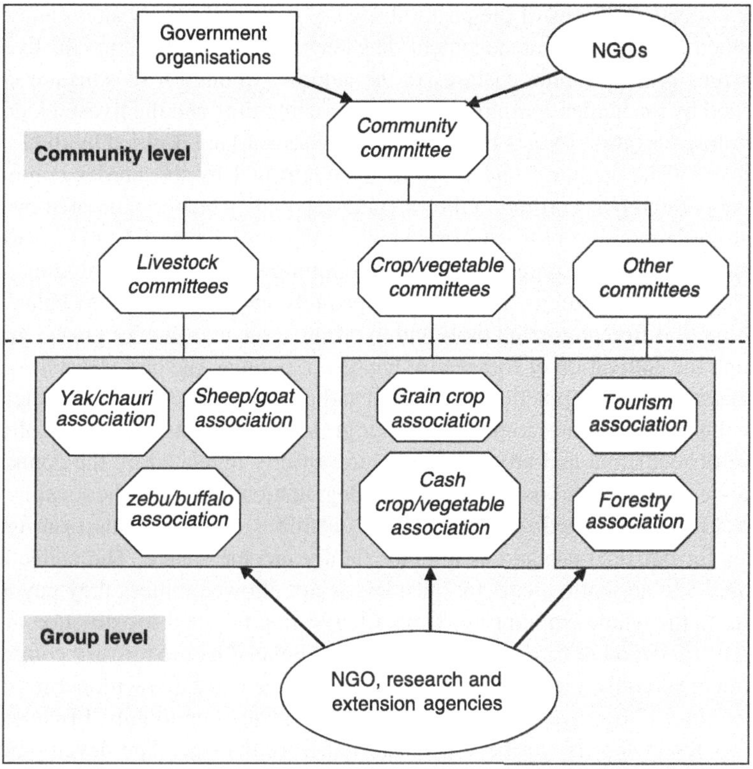

Fig. 3.6 Local pastoral institution arrangements and their linkage with other institutions

well-defined and mutually agreed rights and rules, backed by various social norms and sanctions. In such a way, the community committee can ensure that all community members (including poor and sociopolitically weaker individuals) have relatively equitable access to the pastoral resources. The community committee can also promote the collective actions of livestock grazing and feed collection by enforcing the primary rules and regulations regarding when and for how long the livestock are grazed on certain pasturelands, and when and where hay may be cut for winter feed. The association of livestock keepers such as yak/chauri associations establish rotational grazing rules, regulate herd movements, and make other decisions specific to shared uses of pastoral resources among herders' groups through negotiation and discussion. It is also responsible for mitigating conflicts arising over shared used of pastoral resources within the same herders' group or among different herders' groups with support from the community committee. The coordination and cooperation of the livestock association with other associations such as a crop association, a forestry association, and a lodging (tourism) association can ensure the collective use of different resources related to pastoral management. These grassroots organizations work much better in social functions than the external administrative and political organizations in sustaining pastoral resource management (Dong et al. 2009).

The collective actions of the pastoral society in the case study sites can be well understood in the grazing management. The length of grazing time and the livestock populations grazing alpine pastures in the summer (monsoon season) are strictly controlled by the herder committees, and the grazing time and the livestock populations in low-elevation forests in the winter (dry season) are decided by the forestry committees. The stocking rate is normally controlled by the herder community itself cooperatively according to the carrying capacity, which is estimated by experienced herders on the basis of grass height and cover. Several herders work in a group to herd a certain type of the whole community's livestock, including sheep, goats, buffalo, yaks, zebus, and chauri separately and rotationally to balance the utilization of different grazing plots and to promote the regrowth of grazed grasses. Although the cultivation of forages for hay is not popular, the pastoralists' communities have developed practices to harvest indigenous grasses or fence patches of land with shrubs or stones to protect winter grazing areas and hay fields. Collection and use of medicinal and aromatic plants are strictly regulated by the community committee to reduce the risk of rangeland degradation, although the local herders' households are permitted to harvest small quantities of such plants from grazing pastures for personal use and as a minor family income source. The collection of medicinal and aromatic plants by outsiders is not allowed, unless they pay a very high tax to the whole community. Timber harvesting for shed construction and the cutting of fuelwood in forest areas are strictly regulated by the forestry community committee, and only a small amount of timber can be harvested with permission of the committee or by payment of a high tax to the forestry community. In such a way, grazing sites in forests can be well protected from damage. The development of ecotourism in recent decades has initiated new uses for livestock as pack animals and jobs for local inhabitants as porters, hotel managers, and grocers for foreign mountaineers approaching the Himalaya, resulting in the diversification of local livelihoods. Therefore, pastoral communities collaborate with tourist communities in a collective way to share the benefits of selling livestock products and serving as guides, porters, restaurant managers, and grocers.

3.2.3.3 Property Rights

In a pastoral society, the property rights are mainly related to the pastureland and livestock resources. In the case study sites, the management of pasturelands depends mostly on resource use rights and land tenure systems. Primary pasture resources in these areas, as in other parts of northern Nepal, were previously recognized as either private assets or communal properties owned by the community. However, the central government (i.e., Ministry of Forests and Soil Conservation) took over the management of natural resources as a public property as a result of the Nepal Nationalization Act of 1957. In most cases, some pastoralist households owned small areas of land for crop and hay production and for house and corral buildings in scattered subsidiary settlements. Although the local people lost ownership of pasture resources, they owned use rights to these resources according to the grazing tradition and administrative domains. In this case, the local pastoral communities had to regulate their access and guard against entry of other communities to the natural resources, including rangelands, and strongly resented infringements. Herders from the same community usually negotiate with each other or depend on the community committee's decision for the sharing of grazing pastures, and they normally mitigate conflicts arising from sharing pastures through self-negotiation. Sometimes, conflicts between different communities over use rights of grazing lands happen because of different interpretations of traditional arrangements of grazing areas and administrative boundaries. For example, herders from Gatlang once fought with herders from Chilime, another Tamang VDC in Rasuwa District, over the sharing of a large grazing pasture, Sanjen pastureland, which has been used by herders from Gatlang for a long time but recently had been grouped into Chilime's administrative domain. Recently, both VDCs claimed use rights for this pasture, and they cannot reach an agreement about sharing this pasture through negotiation. Gatlang herders stated that this problem has negatively affected their pastoral production levels and livelihoods. In this case, the local herders had to depend on district or regional governments to make decisions about the pasture resource utilization.

Livestock and their products are privately owned by the individual household, although the livestock from a community or an association are collectively herded on the basis of oral or written agreements of pasture sharing. However, there are some differences in herding management among these three case study sites because of different geographical locations and herding traditions. In Dunche, there are three types of grazing lands: "high-altitude pastures" used normally for summer grazing; "village pasture area" set aside by the community for grazing livestock kept at home for draft and manure production; and "forest edge pasture" for National Park buffer-zone residents to graze a small number of livestock for short periods of time or to collect fodders after paying a fee. In Gatlang, there are two major grazing lands for livestock: summer pastures at high altitude and winter grazing lands in lowland community forests. A group of households (community or association) share the same grazing lands at both locations on the basis of standard animal numbers and specific grazing periods as fixed by the group. In Langtang, seasonal movement between summer and winter grazing lands is regulated by the grazing rights, which are inherited permanently through matrilineal relations or are obtained trough a temporary contract or agreement. Movement to the summer grazing lands involves all livestock owners as a group, and an individual livestock keeper is not allowed to break up grazing lands and make a separate camp. Comparatively, the access to and use of winter grazing lands near their settlements are more flexible; that is, the individual livestock owner in a community can freely graze animals on communal meadow–shrub pastures surrounding the community's settlements.

3.2.4 Implications of the Case Study

These cases testified the importance of indigenous practices and local institutions for natural resource management in northern Nepal. Similar results have been reported across the world (Chapagain 1986; Gilmour 1990; Gadgil et al. 1993; Gill 1993; Rai and Thapa 1993; Farooquee and Saxena 1994; Wu 1997; Chan 2002; Farooquee et al. 2004; Tesfay and Tafere 2004). Although indigenous natural resource management systems may have some shortcomings, the flexibility of these systems to changes and the ability of these systems to adapt demonstrate a major strength in Nepal (Gill 1993), where public support from the government for pastoral development is lacking (Dong et al. 2009). Local pastoralists have extensive experience and knowledge of the local conditions and natural resource use history in this area (Tamang 1993), so they can overcome the physical, climatic, and biological difficulties and utilize the rangeland resources efficiently (Dong et al. 2007). Therefore, effective and appropriate strategies for developing sustainable pastoral management systems in this region require both a clear recognition of indigenous knowledge of pastoral resource management, which has been practiced by local pastoralists for centuries, and integration of the indigenous knowledge with modern technologies.

As mentioned earlier, the strong linkages between property rights, collective action, and natural resource management are very important for technology adoption, economic growth, and environmental sustainability. Although it was previously believed that a resource held under a common property resource regime was inherently inefficient since individuals could not have proper incentives to act in a efficient way (Gordon 1954; Scott 1955; Hardin 1968), it is evident from the case study that clearly recognized pasture use rights and grazing land tenure in traditional pastoral management systems, together with well-defined rules within local institutions, promote the efficient utilization and sustainable development of pastoral resources in northern Nepal. The efficiency of resource utilization under common property resource regimes has been debated for a long time, but it is generally agreed that until collective management under common property institutions is the most viable option for long-term economic and ecological sustainability of the common pool resources. Many studies on the foundation of common property resource regimes in the developing world have shown that local institution arrangements, including customs and social norms, designed to induce cooperative solutions can overcome the collective action problem and help achieve efficient use of common pool resources such as pastoral resources (Gibbs and Bromley 1989; Ostrom 1990). Therefore, local institutions of collective action and indigenous property right systems for pastoral resource management need to be highlighted in the facilitation of rangeland legislation covering traditional rights and customary tenure and cooperation and collaboration between the government and the pastoral society in northern Nepal.

3.3 Case Study from the Indian Himalaya: Importance of Local Adaptations to Climate and Social Changes

3.3.1 Background

Pastoralism contributes a big share to Indian livestock production, which accounts for 25 % of the nation's agricultural GDP and makes India one of the world's largest livestock producers (Bhasin 2011). Pastoralists rear indigenous animal breeds, maintaining the rich genetic variety of livestock. As a result of historical and cultural influences as well as resource availability, various types of pastoral systems, from nomadic to transhumant to agropastoral, can be found across the nation. India's pastoralism is often combined with sociopolitical forms of organization that can be considered tribal (Bhasin 2011). It is estimated that more than 200 tribes with about 6 % of the nation's population are involved in pastoralism in the whole of India (Sharma et al. 2003). Pastoralism exists prevalently among the ethnic tribes living in the drylands of western India, the Deccan Plateau, and in high-altitude regions of the Indian Himalaya (Bhasin 2011): the Gollas and Kurumas of Andhra Pradesh are mostly involved in cattle and sheep rearing; the Rabaris and Bharwads from Gujarat are normally engaged in raising sheep, goats, cattle, and small livestock; the Kurubas and Dhangars from Karnataka usually raise sheep; the Raikas/Rabaris and Gujjars from Rajasthan and western India generally raise camels, sheep and goats; the Gaddis, Gujjars, and Bakarwals from Himachal Pradesh and the western Himalaya normally herd sheep, goats, and buffalo; the Bhutias of Sikkim in the eastern Himalaya and the Changpas of Ladakh in the western Himalaya usually raise yaks (Table 3.2).

The Himalaya cover only 18 % of the territorial lands of India, but they accounts for 50 % of India's forest cover (including rangelands) and 40 % of the species endemic to the Indian subcontinent (Maikhuri et al. 2000). In the Indian Himalaya, the rangelands are represented by warm temperate grasslands, subalpine and cool temperate grassy slopes, alpine meadows of the high mountains and the alpine steppe, cold arid regions, or alpine dry scrub, occupying nearly 35 % of its geographical area (Rawat 1998). These rangelands differ in their climatic and geographical features, as well as in the supporting pastoral communities. Livestock rearing on the rangelands is an integral component of the economy in the Indian Himalaya, and dependence on livestock rearing increases with an increase in altitude (Sundriyal 1995). Over centuries, the alpine grasslands at high elevation have been used as the grazing pastures by migratory livestock of nomads as well as animals from lower valleys during summer (Farooquee 1994). Although the livestock grazing in the upper mountains is dominated in mobile pastoral societies, scholars argued that nobody really knows the exact extent of animal husbandry in the Indian Himalaya (Sharma et al. 2003). On the basis of the estimation from an overview publication, the Indian Himalaya are home to about 50 million domesticated animals, which are mostly kept in systems of combined mountain agriculture (Kreutzmann 2012). The pastoral communities of the Indian Himalaya use the pastoral resources efficiently

Table 3.2 General information about major pastoral groups in the Indian Himalaya (*Source*: Singh 1996)

Pastoral groups	Size	Location and species	Ethnic identities	Outline migration pattern
Bakarwals	NA	Kashmir (mainly goats)	Muslims. Speak Kashmiri and sometimes Hindi	They move to Jammu and Punjab plains in winter and to Kishtwar and other higher alpine valleys of the Kashmir Himalaya in summer months
Gujjars	2,038,692 (1931 census)	Jammu, Himachal Pradesh, and Uttarakhand (mainly buffalo)	Hindu and Muslim. Speak a mix of Gujarati, Urdu, Dogri, and broken Hindi with a Perso-Arabic script	Winters are spent in the regions of Jammu, Punjab, and lower districts of Himachal Pradesh and Uttar Pradesh, Saharanpur regions, and in the areas adjoining Rajaji National Park. They migrate to higher (nonalpine) regions of Himachal Pradesh and Uttarakhand in summer
Changpas	NA	Southeast Ladakh (yaks)	Follow a primitive form of Buddhism. Speak a mix of Ladakhi and Tibetan, with a Tibetan script	Their migration cycle is around the various high-altitude pastures of Rupshu plains in the Changthand region of Ladakh
Gaddis	1,26,300 (2001 census)	Kangra and Dharamsala regions of Himachal Pradesh, parts of Uttar Pradesh and Punjab (sheep and goats)	Hindu Rajputs. Speak Hindi with a Devangri script, and Pahari	Punjab plains and lower districts of Himachal Pradesh during winter months and occupy Lahaul and Dhauladhar pastures in summer months
Bhotias	NA	Upper regions of Garhwal and Kumaon of Uttarakhand (sheep, goats, cattle)	Hindu. Speak the Pahari group of languages with a Devanagiri script	They occupy lower districts of Uttarakhand such as Dehradun and the Bhabhar valley in winter months and move to higher pastures of the Garhwal and Kumaon Himalaya toward Nanda Devi, Gwaldam, Mana pastures, and adjoining regions

(continued)

Table 3.2 (continued)

Pastoral groups	Size	Location and species	Ethnic identities	Outline migration pattern
Bhuttias	21,259 (1981 censu)	North district of Sikkim	Buddhists. Speak a Tibetan dialect	Alpine regions of Lachung and Lachen valleys of the north district of Sikkim and move to lower forest below Mangan in summer
Monpas	34,469 (1981 censu)	Tawang and West Kameng districts of Arunachal Pradesh	Buddhists: their language belongs to the Bodic group of the Tibeto-Burman family	Higher reaches of East Kameng and Tawang districts of Arunachal Pradesh in the summer season and migrate to lowlands around Tawang in the winter months
Kinnauras	59,547 (1981 census)	Kinnaur district of Himachal Pradesh	Rajputs or Khosias and the Berus include both Hindus and Buddhists	In summer, sheep and goat flocks are driven to higher parts of Himachal Pradesh and in winter the flocks are driven to foothills of Uttarakhand and Himachal Pradesh

NA not available

by different means of mobility patterns, socioeconomic organizations, and property rights (Table 3.2). All forms of pastoralism may be considered as different forms of adaptation determined by ecological conditions and technological development levels, making pastoralism critically important in the Indian Himalaya from social, economic, cultural, and environmental dimensions.

Unfortunately, threats and pressures associated with climate change, economic development, and political marginalization have been challenging the sustainability of the traditional pastoral system, including migratory cycles, local economy, and social organization (Bhasin 2011). Some studies showed a decrease in rainfall and unpredictable onset of the monsoon, longer dry spells, higher temperatures linked to decreased water availability, and warmer winters with significantly less snowfall are the major features of climate change in the rural Himalaya (Macchi et al. 2011). The ecosystem services provided by the Himalayan rangelands such as rich biodiversity and food production may become vulnerable to climate change and the large-scale socioeconomic forces (Dong et al. 2010a, b). Extreme weather conditions, drought, epidemics, and predators associated with climate change can result in reduction of animal production (Bhasin 2011). As a result of new threats emerging to water and food security, pastoral production, nutrition, and public health in vulnerable areas such as the Himalaya, hard-fought progress has been made in achieving the Millennium Development Goals on development and poverty alleviation but this may be slowed down or even reversed by climate change (El-Ashry 2009). Social, political, and economic changes are also challenging the sustainability of Indian pastoralism, one of the important human–natural systems in India. As stated by Bhasin (2011): "Currently, the trend towards globalization of the market, with pastoral lands increasingly being commercialized and/or turned in to national parks has created problems for the pastoralists. Due to neglect by officials and policy makers, pastoralists face deprivation from their traditional and customary rights to these grazing areas. The political marginalization of pastoral communities paved the way for forcible eviction from their land and/or restriction of their movements. Many of them left their traditional transhumant way of life and settled along valleys. Some have settled in urban areas others stick to the pastoral activities by changing the composition of livestock by increasing number of goats and decreasing number of yaks. State policies regarding forests, agriculture, irrigation, fodder, famine, pastoral rights and migration are some of the mechanisms that contribute to the alteration of pastoral life-style. Likewise, social crisis, such as phases in domestic developmental cycle and work force shortage in herding groups cause concern in the community." This is not the solely specific case in any Indian districts, and almost all pastoral groups in the Indian Himalaya are facing similar constraints and problems (Bhasin 2011). Therefore, it is necessary to conduct a case study to examine the challenges and problems faced by pastoral communities in the India Himalaya and their adaptation strategies to cope with these difficulties.

3.3.2 Methods

The case study was conducted in the Indian state of Himachal Pradesh (Fig. 3.7), which is located between latitude 30°23′02″N to 33°15′34″N and longitude 75°36′41″E to 79°01′51″E with the altitude ranging from 350 m at its boundary along the Punjab plains to 6816 m at Reo Purgyal in the Zanskar Range (Singh et al. 2009). In the light of regional variations in rainfall, temperature, and humidity, the state can be divided into five climatic zones (Singh et al. 2009): a subtropical zone (below 900 m), a warm temperature zone (900–1800 m), a cool temperature zone (1800–2400 m), a cold high mountain zone (2400–4000 m), and a snow frigid zone (above 4000 m). There are 12 districts, 115 tehsil/subtehsil (similar to VDCs in Nepal), and more than 20,000 villages. According to a recent population census report, the human population of Himachal Pradesh is 6,077,248, with a decadal growth rate of 17.53 %. About 90 % of the population in Himachal Pradesh is rural and belongs to three sociological population groups: the Rajputs and Brahmins, the Scheduled Tribes, and the Scheduled Castes. The Scheduled Tribes include the Gaddis, the Gujjars, and the Bholts, which are mainly engaged in animal husbandry, with many of them practicing migratory grazing (Singh et al. 2009), although livestock rearing is an integral component of the economy, and is inseparable from the agricultural component of every household in rural areas of Himachal Pradesh.

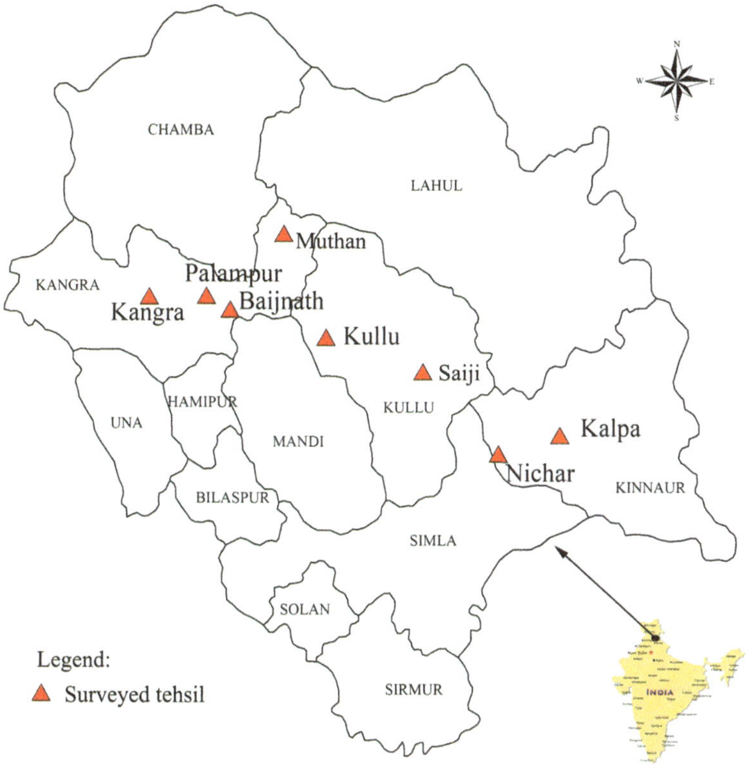

Fig. 3.7 Map of surveyed tehsils/subtehsils in Himachal Pradesh, India

In this study, eight tehsil/subtehsils were selected for the field survey on the basis of information collected from the on-the-desk work (Table 3.3). Sixty-six households were randomly sampled from the selected tehsil/subtehsils for the survey, which was conducted between December 2010 and February 2011. These households represent the general situations of pastoral groups in Himachal Pradesh (Table 3.3). Methods similar to those used in the Nepalese case study, including the collection toolkit of participatory rural appraisal, open-ended questions and pre-tested questionnaires, key-person interviews, and group discussion, were applied in this survey. Herders/farmers were interviewed by face-to-face survey based on a questionnaire mainly on rural animal husbandry practices and problems faced by them. The quality of the results of the questionnaires was controlled through careful checks on the errors in the completed questionnaires. Supplementary information was collected and updated with use of various data sources, including research publications, reports, newsletters, and personal communications. Systematic qualitative techniques recommended by Patton (1990) and Miles and Huberman (1994) were used to analyze all the data.

Table 3.3 Information about rangeland and livestock in the case study sites (data from Singh et al. 2009)

Name of tehsil/ subtehsil	Land area (ha)	Rangeland area (ha)	Grazing livestock numbers (ACU)				
			Cattle	Buffalo	Sheep	Goats	Total
Baijnath	21,325.19	1529.77	21,154	2028	2488	2235	27,904
Kangra	28,429.39	5803.18	30,729	12,810	920	3560	48,020
Multhan	94,693.05	74.17	4458	0	3241	2132	9831
Palampur	44,426.40	5946.24	39,192	6225	3378	3136	51,931
Kalpa	32,678.59	11,062.68	2540	5	1181	361	4087
Nichar	104,414.55	25,804.37	8112	0	4488	2331	14,931
Kullu	290,046.43	29,143.83	67,474	172	13,516	7669	88,831
Saini	27,192.94	3999.74	13,162	19	1496	1357	16,033

ACU adult cattle unit

3.3.3 *Results*

3.3.3.1 Indigenous Practices

The survey indicates that the Gaddi, Kanet, Kauli, and Kinnaura pastoralists in Himachal Pradesh have adopted transhumant grazing practices, which involve cyclical movements from lowlands in winter to highlands in summer to take advantage of the availability of pasture resources varying with seasonal climate change at different elevations. Unlike the nomadic Changpa pastoralists in Changthang of southeast Ladakh, the Gaddi, Kanet, Kauli, and Kinnaura pastoralists in Himachal Pradesh are transhumant shepherd groups who have dwellings in the valleys between mountains and practice long-distance herding of livestock. The pastoralists herd small livestock such as sheep and goats in a vertical migration, in addition to keeping a small number of nonmigratory large livestock such as bulls and cows in their dwellings. They raise sheep for the production of wool, which is woven into rain-resistant blankets, snowshoes for the shepherds, and carpets for family use or sale. They raise goats for the production of milk, which is the staple diet of the shepherd's during migration, and meat, which is mostly sold for family income. Bulls and cows are kept for draft power of plowing cultivated croplands or family drinking milk during the time of year when they stay at home. There are many fixed migratory routes from the highland peaks to the lowland plains, with numerous passes in the Himalayan ranges (Fig. 3.8). Through a year-round movement, the herders can obtain a consistent supply of feed for maintenance, movement, growth, production, and reproduction of the livestock. This traditional transhumant grazing system has capitalized on the physical and climatic characters and the plant communities, and has converted many constraints into opportunities in the fragile environments of the highlands of the Indian Himalaya.

Balance between availability of water and fodders and the requirement of the livestock in different seasons is indispensable to pastoral groups for adaption to migratory grazing in Himachal Pradesh. In early April, the pastoralists begin to

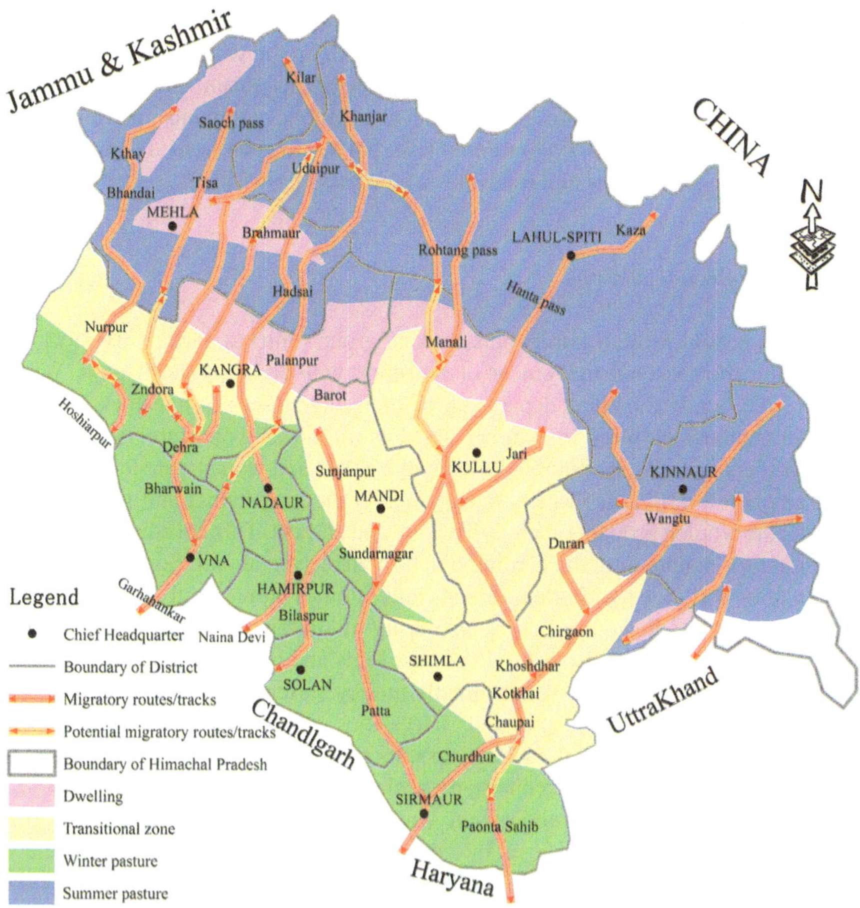

Fig. 3.8 Migratory routes of herding groups in Himachal Pradesh, India

move their flocks of sheep and goats upward (northward) along the low mountain ranges to their dwelling villages by early May. Then they graze their livestock in village forests and middle hill forests in May and June, and harvest the winter crop and prepare the crop fields for the monsoon. Moreover, they pen the sheep and goats on the newly harvested fields for a couple of nights to provide manure as fertilizer for the next crop. By late June, the partial melting of the snow allows the pastoralists to cross over mountain passes into the alpine meadows at high altitude to feed their livestock on the nutritious forages throughout the summer (July, August, and September). By middle-late September, decreased forage availability forces the pastoralists to move their flocks of sheep and goats downward (southward) by recrossing high mountain passes to their own dwelling villages. In the next 2 months, they graze their animals in village forests and middle hill forests, and plow the land and cultivate the winter crops. By the middle of November, the pastoralists start to move down to the winter grazing lands in the lowland forests or the plain pastures within

1 month. By the end of December, the pastoralists reach the winter grazing lands, where they spend 3 months herding their flocks. Along the winter migration routes and on the winter grazing grounds, the pastoralists can obtain forages in various places, such as village pastures, which tend to be public scrublands, private grasslands from which cultivator communities have harvested grass to feed their animals, in stalls, or the streambeds and roadsides where grass grows well.

3.3.3.2 Collective Actions

At the household level, a family acts a social group (unit) to herd its livestock by combining the seasonal movement of livestock with seasonal cultivation. When most of the family members depart to herd their large flocks of livestock on high-altitude meadow pastures in summer and low-altitude silvopastures in winter, a small percentage of the population (mostly the elders and women) are left behind to look after the draft cattle (which subsist mainly on corn stalks) and fields and to process the woolen products such as blankets and carpets. This coordination is very important for maintaining the herding unit as a viable social unit. Some pastoralists live in tents and move with their animals and families along fixed routes, whereas other pastoral families do not use tents during migration and they prefer to move lightly. For those pastoralists who do not use tents and bring tings with them, they normally obtain livestock products directly or they barter animal products for grain with agriculturalists and for other daily necessities with retailers.

At the community level, the village appears as a clear social division. The members of the villages/hamlets consist of a clearly bounded social group. An *al*-association is a form of cooperation and mutual insurance through which a pastoralist can maintain significant interpersonal relations within a broader society. The *al*-association is divided into the *khinds,* named after the ancestors where the split is supposed to have taken place. A *khind is* composed of numbers of *tols*, which are two to three generations in depth and may consist of one or more brothers and share a common hearth. Normally, all the families of a *tol* herd their livestock together under the supervision of two hired shepherds and two *tol* members (belonging to any family who can spare two male members at that time). As labor cannot be purchased in a pastoral community, the pastoralists ensure a stable labor supply through *barton* (obligatory assistance) and cooperation between families. Local governing bodies are necessary to control ownership and transfer of property as well as to adjudicate conflicts. The *Pradhan* (representatives from a group of households) is in charge of settling local disputes within a village, whereas the *Panchayat* (representatives from a group of villages) is responsible for settling disputes between villages. These local institutions are also responsible for serving the social, economic, security, and development needs of all member pastoralists. Besides, local institutions make norms or rules for all the pastoralists to regulate herd movement, information sharing, risk pooling, aggregation, and dispersal of herders across the region. On the other hand, a variety of social and cultural mechanisms, such as religion, folklore, and traditions, support the local institutions to regulate the sustainable migratory herding in Himachal Pradesh.

3.3.3.3 Local Adaptations

Climate change, growing human population, increased infrastructure and economic development, and government policies regarding forests, agriculture, fodder, pastoral rights, and migration are the main driving forces that lead to the alteration of a pastoral lifestyle and threaten the sustainable development of pastoralism in Himachal Pradesh. According to the survey, climate change has been greatly influencing the grazing systems. Most respondents in the survey claimed that they could not find enough forages and drinking water for the grazing livestock on the high-altitude alpine meadow pastures in summer because of glacial retreat and rainfall decline. Because of frequent climate uncertainty and natural disasters, the pastoralists are facing many more difficulties in allocating the grazing time on winter and summer pastures and in planning the balanced year-round feed supplies for the grazing livestock (Table 3.4). With the increase of tree line/timberline associated with warm temperature on the high mountains, the grazing pastures of subalpine forests are shrinking and declining in both area and production. In middle hill there are outbreaks of noxious weeds such as Crofton weed (*Eupatorium adenophorum*), which is known locally as *Kali Basauti*, because of dryness and warmness. Some respondents said that this exotic plant has invaded into subalpine meadows and lowland silvopastures (agroforestry). The spread and dispersal of this invasive plant have led to not only reduced forage production but also decreased animal production (milk and meat), even resulting in the loss of livestock because of its being poisonous. Moreover, some foresters and farmers in the lowland plains and valleys blame the mobile herders for the spread and outbreak of Crofton weed as they think that the migratory livestock carry the seeds of this invasive plant in their hair during middle hill grazing and disperse them in the lowland farms and forests when they migrate there for winter grazing. To cope with these difficulties, the local pastoralists have developed some adaptive strategies; that is, quite a number of the respondents move their livestock earlier in summer to high-altitude pastures for grazing to take advantage of early growth of forages associated with increased temperatures. Some pastoralists raise more goats to control the Crofton weed as goats can eat and digest this noxious plant without great problems. Some pastoralists store the crop

Table 3.4 Pastoralists' responses and adaptations to climate change in the Indian Himalaya

Impacts and mitigation of climate change	Proportion of respondents
Lack of forages	86.1 %
Lack of drinking water	100 %
More livestock loss because of floods	2.8 %
Early grazing to summer pastures	65.7 %
Farmers' organization raises funds for collective actions	12.3 %
Local NGOs help to cope with climate change	Rare
Local governments help to cope with climate change	Rare

residues and cultivated feeds for balancing of the animal requirements during feed-deficient times.

In line with growing human population, increased infrastructure, economic development, and the subdivision and fragmentation of agricultural land, the availability of grazing lands for pastoral communities is dramatically decreasing. As stated by the pastoralist respondents, land use in lower altitudes of the Himachal Pradesh has been greatly altered in recent decades by a number of driving factors, such as afforestation activities, road construction, agroforestry development, and agricultural expansion (Fig. 3.9). These land use changes have seriously reduced the size of available winter pastures for pastoralists and have also disturbed or blocked their migratory patterns. As a result, local pastoralists have to change their migratory routes and have faced serious problems of livestock loss due to roadkill, theft, and tiredness. Especially, agroforestry development and agricultural expansion have increased the heavy tensions between mobile herders and local farmers. In the past, the agricultural or agroforestry cultivators paid the pastoralists to manure their fields, but now the cultivators do not allow the pastoralists to herd their livestock on the agricultural or horticultural fields as they have replaced the livestock manure with chemical fertilizer. The reduced access to winter pastures in plains or low-valley areas has pushed the local pastoralists to adopt the early movement of the herds up through the middle

Fig. 3.9 Development of agroforestry is blocking migratory routes of grazing animals in Himachal Pradesh, India. (Photo by Shikui Dong, 2011)

hill forest belt to high-altitude pastures with prolonged summer grazing. As a result of Forest Department not issuing winter grazing permits and increased grazing taxes charged by local residents, the economy in low-altitude areas has changed from a mixed agropastoral system to an agricultural- or horticultural-based economy. In coping with these social changes, the pastoralists have diversified their livelihoods as agropastoralists, cultivators, and migratory labor. For example, according to the survey, 25 households out of 55 total pastoralist households in the village of Nawal in the Palampur tehsil have shifted their livelihood to agricultural production. The survey indicates the diversification of livelihood is a prevalent trend in a pastoral society to adapt to the socieconomic changes in Himachal Pradesh, even in the Indian Himalaya.

In addition, increased political marginalization is presently playing an important role in accelerating the decrease of pastoral activities in Himachal Pradesh, even in the whole of the Indian Himalaya. Bcause of the small population and migratory lifestyle, pastoralists in the Indian Himalaya have often been overlooked in the policy decisions at various levels. According to pastoralist respondents, nonparticipation in policymaking and ignorance of their rights and status by the local and central governments in India have seriously marginalized these communities. Some newly implemented policies related to farming, forest and animal husbandry, and poverty alleviation have been directly or indirectly influencing pastoral production in the region. Various development schemes for the pastoral population have an agricultural preference, and pastoralism is considered to be an activity supplementary to agriculture, which has resulted in a bias against pastoralists. As one of the priorities for regional development, sedentarization of pastoralists is now widely supported by the governments in Indian Himalayan states such as Himachal Pradesh. The establishment of national parks and protected areas, along with the expansion of agriculture/agroforestry into marginal pasture areas, has suppressed the pastoral production systems in the region. The records of the herder respondents indicate that the recent establishment of the Great Himalayan National Park in Himachal Pradesh has deprived them of access to about 300 km^2 of summer pastures without their having been allotted grazing rights in any alternative regions. The afforestation program for carbon sequestration in lowland forestry areas initiated by the Forestry Department has altered or fragmented the traditional migration routes, forcing the pastoralists to shorten the grazing time on the winter pastures or to find other alternatives to compensate for the winter grazing.

3.3.4 Implications of the Case Study

Similarly to what was found by previous researchers (Bhasin 1988, 2011; Chakravarty-Kaul 1998; Ives and Messereli 1989), transhumant grazing accompanied by vertical movement of livestock in a cyclic manner is a very important indigenous grazing practice in the Indian Himalaya. This practice adopted by pastoral communities living in the high-altitude areas of the Indian Himalaya exploits the seasonal abundance of grazing areas on both temporal and spatial scales. In most cases, mobile grazing in the Indian Himalaya, as in other Himalayan areas such as the Nepalese Himalaya, is guided by indigenous rules and institutions (Chakravarty-Kaul 1998; Dong et al. 2009). Mobility is one of the most important adaptations in the pastoral society of the Indian Himalaya, through which pastoralists can successfully manage their environment with a high degree of diversity. The mobility characterized by seasonal movement promotes the capability of the opportunistic migratory pastoralism to balance the relationships between livestock, pastures, and the human population by creating the possibility for marginal regions to support livestock and human life. Migration to different locations with a combination of seasonal and ecological variables in pasture and water is the basis for the survival strategy of pastoralists in the harsh environments in this region. As stated by Janzen 1993, "mobile livestock keeping is a best active human adaptation to the harsh environment and is probably the only way of putting the pastures to economic use without a huge expenditure of capital." The indigenous mobile grazing system allows pastoralists to take advantage of pasture resources with low productivity and water resources with an irregular spatial distribution.

The traditional pastoral system, including a migratory cycle, local economy, and social organization, is based on efficient use of seasonally varied resources through collective actions in marginal environments of the Indian Himalaya. Well-organized local institutions have strengthened the collective actions of pastoral communities at both the household level and the community level. The commonly agreed norms and rules among the pastoral societies have promoted the efficient use of pastoral resources in the Indian Himalaya. Although climatic, socioeconomic, and political changes have brought pressures and threats to sustainable development of pastoralism in the Indian Himalaya, local pastoralists have adjusted accordingly in numerous ways. As mentioned earlier, adaptation is about building resilience and reducing vulnerability of coupled social–ecological systems such as pastoral systems to absorb disturbance and still retain their basic function and structure (Walker et al. 2004; Folke 2006; Walker and Salt 2006; Kassam 2010). This case study shows that local pastoralists in the Indian Himalaya have developed adaptive management systems in their traditions of pastoral resource use. The flexibility of these management systems to climatic, socioeconomic, and political change seems to be a key strength. Effective and appropriate strategies for sustaining the pastoral development in the Indian Himalaya require a comprehensive understanding of traditional pastoral systems as the local people have practiced them over centuries. The autonomous adaptations adopted by local pastoralists require an enabling policy environment in the context of the complexity of these changes, especially for the pastoral societies in the Indian Himalaya, which are generally politically marginalized. Deep understanding of traditional production systems and indigenous knowledge, strategies, and practices can empower pastoralists and enhance their capacity to maintain the subsistence economy in fragile environments

3.4 Case Study from the Qinghai–Tibetan Plateau of China: Social Dynamics of Pastoral Communities To Cope with Rangeland Degradation

3.4.1 Background

The Qinghai–Tibetan Plateau (QTP) is located in western China and is known as the "Roof of the World" because of its high altitude. The QTP is also called the "Water Tower in Asia." because it is the source area of many of Asia's major rivers. Rangelands/grasslands cover about 60 % of this vast land mass and provide critical ecosystem services to humans on local, regional, and global scales (Foggin 2008). These rangelands have served as the dominant grazing pastures for indigenous livestock and are regarded as one of the major pastoral production bases in China (Dong et al. 2012). Extensive grazing has been the major land use for the QTP's rangeland in throughout history, and pastoralism has played key roles in local livelihood and the regional economy (Dong et al. 2011). Tibetan-dominated communities have lived on these rangelands as pastoralists to raise yaks and Tibetan sheep for meat, milk, wool/hair, and hide for centuries (Long et al. 2008). The QTP supports more than 13 million grazing yaks (more than 90 % of the world's total population) and about 42 million grazing Tibetan sheep (Long et al. 1999; Dong et al. 2012). For thousands of years, the QTP's pastoralists have lived harmoniously with nature through their indigenous grazing practices (Dong et al. 2007; Cai and Zhang 2013). However, rangeland degradation associated with climate change and human disturbance is threatening the pastoral production system in the fragile and vulnerable landscapes of the QTP (Dong et al. 2010a, b). The rangeland degradation can be the cause and the effect of sociopolitical changes and will negatively affect the productivity and sustainability of pastoral systems, limiting the sustainable development of a coupled human–natural system of pastoralism on the local scale and ecological, social, and economic systems on the regional scale (Dong et al. 2012).

Although the analysis with three-dimensional "vulnerability/resilience" coordination framework in Chap. 2 shows that that rangeland degradation has increased the vulnerability of the QTP's pastoralism in all three dimensions of livelihood, institutions, and ecosystems, it is critically important to examine the potential of pastoral societies to maintain the human–natural system of the QTP's pastoralism, particularly from the perspectives of local people's perceptions, actions, and behaviors, as well as local institutions' responses and reactions. Much historical and present evidence regarding pastoral societies across the QTP has shown that local pastoralists do have the wisdom to cope with the changes resulting from inside and outside drivers and to adapt to new situations through mobility, diversification, preparedness, exchange, and local specificity. For example, the local pastoralists have used the strategy of mobile grazing for generations to cope with uncertain environmental changes (such as drought, snowstorms, landslides, diseases, pest outbreaks) to secure the supply of feed and water resources. Over centuries, the adaptive management applied by local pastoralists has capitalized on the physical and climatic characteristics and the plant communities and has converted many physical and ecological constraints into socioeconomic opportunities (Cai and Zhang 2013). Therefore, this case study was conducted to documents the local pastoralists' actions to cope with rangeland degradation resulting from climate change and human disturbance, as well as their strategies regarding sustainable pastoral management on the QTP.

3.4.2 Methods

The case study was conducted between 2010 and 2012 at two pastoral sites in Gansu Province, Zhuaxixiulong Township of Tianzhu Tibetan Autonomous County, and Huangcheng Township of Sunan Yugur Autonomous County (Fig. 3.10). Both sites are located in the Qiliang Range, the eastern edges of the QTP with the an average elevation above 3000 m. Tianzhu was the first Tibetan autonomous region formed after the People's Republic of China was founded in 1949. With 230,000 people, Tianzhu has the highest population density in all pastoral counties in Gansu Province, even in the whole of China. About 35 % of the population of Tianzhu are Tibetans, whose livelihood is mainly pastoralism. Sunan is home to the Yugur ethnic minority groups, who have practiced traditional transhumant grazing for centuries. There are 15,000 Yugurs in China, and 90 % of them live in Sunan as pastoralists, whose ethnic traditions are somewhat similar to Tibetan, Mongolian, and Han Chinese traditions. In addition to cultural–ethnic differences, there are significant

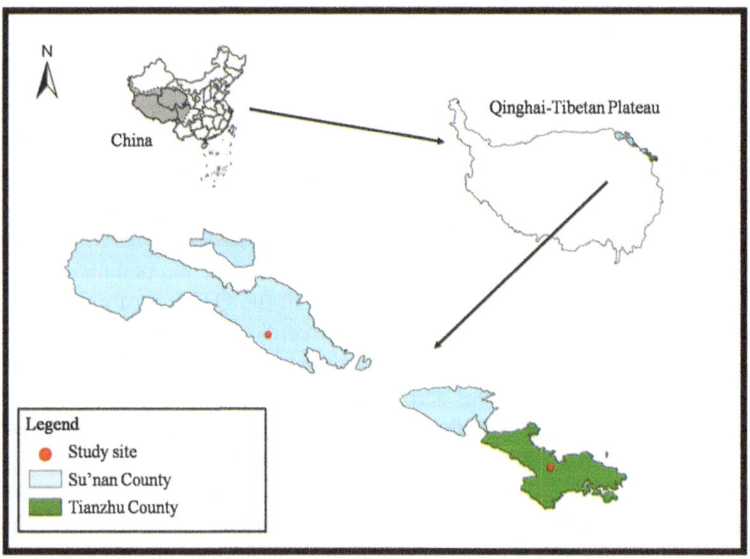

Fig. 3.10 Location of case study sites, Tianzhu County and Sunan County of Gansu Province, China

differences ingrazing pastures in terms of rangeland vegetation; that is, the humid alpine meadow mainly grazed for yak farming in Tianzhu and the dry alpine steppe mainly grazed for sheep farming in Sunan. At both sites, rangeland degradation associated with climate change and overgrazing is a key limit for sustainable pastoral production. Although Sunan's rangeland is much healthier than Tianzhu's rangeland because of better management and lower grazing pressure, both sites are being greatly affected by climate change.

To collect the general information about the indigenous grazing practices, traditional pasture management, pastoral institutions, local perception, and response and adaptation to environmental changes, the field surveys were performed by use of integrated approaches including participatory rural appraisal, open-ended questions and pretested questionnaires (14 and 10 households in Tianzhu and Sunan respectively), key-person interviews (three and four individuals in Tianzhu and Sunan respectively), and group discussion (24 and 11 participants in Tianzhu and Sunan respectively). Supplementary information about problems, constraints, challenges, opportunities, and changes in pastoral management systems, external public support, and partnerships was collected and recorded through group discussion and personal communications (including both professionals and practitioners). All the primary information of the survey was documented by transcription of sound recordings or film recordings. Secondary information regarding pasture management, pastoral development, and government policies was collected and updated with use of various data sources, including research publications, reports, newsletters, and yearbooks. The data quality was controlled by careful investigation and cross-checks with different sources. Systematic qualitative techniques recommended by Patton (1990) and Miles and Huberman (1994) were used to analyze all the data.

3.4.3 Results

3.4.3.1 Local Perception and Preparedness

The case study indicates that local people have a clear perception about environmental changes, socioeconomic transformation, and political dynamics. Simultaneously, the local people have to prepare to adapt to or cope with all these changes and dynamics. Although there are no experimental instruments to record the real scenarios of precipitation and temperature dynamics, the local pastoralists can give a clear description of climate change from their personal experiences. Almost all the interviewees in the case study reported a general trend of temperature change, much hotter summers and autumns but colder springs and colder winters than before. The pastoralist respondents felt that there is a 15- to 30-day delay in the start of spring and that summer starts 15–30 days earlier in comparison with the past. The dryness is a significant change in the patterns of precipitation and water supply. Most of the interviewees reported that both rainfall and snowfall have declined significantly in recent years and water scarcity is a big problem for forage and livestock production (Table 3.5). A male herder in Tianzhu County said: "Decades ago, there were many wells on the summer pastures in the surrounding mountains; we had to wear rubber boots to herd the livestock on the dense grasses, which were as high as my knees. However, currently, as most of the wells have dried up, livestock cannot obtain enough drinking water and spend a long time walking to obtain enough grass for grazing. If this situation continues, the drought and warming continues in the coming years, the springs will disappear. Grazing will face a big problem." Similarly, a female herder in Sunan County stated: "As the herders, we are really afraid of drought on the rangeland. Most of the time, the livestock drink water from springs, which dry up in dry years. In the past 2 years, there has been a continuous drought. Recently, there are many places (in pastoral areas) where drinking water is hardly available. We have to go a few kilometers to find drinking water. The drought directly affects our forage production. The damage caused by drought is really severe in pastoral areas."

Rangeland degradation is another evident environmental change, which has been generally noticed by the local people. More than 80 % of respondents in this case study felt that rangeland degradation, including decreased grass production and forage quality and decreased grass coverage and height, had happened over the past few decades and is continuing. As described by a rangeland extension agent in Tianzhu County "the rangeland condition has indeed changed. Compared with the 1970s and 1980s, grass height here (Zhuaxixiulong Township) has decreased around 10 %, and plant cover has decreased about 20 %." A similar statement was obtained from a herder in Tianzhu County: "Although the government has invested a lot of subsidies to restore and protect the grasslands, rangeland degradation has not been completely mitigated. Our yaks were larger in the 1970s and 1980s than the present when you compare animals of the same age.

Table 3.5 Local pastoralists' perceptions and adaptation strategies regarding climate change

Items	Tianzhu County	Sunan County
Temperature change	Colder winter and spring, hotter summer and autumn	Colder winter and spring, hotter summer and autumn
Precipitation change	Less snowfall in winter and spring, less rainfall in summer and autumn	Less rainfall in summer and autumn
Feelings about climate change	Spring ends earlier, summer starts earlier, mountain wells decline, river flow decreases	Spring ends earlier, summer starts earlier, mountain wells decline
Impacts of climate change on grazing systems	Shortage of forage, lack of drinking water for livestock	Shortage of forage, lack of drinking water for livestock
Changes in grazing strategies	Early movement to summer pasture and lengthening of summer grazing	Early movement to summer pasture and lengthening of summer grazing
Institutional responses	More collective actions, enhancing partnerships and collaboration with professionals	More collective actions, enhancing partnerships and collaboration with professionals

Because of declining rangeland quality and water shortage, the yaks become smaller. For example, in the past a 6-year-old yak produced around 150 kg of beef, but nowadays they produce just 120 kg of beef." As for the cause of rangeland degradation, both the professionals and the pastoralists stressed climate change and overgrazing. For example, the rangeland extension agent said: "Grassland degradation is partly due to global warming, but a main reason is overgrazing." A herder in Sunan said: "Personally, I think overgrazing is happening here." In Sunan's dry alpine steppe (a type of rangeland), a grasshopper disaster associated with a drier climate is becoming an environmental problem according to local pastoralists and professionals. To mitigate rangeland degradation, the local pastoralists and professionals cared more about the carrying capacity of rangelands for long-term development. Most of the herders were willing to prevent rangelands from degrading in collaboration with professionals.

Growing population, increased living standard, and diversified livelihoods are some socioeconomic transformations experienced by the local people. Through public survey and interviews, it was found that the present (human) populations in Zhuanxixiulong Township of Tianzhu County and Huangcheng Township of Sunan County are 1.4 times and 1.8 times as high as they were 3 decades ago, and there are growing pressures on grazing pastures because of the increased population. As a female herder in Sunan said "Our population has increased. Each family may have three to four children. If a family has many children, the pasture is divided between the children; the pasture becomes smaller for each family. The rangeland has more pressure now than before, as the size of the pastures cannot be increased with the human population. Moreover, the survey of this case study shows that the increased living standard is threatening pastoral production." A male herder in Tianzhu County stated: "Our living expense is quite high here. If we raise fewer animals, our income will not be increased and our living standard will not be improved. So we need to raise more animals. But if we raise more animals, it's bad for the rangeland. We are falling into a trap (between rangeland protection and income generation)." Clearly, the local pastoralists are concerned more about social and environmental problems such as population growth, increased cost of living, overgrazing, and rangeland degradation, although their living standards were greatly improved. Therefore, they are ready to cope with them through local preparedness(e.g., depopulation through migration and livelihood diversification). In Tianzhu, as stated by a male herder, "a few pastoral people have moved into cities for migratory labor."

Political changes and their impacts are widely sensed by the local people. The survey of the case study indicates that the Rangeland Individualization and Herder Settlement policies initiated by the central government in the 1990s have been gradually landed in both counties. Conflicting responses to the impacts of these policies were obtained from the interviewed pastoralists. Some of them thought that the implementation of these policies has promoted the effective utilization of rangeland resources and improved the herders' livelihoods, whereas some of them believed the implementation of these policies has accelerated rangeland degradation by changing indigenous grazing practices. With the implementation of "Grassland

Ban" (or "Retire Livestock, Restore Grassland") aiming at protecting rangeland ecosystems and restoring degraded rangeland since 2002, the pastoralists are encouraged to fence their individual pastures for rotational grazing or fallow and resettle themselves in towns or cities for other livelihood. As a result, both pastoral livelihood and rangeland conditions have been greatly changed. In Sunan, a local official stated: "In recent years, our herders have moved from tents in the rangelands to apartments in town built by the government (as ecological immigrants to other livelihood). This is a big change." However, most pastoralists have to struggle with preparing for unexpected difficulties, such as building new social networks and learning new technologies of grazing management.

3.4.3.2 Adaptive Action and Management

To cope with environmental, socioeconomic, and political changes, local people have developed adaptive strategies in addition to preparedness. Although local pastoralists are still maintaining transhumant pastoralism, a cycling movement of livestock between summer and winter pastures, they have made some adjustments according to the changes in the climate conditions and rangeland health on temporal and spatial scales to maintain sustainable grazing. To overcome the problems of drinking water shortage and forage production reduction associated with climatic dryness and water scarcity, the local pastoralists move their livestock upward to summer pastures at high altitude earlier than before, and they move their livestock downward to winter pastures at low altitude later than before (Fig. 3.11). To take advantage of the early start of summer, the local pastoralists adopt a prolonged grazing time with higher livestock densities on summer pastures and a shortened grazing time with lower livestock densities on winter pastures. As such, rangeland degradation of winter pastures, which are normally overgrazed by high livestock populations with long winter grazing, can be somewhat mitigated. In addition, the local pastoralists strived to mitigate rangeland degradation through "frequent movement" and "reseeding of the campsite on departure" when they moved their livestock along the grazing routes.

With the implementation of government's Rangeland Individualization and Grassland Ban policies, the pastoralists have translated their indigenous transhumant grazing practices into rotational grazing management. They fence their individualized pastures (mostly winter pastures) into different paddocks, which are rotationally grazed by the appropriate amount of the livestock according to the carrying capacity of the rangelands that are set by local extension agents from their measurements or by the pastoralists themselves from their own estimation. Under the supervision of professionals (local extension agents or researchers), the local pastoralists plant some fodder crops such as oats in their yards or livestock pens to increase supplemental feed for livestock during longer and colder winters. In such a way, they can control the grazing pressures on the native rangelands and reduce the risk of rangeland overgrazing and degradation. In collaboration with the professionals, the pastoralists adopt innovative strategies to combat natural disasters and miti-

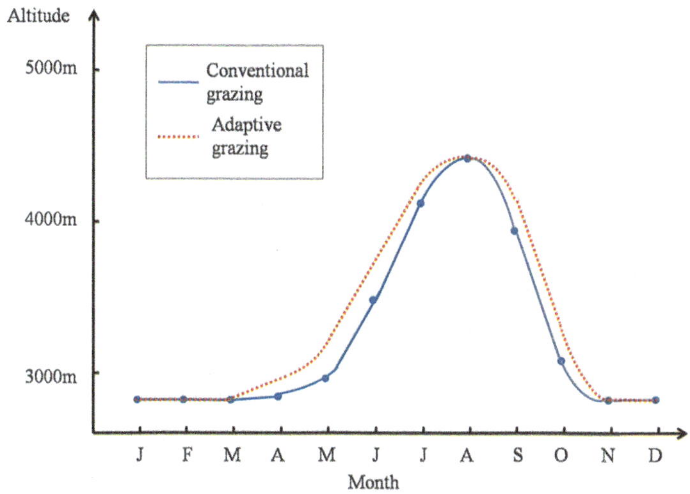

Fig. 3.11 Adaptive grazing strategies used by local pastoralists in coping with climate change

gate rangeland degradation. For example, the herders in Sunan County have been working with researchers to bring grazing chicken to control the grasshopper plague in an experimental project (Fig. 3.12); the herders in Tianzhu County have been working with local extension agents to control the rodent damage through the application of pesticides and to restore the degraded pasturelands through fencing and reseeding in a pilot project.

3.4.3.3 Multiple Partnerships and Networks

Collaboration among different stakeholders, pastoralists, researchers, extension agents, and government officials through social networks or political partnerships is a good way to promote local adaptations and innovations in political, technological, and social dimensions for sustainable pastoralism in the case study sites. The researchers have done on-the-ground work closely with local pastoralists to solve the real-world problems that are challenging and threatening their livelihood. As an interviewed researcher said: "The basis of doing research is to get the research questions from the herders. If they have some problems that need to be solved, we help them find solutions or develop research projects. So our research is really from the grassroots, we get support from the herders." This can be verified in a separate interview by the statement from one female herder in Sunan County: "The researchers come here to do some (scientific) analysis. They really help us and benefit our pastures. If it benefits our pasture, then it also benefits our income. So we welcome them (scientific researchers) from our hearts and support them." The researchers also work closely with local extension agents and government officials to translate their research findings into pastoral practices. For example, In Zhuaxixiulong

Fig. 3.12 Chicken grazing on alpine steppe in Sunan County initiated by researchers at Lanzhou University, China (photo by Kiran Goldman, 2010)

Township of Tianzhu County, the scientific researchers work with local pastoralists to experiment on the cultivated perennial grasslands with two goals: to reduce the grazing pressure on native rangelands by rearing livestock on productive cultivated pastures, and to provide alternative options for restoring degraded rangelands. The local extension organization (Grassland Station of Tianzhu County) is responsible for translating the research findings from these experimental studies into practical application through a pilot project focusing on reseeding the lands that were damaged by rodents. From our on-the-ground observations and local professionals' assessment results, it is evident that the ecological function and productivity of the rangeland as well as livestock have been greatly improved.

The survey in this case study showed that the local pastoralists have also built partnerships with the government, which provide financial support and policy instrument measures for the local pastoralists, such as funding for building stalls and fences and an eco-compensation policy for reducing livestock numbers on the rangelands. One interviewed male herder in Tianzhu County said: "The government's policies and investments have benefitted us greatly. For example, we would not be able to afford the pesticide for controlling rodents or the materials for building fences (for rotational grazing and fallow). In recent years, the government has provided a lot of subsidies for fencing. This has been very useful for rangeland protection. Without the government's support and investment, we can do nothing to combat the rangeland degradation." One interviewed female herder in Sunan County stated: "As herders, we see the rangelands as our life. In recent years, we have protected our rangelands very well. We have built fences (to protect rangelands) and improved our forage species (on cultivated pastures) with the government's policy and financial support."

From the survey, it can be seen there are some emerging partnerships among the pastoralists, although the Rangeland Individualization policy has broken their traditional community-based pastoral management systems. Some households in the case study sites come together voluntarily to collectively graze their livestock, plant their fodder crop, and harvest and store forages to copr with the uncertainties of climate change. The household collectives have made rules or norms on their own as oral or written agreements to regulate the pastoral activities, such as setting the migratory time, finding the drinking water resource, and fixing grazing. Additionally, the local pastoralists have built close networks among relatives, friends, and neighbors, through which they can learn new practices, share experiences, and communicate information. The survey indicates that the herders try to build adaptive capacities by themselves with the support obtained through the internal and networks as well as multistakeholder partnerships.

3.4.4 Implications of the Case Study

It is widely recognized that rangeland degradation associated with climate warming, overgrazing, and policy changes has threatened the pastoral production systems in the fragile and vulnerable areas of the QTP (Yeh 2003, 2010; Xu and Liu 2007; Wang et al. 2007; Harris 2010; Dong et al. 2010a, b, 2011, 2012). Klein et al. (2004) found from a simulation experiment that climate warming resulted in the decline of species richness in the QTP's alpine meadow and shrubland. They also observed that experimental warming led to a decline of the net productivity of alpine plants, particularly palatable grasses during the growing season (Klein et al. 2007). Zhang et al. (2015) found that experimental warming significantly reduced the vegetation living state of the QTP's alpine steppe. In contrast, Wang et al. (2012) concluded from a comparative experiment that heavy grazing rather than warming causes degradation of the QTP's alpine meadows. This viewpoint was supported by a great number of scholars who insist that the dominant drivers of alpine degradation in the QTP are overgrazing (Ma et al. 1999; Shang and Long 2005; Wu and Du 2007). In addition, some scholars stressed that population growth (Zhang et al. 2004), rangeland individualization, and fencing facilitated by government policies (Yan et al. 2005) have resulted in overstocking and rangeland degradation. However, Harris (2010) stated that the major causes of the QTP's rangeland degradation remained uncertain because of the vaguely tested hypotheses (e.g., overstocking is a clear driver of rangeland degradation, although policy initiatives aimed at sustainability may lead to overstocking because of insufficient understanding of current social–ecological systems of pastoralism). Therefore, Dong et al. (2012) argued that breaking human–natural systems (social–ecological systems) is greatly associated with rangeland degradation in the QTP and other areas of the developing world, where policy instruments do not achieve the objectives of promoting sustainability of rangeland production systems mainly because of overlooking emergent issues at the interface between the ecological, economic, and social perspectives (Fig. 3.13). There are inextricable linkages between the drivers of change and the adaptive responses in terms of the social, institutional, and biophysical constraints and challenges faced by local pastoral society today (Wu et al. 2015).

As the real receptors and reactors to all these drivers of environmental, socio-economic, and political changes in the pastoral realm, the local pastoralists have evolved their own perceptions and are well prepared to cope with all these changes. Most importantly, they have developed adaptive action strategies based on their own knowledge and wisdom to maximize the positive impacts of these changes and minimize their negative effects, and even to convert many limitations of these changes into opportunities in sustainable pastoralism. These adaptive action strategies, according to this case study, mainly include mobility, specificity, preparedness, diversification, exchange, collaboration, and partnership. All these strategies can promote practical applications of newly developed natural resource management framework; that is, a three dimensional framework of adaptive management, social learning, and resilient thinking (complex adaptive system) in the pastoral system of the QTP. In this three dimensional framework, as summarized by Tyler (2008), "adaptive management typically emphasizes natural science and ecological systems, social learning emphasizes

Fig. 3.13 Rangeland degradation of Gonghe County where the "Grazing Ban" policy is advocated using a big board (Photo by Shikui Dong, 2012)

human agency and interaction, and resilience thinking addresses social-ecological systems as complex entities that behave in dynamic and cyclical fashion. They (these three dimensions) can offer insights into practices that support learning, adaptation and sustainability." The results of this case study suggest that new institutions are needed to foster the adaptive action strategies of the QTP's pastoral systems in the era of environmental degradation, socioeconomic transformation, and political dynamics. The new institutions need to foster more widespread interactions among pastoralists and other stakeholders through networks, partnerships, consultative bodies, and collective actions to enhance the resilience of coupled human (e.g., indigenous practices and tradition) and natural (e.g., physical conditions) systems on the QTP.

3.5 Strengthening the Resilience of Human–Natural Systems of Pastoralism in the Himalayas

Although these three case studies on Himalayan pastoralism differ in socioeconomic, political, demographic, and cultural settings, they have addressed similar issues regarding the causes and effects of environmental, socioeconomic, and political changes in pastoral systems. Moreover, they have commonly highlighted the complicated interactions and feedbacks between human and natural systems of pastoralism in coping with all these changes, and the integration of various tools and strategies from the ecological and social sciences as well as other disciplines in sustainable pastoral development. As such, these three case studies have offered unique interdisciplinary insights into human–natural systems of pastoral management practices that support learning, adaptation, and sustainability. Moreover, the three case studies have highlighted the importance of human–natural systems in formulating a more integrated understanding of nature and society to promote the resilience of pastoral systems in the Himalaya. As stated by Liu et al. (2007a), coupled human–natural systems challenge traditional planning and management assumptions and strategies for natural resources and the environment. The success or failure of many policies and management practices is based on their ability to take into account the complexities of human–natural systems (Liu et al. 2007b).

The implications of the coupled human–natural system approaches for sustainable pastoral development in the Himalayas can be oriented specifically to policy decision making. Local pastoralists in a wide area of the Himalaya (Nepalese Himalaya, Indian Himalaya, and Chinese QTP) represent a repository of rich indigenous knowledge essential to sustain pastoral management, and underscore the need to integrate local adaptations and collective actions coping with climate changes into modern technological development and public decision making. Property rights and local institutions are vital components of political instruments in rational sharing of rangeland resources among the pastoral communities and need to be considered in the common decision analysis of pastoral conservation and production in the Nepalese Himalaya and the Indian Himalaya. Partnerships and collaboration through social networks among different stakeholders are critically important for maintaining the stability of pastoral production systems in the QTP of China, and they need to be integrated in the pastoral resource research, planning, and government.

Management practitioners and policymakers responsible for enhancing resilient pastoralism in the Himalaya are expected to become experimental learners by systematically applying scientific knowledge and approaches of coupled human and natural systems to their management and decision-making practices. Policy decisions must balance the needs of society with the best scientific knowledge of coupled human and natural systems. To facilitate this, the interfaces between social, economic, physical–biological, and ecological components in resilient human–natural systems of pastoralism must be improved. There is a pressing demand to collect and integrate new and existing research results into packages that can be used by management practitioners and decision makers. Socioeconomic and cultural components of human–natural systems for sustainable pastoralism need to be stressed and well integrated with scientific objectives and policy priorities to equitably balance local pastoralists' needs with national or regional pastoral management policies and strategies. Comprehensive programs of integrated ecological, social, and economic research should be facilitated to provide a sound foundation for decision making. Increased support and funding for researches into human–natural systems are critical to the future of Himalayan pastoralism, and must include interdisciplinary investigations of pastoral resource use and management systems, complex adaptive systems, and syntheses of the state of current knowledge.

Acknowledgments The authors acknowledge the funding organizations including Ministry of Science and Technology of the People's Republic of China, National Natural Science Foundation of China, and Asian Scholar Foundation, India China Institute of New School, Beijing Normal University for their financial supports. The authors are grateful to the scholars whose articles, tables and figures are cited in this chapter. The authors are most grateful to the farmers, herders and professionals who have been interviewed.

References

Alirol P (1979) Transhumaning animal husbandry systems in Kaligchowk Region (central Nepal). Swiss Association for Technical Assistance/Integrated Hill Development Project, Kathmandu

Bhasin V (1988) Himalayan ecology: transhumance and social organization of Gaddis in Himachal Pradesh. Kamal Raj Enterprise, New Delhi

Bhasin V (2011) Pastoralists of Himalayas. J Hum Ecol 33(3):147–177

Cai H, Zhang XF (2013) A participatory survey on status and local pastoralists' consciousness to animal husbandry in alpine grassland. Grassl Turf 33:66–72

Chakravarty-Kaul M (1998) Transhumance and customary pastoral rights in Himachal Pradesh: claiming the high pastures. In: Mountain research and development. International Mountain Society and United Nations University

Chan YN (2002). Livestock and rangeland management in Shangri-La Gorge, Shangri-La County, Northwest Yunnan, China. MSc thesis, Lund University

Chand SP, Gurung BD, Rood PG (1991) Farmers' traditional wisdom: where does it stand within the present agricultural research systems of Nepal? Pakhribas Agricultural Centre (PAC) occasional paper no 4, Dhankuta

Chapagain DP (1986). Managing public lands as a common property resources; a village case study in Nepal. PhD thesis, University of Wisconsin, Madison

Derville M, Bonnemaire J (2010) Marginalization of yak herders in Bhutan: can public policy generate new stabilities that can support the transformation of their skills and organization? Innovation and sustainable development in agriculture and food. ISDA, Montpellier, pp 1–10

Dong SK, Lassoie JP, Yan ZL, Sharma E, Shretha KK, Pariya D (2007) Indigenous rangeland resource management in the mountainous areas of northern Nepal: a case study from the Rasuwa District. Rangeland J 29:149–160

Dong SK, Lassoie JP, Shrestha KK, Yan Z, Sharma E, Pariya D (2009) Institutional development for sustainable rangeland resource and ecosystem management in mountainous areas of northern Nepal. J Environ Manage 90:994–1003

Dong SK, Li JP, Li XY, Wen L, Zhu L, Li YY, Ma YS, Shi JJ, Dong QM, Wang YL (2010a) Application of design theory for restoring the "black beach" degraded rangeland at the head water areas of the Qinghai-Tibetan Plateau. Afr J Agric Res 5:3542–3552

Dong SK, Wen L, Zhu L, Li XY (2010b) Implication of coupled natural and human systems in sustainable rangeland ecosystem management in HKH region. Front Earth Sci China 4(1):42–50

Dong SK, Wen L, Yang ZF, Liu SL, Lassoie JP, Zhang XF, Yi SL, Li JP (2011) Vulnerability of worldwide pastoralism to global changes and interdisciplinary strategies for sustainable pastoralism. Ecol Soc 16(2):10, http://www.ecologyandsociety.org/vol16/iss2/art10/

Dong SK, Lassoie JP, Wen L, Zhu L, Li XY, Li JP, Li YY (2012) Degradation of rangeland ecosystems in the developing world: tragedy of breaking coupled human-natural systems. Int J Sustain Soc 4:357–371

El-Ashry M (2009) Adaptation to climate change: building resilience and reducing vulnerability. Recommendations from the 2009 Brookings Blum Roundtable. United Nations Foundation. http://www.brookings.edu/~/media/Research/Files/Papers/2009/9/climate%20change%20poverty/09_climate_change_poverty_el_ashry.PDF

Farooquee NA (1994) Transhumance in the central Himalaya: a study of its impact on environment, Ph.D. Thesis, HNB Garhwal University, Srinagar, India.

Farooquee NA, Saxena KG (1994) Adaptation, conservation and livestock management in high hills of central Himalaya: indigenous knowledge and practice. In: Somvanshi R, Lokeshwar RR (eds) Current advances in veterinary science and animal production in India. International Book Distribution Company, Lucknow, pp 123–129

Farooquee NA, Majila BS, Kala CP (2004) Indigenous knowledge systems and sustainable management of natural resources in a high altitude society in Kumaun Himalaya, India. J Hum Ecol 16:33–42

Foggin JM (2008) Depopulating the Tibetan grasslands. Mt Res Dev 28:26–31

Folke C (2006) Resilience: the emergence of a perspective for social-ecological systems analyses. Glob Environ Change 16:253–267

Gadgil M, Berkes F, Folke C (1993) Indigenous knowledge for biodiversity conservation. Ambio 22:151–156

Gibbs JN, Bromley DW (1989) Institutional arrangements for management of rural resources: common property regimes. In: Berkes F (ed) Common property resources: ecology and community-based sustainable development. Belhaven, London, pp 22–33

Gill GJ (1993) Indigenous system in agriculture and natural resource management: the policy dimension. In: Tamang D, Gill GJ, Thapa GB (eds) Indigenous management of natural resources in Nepal Tamang. Proceeding of the workshop on indigenous management of agriculture and natural resources, Dhulikhel, Nepal, 8–9 June 1992. Winrock International, Kathmandu, pp 3–12

Gilmour DA (1990) Resource availability and indigenous forest management systems in Nepal. Soc Nat Resour 3:145–158

Gordon HS (1954) The economic theory of a common pool resource: the fishery. J Polit Econ 62:124–142

Hardin G (1968) The tragedy of commons. Science 162:1243–1248

Harris RB (2010) Rangeland degradation on the Qinghai-Tibetan plateau: a review of the evidence of its magnitude and causes. J Arid Environ 74:1–12

Ives JD, Messereli B (1989) The Himalayan dilemma: reconciling development and conservation. Routledge, New York

Janzen J (1993) Mobile Livestock Keeping in Somalia - General Situation and Prospects of a Way of Life Undergoing Fundamental Change. In: Baumann M. P. 0., Janzen J. and Schwartz H. J. (eds.) Pastoral Production in Central Somalia, . OTZ, Eschborn

Kassam KS (2010) Pluralism, resilience, and the ecology of survival: case studies from the Pamir mountains of Afghanistan. Ecol Soc 15(2):8, http://www.ecologyandsociety.org/vol15/iss2/art8/

Klein JA, Harte J, Zhao XQ (2004) Experimental warming causes large and rapid species loss, dampened by simulated grazing, on the Tibetan Plateau. Ecol Lett 7:1170–1179

Klein JA, Harte J, Zhao XQ (2007) Experimental warming, not grazing, decreases rangeland quality on the Tibetan Plateau. Ecol Appl 17:541–557

Kreutzmann H (2012) Pastoralism—a way forward or back? In: Kreutzmann H (ed) Pastoral practices in High Asia. Agency of 'development' effected by modernisation, resettlement and transformation. Springer, Dordrecht, pp 323–336

Land Resource Mapping Project (1986). Summary report of LRMP. His Majesty's Government of Nepal and Government of Canada. Land Resource Mapping Project, Kathmandu

Liu, J.G., Dietz T., Carpenter S.R, Alberti M., Folke C., Moran E., Pell A.N., Deadman P., Kratz T., Lubchenco J., Ostrom E., Ouyang Z.Y., Provencher W., Redman C.L., Schneider S.H., and Taylor W.W. (2007a). Complexity of coupled human and natural systems. Science 317(5844):1513-1516.

Liu, J.G., Dietz, T., Carpenter, S.R., Folke, C., Alberti, M., Redman, C.L., Schneider, S.H., Ostrom, E., Pell, A.N., Lubchenco, J., Taylor, W.W., Ouyang, Z.Y., Deadman, P., Kratz, T. and Provencher, W. (2007b) Coupled human and natural systems, Ambio, 36: 639–649.

Long RJ, Zhang DG, Wang X, Hu ZZ, Dong SK (1999) Effect of strategic feed supplementation on productive and reproductive performance in yak cows. Prev Vet Med 38:195–206

Long RJ, Ding LM, Shang ZH, Guo XH (2008) The yak grazing system on the Qinghai-Tibetan plateau and its status. Rangeland J 30:241–246

Ma YS, Lang BN, Li QY, Li YF, Li FJ (1999) The present status of the grassland eco-environment at the headwater areas of Qinghai-Tibetan Plateau and resume strategies of degraded grassland. Grassl China 55:59–61

Macchi M, Gurung AM, Hoermann B, Choudhury D (2011) Climate variability and change in the Himalayas: community perceptions and responses. International Centre for Integrated Mountain Development, Kathmandu,

Maikhuri RK, Nautiyal S, Rao KS, Chandrasekhar K, Gavali R, Saxena KG (2000) Analysis and resolution of protected area-people conflicts in Nanda Devi Biosphere Reserve, India. Environ Conserv 27(1):43–53

Meinzen-Dick R, Gregorio MD (2004) Collective action and property rights for sustainable development: overview. In: Meinzen-Dick R, Gregorio, MD (eds) Collective action and property rights for sustainable development: 2020 vision for food, agriculture and environment. 2020 focus II, brief 1 of 16'. International Food Policy Research Institute, Washington, pp 1–2

Miles MB, Huberman AM (1994) Qualitative data analysis: an expanded sourcebook, 2nd edn. Sage, Thousand Oaks

Ostrom E (1990) Governing the commons: the evolution of institutions for collective action. Cambridge University Press, London

Patton MQ (1990) Qualitative evaluation and research methods, 2nd edn. Sage, Newbury Park

Rai N, Thapa MB (1993). Indigenous pasture management systems in high-altitude Nepal: a review. HMG Ministry of Agriculture/Winrock International Policy Analysis in Agriculture and Related Resource Management, research report series, no 22. Winrock International, Kathmandu,

Rajbhandary HB, Pradhan SL (1990). Livestock development and pasture management. In: National Planning Commission/The World Conservation Union (eds) Background papers to the national conservation strategy for Nepal. The World Conservation Union, Kathmandu, pp 1–10

Rawat GS (1998) Temperate and alpine grasslands of the Himalaya: ecology and conservation. Parks 8:27–36
Scott A (1955) The fishery: the objectives of sole ownership. J Polit Econ 63:116–124
Shang ZH, Long RJ (2005) Formation reason and recovering problem of the "black soil type" degraded alpine grassland in Qinghai-Tibetan Plateau. Chi J Ecol 24(6):652–656
Sharma VP, Rollefson IK, Morton J (2003) Pastoralism in India: a scoping study. Centre for Management on Agriculture, IIM, Ahmadabad
Shrestha TB (2001). Status review national strategies for sustainable development forestry/rangeland/biodiversity. The World Conservation Union, Kathmandu
Singh KS (1996) Peoples of India series. Anthropological Survey of India, New Delhi
Singh JP, Roy MM, Radotra S (2009) Grasslands of Himachal Pradesh. Indian Grassland and Fodder Research Institute , Jhansi
Sundriyal RC (1995) Grassland forage production and management in the Himalaya: a review. J Hill Res 8(2):135–150
Tamang D (1993) Challenges and opportunities in farm and community resource management in Nepal. In: Tamang D, Gill GJ, Thapa GB (eds) Indigenous management of natural resources in Nepal Tamang. Proceedings of the workshop on indigenous management of agriculture and natural resources, Dhulikhel, Nepal, 8–9 June 1992. Winrock International, Kathmandu, pp 13–23
Tesfay Y, Tafere K (2004). A pastoral forum organized by the Dryland Coordination Group (DCG) in Ethiopia, 27–28 June 2003. DCG report no 31, November 2004. Dryland Coordination Group, Mekelle
Tourism for Rural Poverty Alleviation Program (2005) Tourism resource mapping profile: Resuwa District. (NPTC limited), Chabahil: Kathmandu, Nepal
Tyler SR (2008) Adaptive learning in natural resource management: three approaches to research. Working paper 22. International Development Research Centre
Walker B, Salt D (2006) Resilience thinking: sustaining ecosystems and people in a changing world. Island, London
Walker B, Holling CS, Carpenter SR, Kinzig A (2004) Resilience, adaptability and transformability in social-ecological systems. Ecol Soc 9(2):5, http://www.ecologyandsociety.org/vol9/iss2/art5/
Wang GX, Wang YB, Li YS, Cheng HY (2007) Influences of alpine ecosystem responses to climatic change on soil properties on the Qinghai-Tibet Plateau, China. Catena 70:506–514
Wang SP, Duan JC, Xu GP, Wang YF, Zhang ZH, Rui YC, Luo CY, Xu BRBY, Zhu XX, Chang XF, Cui XY, Niu HS, Zhaio YQ (2012) Effects of warming and grazing on soil N availability, species composition and ANPP in alpine meadow. Ecology 93:2365–2376.
Wu N (1997) Indigenous knowledge and sustainable approaches for the maintenance of biodiversity in nomadic society: experiences from eastern Tibetan Plateau. Erde 128:67–80
Wu X.Y., Zhang X.F., Dong S.K., Cai, Zhao T.R., Yang W.J., Jiang R., Shi Y.D., Shao J.L. (2015) Local perceptions of rangeland degradation and climate change in the pastoral society of Qinghai-Tibetan Plateau. The Rangeland Journal, 37: 11–19.
Xu WX, Liu XD (2007) Response of vegetation in the Qinghai-Tibet Plateau to global warming. Chin Geogr Sci 17:151–159
Yan ZL, Wu N, Yeshi D, Ru J (2005) A review of rangeland privatization and its implications in the Tibetan Plateau, China. Nomadic Peoples 9:31–51
Yeh ET (2003) Tibetan range wars: spatial politics and authority on the grasslands of Amdo. Dev Change 34:499–523
Yeh E (2010) Restoring the grasslands? Chinadialogue, 26 January
Zhang HF, Liu FG, Zhou Q, Duo HR (2004) Degradation mechanism of the grass in Qinghai Plateau and its prevention and control countermeasures. J Nat Disasters 13:115–120 (in Chinese)
Zhang Y., Gao Q.Z, Dong S.K., Sherman R., Wang X.X., Li,Y.Y., Tang L., Wu X.Y. (2015) Effects of grazing and climate warming on plant diversity, productivity and living state in the alpine rangelands and cultivated grasslands of the Qinghai-Tibetan Plateau. The Rangeland Journal 37: 57–65.

Chapter 4
Sociocultural and Ecological Systems of Pastoralism in Inner Asia: Cases from Xinjiang and Inner Mongolia in China and the Pamirs of Badakhshan, Afghanistan

Karim-Aly S. Kassam, Chuan Liao, and Shikui Dong

Abstract In pastoral societies, economic and ecological aims are not necessarily in conflict. These societies, through mobility, engage different ecological niches as a livelihood strategy. Specific case studies from Inner Asia indicate that instead of seeking to replace pastoralism as an ecological profession through forced sedentarization, governments should seek to enhance its historically proven potential for food and livelihood security. The case from the Altay Mountains and the Tian Shan documents the effect of sedentarizing pastoral communities, resulting in the removal of sociocultural and ecological diversity, with profound consequences on income. It is an example of the central government asserting administrative authority in the name of ecological restoration while pursuing strictly an instrumental agenda of economic extraction of key renewable and nonrenewable resources. The case from Inner Mongolia shows increased economic and ecological vulnerability of pastoral societies caused by government-induced sedentarization programs but also illustrates the adaptive capacity of pastoral institutions under such policies. The final case, from the Pamirs, shows that under conditions of political and economic stress, interactions between diverse ecological professions such as farmers and herders is central to livelihood and food security through mutual dependence. It is the basis for survival.

K.-A.S. Kassam (✉)
Department of Natural Resources and American Indian and Indigenous Studies Program, Cornell University, Ithaca, NY 14853-3001, USA
e-mail: ksk28@cornell.edu

C. Liao
School of Natural Resources and Environment, University of Michigan - Ann Arbor, Ann Arbor, MI, USA

S. Dong
School of Environment, Beijing Normal University, Beijing, China

S. Dong et al. (eds.), *Building Resilience of Human-Natural Systems of Pastoralism in the Developing World: Interdisciplinary Perspectives*, DOI 10.1007/978-3-319-30732-9_4

4.1 Introduction

Variation and difference are the hallmarks of pastoralism. Pastoralism is not merely a livelihood strategy but a way of life that is fundamentally based on adaptation to changing seasonal and, therefore, climatic conditions in different ecological contexts. Mobility through pastoral activities and the subsequent food security arising from those undertakings are not only a necessity but are a recognized behavioral norm with sociocultural significance. Thus, pastoralism is not only an ecological profession strategic to securing human survival, but in turn, generates a mutually reinforcing sociocultural identity that draws primarily from connectivity with the ecosystems in which humans seasonally dwell. The cultural values and social institutions, in turn, facilitate pastoral activities. The relationship is neither linear nor deterministic. Pastoralists are not hemmed into an ecological niche but rather engage in complex connectivity with diverse habitats. Environments simultaneously shape and are a product of human actions. Complexity and uncertainty effect pastoralism as a livelihood strategy and a way of life; the system is dynamic.

Central to understanding pastoralism is recognition of the mutual relationship between cultural and ecological diversity. Drawing on already published applied research on Inner Asia, specifically the Altay Mountains and the Tian Shan (Liao et al. 2014a, b), Inner Mongolia (Dong et al. 2007; Dong and Ren 2015), and the Pamirs (Kassam 2010), we will explore the implications of externally induced perturbations to pastoral systems as livelihood strategies. The first case study examines the implications of decades of centralized planning through collectivization, then decollectivization, and now sedentarization policies on Kazakh pastoralists and their livelihoods in Xinjiang (northwestern China). The second case study draws evidence from Inner Mongolia (China) to illustrate the impacts of institutional arrangements driven by privatization and their effects on pastoral livelihoods as well as adaptive responses to government policies. The third case study examines the relationship between pastoralists and farmers in mutually securing each other's food security and survival under conditions of war in Badakhshan (northern Afghanistan). We will conclude with a discussion of the diversity and potential insights the three case studies reveal.

4.2 Case Study 1: Livelihood Diversity and Pastoralism in the Altay Mountains and the Tian Shan of Xinjiang, China

4.2.1 Context

The Xinjiang Uyghur Autonomous Region is located in northwestern China, and lies in the center of the Eurasian landmass (Fig. 4.1). It spans more than 1.6 million square kilometers. Situated in the middle of the ancient Silk Road, Xinjiang has a

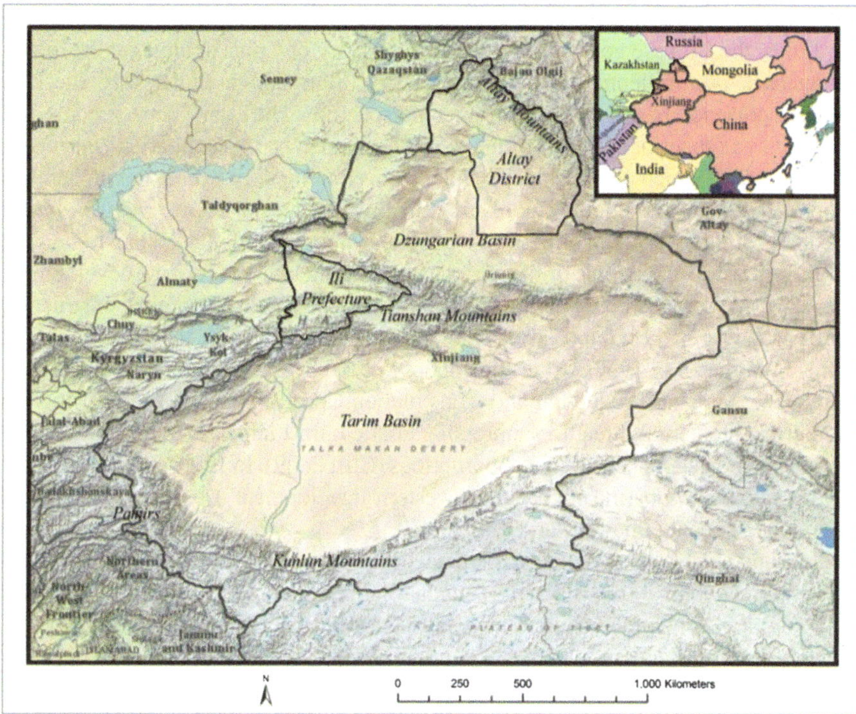

Fig. 4.1 Altay District and Ili Prefecture in Xinjiang, China

border of more than 5600 km, neighboring eight countries from the northeast to the southwest, including Mongolia, Russia, Kazakhstan, Kyrgyzstan, Tajikistan, Afghanistan, Pakistan, and India. The region is characterized by biophysical diversity: movement from the south to the north involves crossing physical landscapes that range from the second highest point (K2, 8611 m) to the second lowest point (Aiding Lake, −154 m) on the planet (Starr 2004).

The physical geography of Xinjiang can be summarized as "two basins within three mountains" (XUAR Chorography Committee 2010). The Tarim Basin is between the Kunlun Mountains in the south and the Tian Shan in the north. The Dzungarian Basin is between the Tian Shan in the south and the Altay Mountains in the north. In the middle of the Tarim Basin lies the Taklimakan desert, where the annual rainfall is less than 30 mm (Li 1991). As the region most remote from oceans in the world, the water vapor from the sea almost disappears because of distance and mountain barriers.

The Han Chinese name for the region reveals a history of repeated conquests, resultant rebellions, and external exploitation of this frontier region (Kassam 2001). In Chinese, the word "Xinjiang" consists of two characters: *xin* meaning "new," and *jiang* meaning "territory." The glyphic components of the character *jiang* consist of the bow, the earth, and the fields, meaning land that needs weapons to protect it.

Although non-Han Chinese sources maintains that Xinjiang was annexed by China in the 1760s, the Chinese government asserts the history of China's rule over the region dating back two millennia to the Han Dynasty (Starr 2004). Either way, the Han Chinese presence is driven by a frontier perspective. The Han Chinese perceived themselves as superior residents of the *core*, surrounded by the "barbarian" *periphery* (Amitai 2005), which also includes Xinjiang. Even in modern China, the indigenous peoples living in the ethnic regions are still given a special name: *shaoshu minzu* (meaning "minority nationalities"). A general perception of the ethnic regions is "backwardness," and people of the periphery require help in the form of development interventions from the core (Cerny 2010). The Xinjiang Production and Construction Corps was originally composed of soldiers who participated in the "liberation" of Xinjiang in the early 1950s. After the collapse of the Soviet Union, the primary mission of the Xinjiang Production and Construction Corps shifted from protecting the frontier from external threat to suppressing ethnic unrest largely due to limited local autonomy and unequal economic opportunities (Cliff 2009). In 1999, China embarked on the *xibu da kaifa* (meaning "great western development") campaign. This has informed its recent policy toward Xinjiang.

Xinjiang's economic structure displays distinct characteristics of periphery and frontier (Becquelin 2004). Since the foundation of the People's Republic of China in 1949, significant natural resources have been extracted to support economic development in the Chinese core, whereas manufactured goods are shipped in the opposite direction. Xinjiang is a major supplier of primary products, including energy, strategic minerals, livestock, and cash crops (Goodman 1989, 2004; Toops 2004). There is little doubt that Xinjiang will become the energy base of China, with reserves of more than 2.5 billion tons of petroleum and 700 billion cubic meters of natural gas (Xinhua 2007a).

Unlike the pastoral cultures of Xinjiang, who maintained complex connectivity with their habitat, the central government's connectivity has been instrumental as it views this region as a frontier for exploitation, and its connectivity with the habitat is purely an instrument, a source of wealth. Before the foundation of the People's Republic of China, livestock herding activities were organized in the unit of tribes. Each tribe had its own winter, spring/fall, and summer pastures, which were exclusive. In addition, each tribe had its own migration route. Although the pastures were shared by all tribal members, the livestock were owned by individual households (Mi'erzhahan 2004). Some wealthy households chose to settle in towns or villages, retaining ties with poorer herdsmen, who raised animals for them in return for a share in the herd (Benson and Svanberg 1998).

Changes started in the 1960s, as pastoralists were forced to "hand in" their livestock and herd for the communes. The pastoral unit (*muye dui*) served as a substitute for tribal institutional structures. In this way, the traditional resource use patterns were preserved, and pastures remained sustainable, until decollectivization spread to these remote areas in the mid-1980s. Subsequently, livestock and pasture lands were assigned to individual households according to their communal herding units. However, inequitable allocation severely limited some households' access to pastures and water resources (Miller 2000). Although individual households were stimulated to acquire wealth under the newly introduced market-oriented economy, increases in livestock production have been largely achieved by exploitation of pasture resources. Although other reasons might have led to pasture degradation, arguably the resource use patterns under the current land tenure have played an important role in exacerbating the situation since its initiation (Longworth 1993).

Despite a frontier perspective, in the past decade, the central government has initiated a series of ecological restoration, sedentarization, and development projects throughout its pastoral areas (Xinhua 2007b). These policies, ironically, were justified on the basis of current resource use patterns having seriously damaged the pasture lands (Harris 2010). Moreover, the Twelfth Five-Year Plan of China further confirmed the determination to "civilize" the pastoralists by settling them down and transforming them into modern ranchers (NDRC 2011). However, a review of these projects indicates further economic disenfranchisement and social marginalization for disadvantaged indigenous peoples, while generating questionable environmental benefits (Yeh 2009). Encroaching interests on the pastures from outside combined with inherent difficulty to manage the semiprivatized common resources have challenged the sustainable use of pasture lands. Given these challenges, pastoralists have been sedentarized, started cultivating crops, tried diversifying income sources, and even emigrated to other countries (Cerny 2010; Fernandez-Gimenez and Le Febre 2006).

4.2.2 Methods

Semistructured interviews were conducted with 159 households in the summer of 2011. Ninety-six of them were in Altay District, covering four counties: Aletai, Fuhai, Buerjin, and Habahe (Fig. 4.2a). Sixty-three of them were in Ili Prefecture, covering six counties: Zhaosu, Tekesi, Gongliu, Xinyuan, Nileke, and Yining (Fig. 4.2b). Although the sampling method was unstructured, we tried to interview respondents who represented diversified perspectives. We visited households on summer pastures, on transitional pastures, in winter villages, and in resettlement villages. Interviews were conducted at individual homes, including houses, huts, yurts, and tents. In sum, the aim of household sampling was to capture the relative variation in the physical environment, migration patterns, livestock structures, and income sources.

In each household, we first recorded the coordinates using a GPS instrument. Then we interviewed the male head of the household, if he was available. We only wrote down the personal characteristics of the major interviewee, but we recorded all comments contributed by other family members. When the head of the household was absent, we interviewed another family member who was willing to participate and talk. The questions were asked in Chinese and translated into Kazakh by a local facilitator, who was fluent in both Chinese and Kazakh. Questions sought to capture a broader perspective of livelihoods, which included household income, livestock and other assets, and subsistence activities.

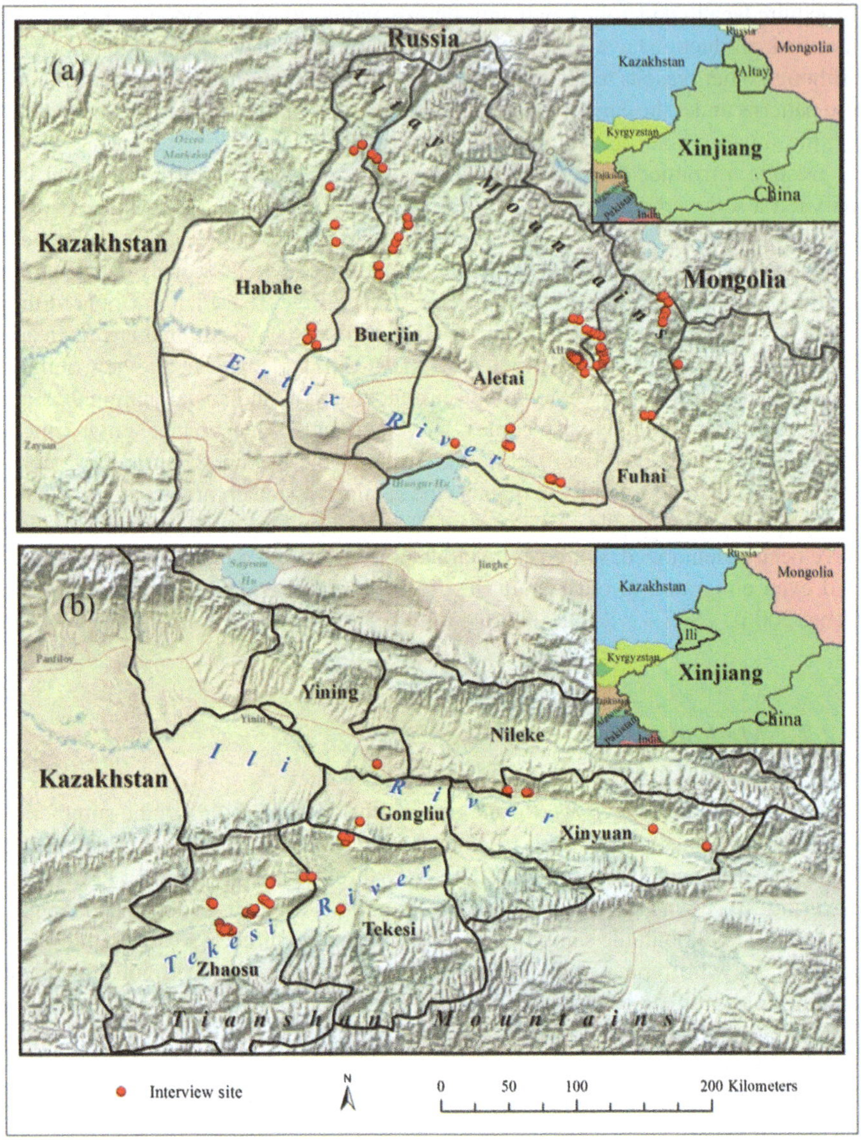

Fig. 4.2 Interview sites in Altay District (**a**) and Ili Prefecture (**b**) of Xinjiang, China

4.2.3 The Role of Livestock

The major livestock raised by pastoralists are cattle, sheep, and goats, but they also keep a small number of horses and camels (Table **4.1**). Each kind of livestock plays different roles. In general, cattle, sheep, and goats are mainly raised for markets, whereas horses and camels are largely used for transportation.

Table 4.1 Number of livestock owned by interviewed households

Livestock	In Kazakh	In Chinese	Median	Mean	Standard deviation	Maximum	Mininium
Cattle	*Sier*	*Niu*	10	12.08	10.11	60	0
Sheep/goats	*Koyi*	*Yang*	40	69.07	80.73	400	0
Horses	*Utt*	*Ma*	3	4.93	6.65	35	0
Camels	*Tuye*	*Luotuo*	0	0.89	2.15	11	0
Livestock units[a]			20.40	28.72	24.50	118.80	0

[a]1 livestock unit = 1 cow = 1 horse = 0.8 camel = 6.5 sheep or goats (Chilonda and Otte 2006)

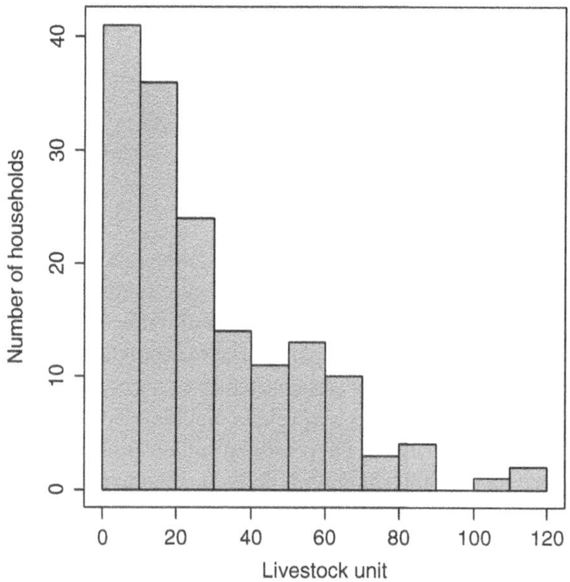

Fig. 4.3 Livestock unit distribution

The distribution of livestock units owned by individual households follows not a normal distribution but a Poisson distribution, with more households at the lower end (Fig. 4.3). Almost 40 % of them have less than 15 livestock units, whereas less than 15 % possess more than 60. This indicates that most of these households are maintaining their livelihoods on the basis of a very limited number of livestock.

A comparison of average livestock numbers in Altay and Ili is shown in Fig. 4.4. Individual households in Altay (32.1) raise significantly more livestock units than those in Ili (23.5). In terms of specific livestock types, the Altay pastoral households keep more cattle, sheep/goats, and camels, but their average horse number is slightly lower than that of their Ili counterparts. Arguably, such livestock structures in these two regions reflect the environmental differences: camels exist only in Altay, where the Gobi desert is prevalent; more horses are raised in Ili, where the pastures are of better quality.

Fig. 4.4 Average livestock numbers in Altay and Ili

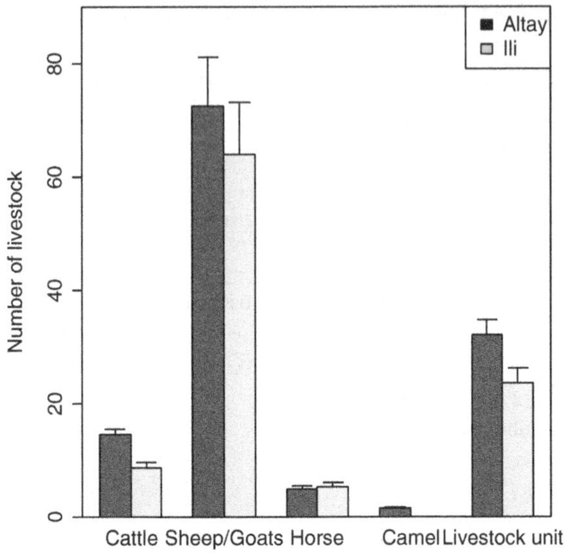

Most pastoralists only sell male calves, and keep females for milk or reproduction. According to the owners of large cattle herds, the proportion of females to males is between 10:1 and 5:1. Compared with other livestock, cattle are more susceptible to the threat of poisonous plants. Four respondents in Altay mentioned that their cattle died after consuming certain species of herbs. According to their description, the proliferation of poisonous species coincides with drought. When rainfall is low, most grass species wither, but poisonous plants prosper. Although cattle appear to know the toxicity of plants, they have no choice but to consume them when they are extremely hungry. Other kinds of livestock move more frequently to avoid the poisonous plants in their search for forage during drought periods.

There is a common word for sheep and goats in Kazakh (*koyi*) and Chinese (*yang*). Pastoralists are fully aware of the difference between sheep and goats, but they tend to use *koyi* to refer to these two species. From fieldwork observations, only 10–20 % of the *koyi* are goats. Kazakh pastoralists think the sheep are more economically valuable than goats because sheep grow much faster in their context. In both Altay and Ili, sheep/goats are the dominant livestock species, and almost 90 % of livestock income is from them.

Although the number of horses is much smaller than that of cattle and sheep/goats, horses play a significant cultural role among Kazakh pastoralists. The Kazakhs are proud of their mobile pastoral culture. Children start to learn horseback riding at the age of 5 years no matter what sex they are. In addition, a variety of sports and entertainment activities on the pastures are based on horseback riding. As a major source of transportation, horses are seldom raised to earn cash except for a few households in Zhaosu County[1] in Ili.

[1] Zhaosu, as the hometown of "heavenly horses" in ancient tales, has a long tradition of horse raising.

Table 4.2 Income of sampled households

Sources of income	Mean income (yuan)	Standard deviation	Proportion of households involved (%)	Mean total income of households involved (yuan)
Livestock	37,612.6	45,612.2	76.7	49,019.7
Crops	6510.7	14,145.3	30.2	21,566.7
Wages	4839.2	18,332.0	17.0	28,497.8
Herding fees	2987.5	11,537.1	26.4	11,310.0
Subsidies	1867.5	9725.5	12.6	14,847.0
Small business	769.8	1757.9	19.5	3948.4

Only 36 of 96 households in Altay own camels, whereas none of the 63 households in Ili do. Although camels are helpful in moving belongings during migration, more and more households are choosing not to keep camels anymore. Instead, they rent a truck to move their belongings. The average truck rental fee was about 500 yuan, which was almost half the price of a sheep in 2010. Given that the median number of sheep was 40, the cost of renting a truck to move back and forth in a year would been 2.5 % of the sheep flock value.

4.2.4 *Diversified Sources of Income Among Pastoral Households*

Household income was either estimated indirectly or reported directly by the interviews, depending on the specific sources. In general, there are six sources of income: livestock, crops, wages, herding fees, subsidies, and a small business (Table **4.2**). Income here is just cash income without consideration of household self-consumption. According to our interviews, most households consume a very small part of their livestock or crop. Meat is considered a luxury that is mainly sold to earn cash, just as crops are aimed at regional markets rather than for local consumption.

Respondents usually reported the number of livestock they sold each year and the size of crop fields they cultivated. On the basis of the local prices[2] of livestock and crops around the fieldwork period, the income from these two sectors could be estimated. Herding fees were calculated according to the number of livestock cared for, the length of time the respondents herd for others, and the herding price for each kind of livestock.[3] Other sources of income such as wages, subsidies, or a small business were directly reported by respondents.

[2] In 2010, the price of a lamb was about1100 yuan, that of a calf was about 2500 yuan, and that of a horse was about 5000 yuan. The average income from a *mu* of crop field is about 800 yuan. 1 *mu* = 666.67 m^2.

[3] In 2010, the price for herding one cattle was 50 yuan per month and the price for herding a sheep/goat was 8 yuan per month.

Quite a number of respondents pointed out that the price of livestock had just increased to a satisfactory level in the previous couple of years. Therefore, the estimation is based on the highest price. Five years ago, the price of a lamb was about 200 yuan, which was less than 20 % of the value in 2010. Since pastoral households largely depend on the sale of livestock to sustain their livelihoods, their welfare is closely linked to the livestock price. This makes them vulnerable to unexpected price fluctuations and disease. In addition, some households mentioned that although they became well off because of higher livestock prices, the cost of other necessities increased accordingly, which offset their increasing income. Therefore, vulnerability continues to be a major concern.

The details of each income source are presented in Table **4.2**. The most import source is livestock. The average income from this sector is about 38,000 yuan, and 77 % of households are more or less dependent on the sale of sale livestock to sustain their livelihoods. For those engaged in this sector, the average total income is more than 49,000 yuan.

The second most important source of income is crop cultivation, in which 30.2 % households are engaged. Cultivation of hay and other crops used for livestock consumption is not counted here. Popular crops cultivated in the study areas are cash crops, which include certain kinds of beans and melons. However, crop cultivation is not Kazakh people's comparative advantage, especially under harsh environmental conditions that require more labor and capital investment. Therefore, quite a number of Kazakh households choose to rent their crop fields to Han Chinese.

Seventeen percent of households are engaged in wage labor. The average income from this sector is 4839 yuan, but for those who are involved in this sector, their average income is about 28,000 yuan. In general, there are two types of wage income. The first type is employment in government organizations. Respondents belonging to this group have a relatively steady income. The second type is temporary seasonal employment, which mainly includes construction and farming work. Some Kazakhs have to seek such employment on a daily basis.

More than a quarter of households take care of livestock owned by other individuals to earn income through a "hired herding fee." This has become prevalent especially in recent years, not only because some newly settled Kazakh pastoralists continue to maintain a substantial amount of livestock, but also because immigrant Han Chinese raise animals for profit and self-consumption. Except for a small proportion of hired herders who take care of the livestock of others throughout the year, most of them only do that during the warm season from May to September. Some hired herders expressed concerns about theft of livestock for which they are responsible. Loss of even one animal requires compensation, which takes them several months.

Households that depend on government subsidies to maintain their livelihoods account for 12.6 % of households. In most cases, subsidies were given to pastoral households as compensation for their giving up land tenures for pasture conservation purposes. As the implementation of pasture fencing is becoming intensive, more households will receive income from this sector soon. Compared with others, households from a community in Kanasi National Park receive a much greater subsidy because of tourism development. This is because they are deprived of the rights to rent their houses to tourists, from which they could earn much more. Conflicts occur every year when it comes to their rights to rent their houses and how much compensation they should get if they give up renting. In addition, some households simply receive a subsidy for poverty relief. However, eligibility for a poverty subsidy is always controversial. Quite a number of respondents complained about the unfairness, because the subsidy was usually allocated to households who maintained a good relationship with the local officials.

About 20 % households run a small business as a source of income. This is practiced by their selling milk and processed milk products, either to middlemen who purchase milk from a number of households or to tourists. Another form of a small business is a small grocery store operated from a yurt, as access to certain grocery items is very limited on pasturelands.

4.2.5 Discussion

Evidence from Altay and the Tian Shan indicates pastoralism continues to be a viable livelihood strategy. We identified six distinct livelihood strategies as the optimal fit in our cluster analysis (Everitt et al. 2011). The summary statistics of the identified strategies are given in Table **4.3**.

Farmers (cluster 1) represents 13.8 % of the entire sample. On average, they receive more than 70 % of income from crops, which is almost four times as much as agropastoralists, for whom crop revenue is the second most important source of income. About 10 % comes from livestock, which is much less than for the agropastoralists, who derive more than 60 % from this sector. Another key distinction between farmers and agropastoralists is the average household income. Farmers earn only 55 % of what agropastoralists do. In addition, farmers' income is also about 40 % less than the overall average. Income from other sources is minimal for this cluster.

The households in cluster 2, mixed smallholders, earn the least income compared with other clusters, only 47 % of the overall average. They rely heavily on government subsidies to maintain their livelihoods. Another feature of this cluster is the reliance on a small business. About 20 % of their income is from selling milk products and grocery items, whereas none of the other clusters derive more than 5 % of their income from this sector. The remaining 20 % of income is either from livestock or herding fees. Mixed smallholders are not engaged in crop cultivation or wage labor at all.

The third livelihood strategy (cluster 3), agropastoralism, is a combination of livestock herding and crop cultivation. Agropastoralists are the second largest group, representing about 20 % of the whole sample. They have the second highest mean

Table 4.3 Livelihood strategies estimated via *k*-means cluster analysis

Variables	Cluster 1	Cluster 2	Cluster 3	Cluster 4	Cluster 5	Cluster 6	Mean
Livestock (%)	10.7	10.7	60.2	0.0	95.1	2.7	57.2
Crops (%)	70.3	0.0	18.0	2.7	0.6	1.9	13.9
Wages (%)	6.1	0.0	9.9	88.2	0.4	0.0	9.7
Herding fees (%)	6.6	11.0	7.3	3.0	2.1	93.2	8.7
Subsidies (%)	2.2	43.9	1.4	2.0	0.1	0.0	4.9
Small business (%)	4.2	21.1	3.2	4.1	1.7	2.2	4.4
Other variables							
Household income	33,179.1	25,593.3	60,125	45,943.3	68,016.1	37,300	54,587.4
No. of households	22	15	32	12	71	7	159
Fraction of household (%)	13.8	9.4	20.1	7.5	44.7	4.4	100.0
Strategy name	Farmer	Mixed smallholder	Agropastoralist	Wage laborer	Pastoralist	Hired herder	Whole sample

income among the six groups. Livestock revenue, as the most important income source for this cluster, constitutes more than 60 % of their income. This is followed by crop revenue, which accounts for almost 20 % of the total.

The distinguishing feature of wage laborers (cluster 4), representing 7.5 % of the sample, is their dominant reliance on wages as a source of income, which accounts for almost 90 % of the total. This cluster is the only one that gains no income from livestock. Their income from other sources is also minimal. Although wage laborers are the third wealthiest group, their income is still about 15 % less than the average.

Cluster 5, pastoralist, the largest group among the six clusters, represents almost 45 % of the entire sample. More than 95 % of the income of pastoralists is from livestock, whereas the other sources are negligible. Their dominant reliance on livestock makes them the wealthiest group. They earn more than 68,000 yuan annually, which is 2.5 times more than the poorest cluster.

Cluster 6 exhibits characteristics that can be best described as hired herder. Households in this cluster derive 93.2 % of their income from herding fees. Their income from the sale of livestock is minimal, but their work is similar to that of pastoralists in terms of tending to livestock. A major difference is that hired herders do not own most of the animals they herd. Although this cluster accounts for less than 5 % of the entire sample, all other clusters are more or less engaged in herding livestock for others. As an emerging source of income, being a hired herder is becoming more prevalent. Hired herders earn a mean income of 37,300 yuan, which is 30 % less than the average.

Although pastoralism is the preferred livelihood strategy, only 45 % of households are currently able to derive a large share of their income from livestock. The ongoing transition, which is from depending heavily on livestock herding to relying on diversified income sources, is exactly what the government wants to achieve in the Twelfth Five-Year Plan. The official policies aim at sedentarizing pastoralists and transforming them into modernized ranchers who are able to produce large quantities of dairy products and meat using an industrialized approach. However, only the first half of this approach is being implemented, and the second part has been abandoned. In the implementation of these policies, new houses ranging from 60 to 90 m^2 with a 3-*mu* (about 2000-m^2) yard are sold to pastoralists with a subsidy. In addition, another 50-*mu* (about 3.33-ha) hayfield is given for free as a bonus. However, almost all respondents complained that a 50-*mu* hayfield is far from being enough to sustain a viable number of livestock. What makes the situation worse is that the quality of the bonus hayfield is much worse than that of the land they owned before. There is little water, and the soil is highly saline and alkaline. Some households also noted that the assigned hayfield is too far from the village, which makes it difficult to manage the land.

In response to a series of socioecological changes and policy pressures, pastoral households are trying to diversify their sources of income. However, such diversification is accompanied by reduced welfare, which is directly reflected in household income; therefore, it is hard to conclude that diversified income sources can always contribute to household welfare. From our analysis, higher income diversity is associated with lower annual household income in the pastoral context (Liao et al. 2015).

In response to the identified socioecological challenges in the pastoral contexts, researchers working in different study areas almost unanimously reached the conclusion that future development activities need to be built on the foundation of the livestock economy instead of seeking other ways to replace it (Behnke 1993; Sandford 1983), especially in the arid and semiarid lands. Research findings in the same context have indicated that mobile livestock herding is inherently diverse and highly adaptive, which involves complex spatial movement, land use patterns, and a livestock portfolio (Liao et al. 2014a, b). It is such that ecological diversity allows pastoralists to make better use of the rangeland resources constantly in disequilibrium.

4.3 Case 2: Adaptation To Mitigate Pastoral Vulnerability Associated with Institutional Transformations in Inner Mongolia, China

4.3.1 Context

Inner Mongolia, covering a total area of approximately 183 million square kilometers and a total distance of about 2400 km from west to east, is located in northern China, bordering Mongolia to the north and Russia to the east. Over 90 % of the territory is covered by rangelands, which can be classified as temperate meadow, temperate typical grassland, temperate desert grassland, and temperate desert from east to west. With the largest rangeland regions, Inner Mongolia is regarded as one of the five pastoral production bases in China. Over centuries, nomadic pastoralism has been practiced as the dominant land use in vast rangeland areas, and history has proved that nomadic pastoralism is the best production model for protecting the rangeland ecosystem of Inner Mongolia (Wu and Du 2008). Petroglyphs in this region indicate that nomadic pastoralism appeared as early as 3000 years ago (Wu and Du 2008).

Historically, the rangelands of Inner Mongolia in China as well those of Mongolia (which was separated from the Chinese Empire in 1919) were alternatively controlled by different pastoral groups, including Huns, Xianbeis, Rourans, Turkics, Uyghurs, Khitans, Jurchens, and Mongolians (Wu and Du 2008). Pastoral production on the rangelands of this region can be categorized into five phases in the administrative systems according to Wu and Du (2008): "phase I, the tribal nomad system before the Genghis Khan' reign; phase II, the subinfeudation nomad system of the Genghis Khan; phase III, the league and banner Zhasake nomad system from the Qing Dynasty to the Republic of China in the twentieth century; phase IV, the small-area nomad system from the founding of the People's Republic of China to 1996; and phase V, the land contract enclosed-stocking system from 1996 (initiated in 1980s) to the present." The fifth phase is the most critical period, with major destruction and degradation of grasslands, decrease of livestock capacity, and decline of herders' income in the entire region of Inner Mongolia (Wu and Du 2008). Implementation of the Livestock and Grassland Double-Contract Responsibility System (LGDCRS) started in the 1980s, and expansion of the Grassland Ecology Protection Projects (GEPP) which is aimed at "retire livestock, restore grassland" initiated in 2000 are two distinct drivers for dramatic changes in socioeconomic institutions in pastoral areas in Inner Mongolia in recent decades (Wang and Zhang 2012).

With the transition from a command economy to a market economy in the early 1980s, the LGDCRS has been implemented in Inner Mongolia with the aim of promoting grassland protection and livestock husbandry development through acceleration of the transition from transhumant grazing to settled living and grazing, enhancing the grassland livestock breeding and increasing planted fodders and forages (Li and Zhang 2009). The year 2000 was the turning point in the government's attention to rangeland ecosystem protection because of frequent sandstorms in northern China and severe floods in southern China, and the GEPP of fencing grassland, decreasing livestock numbers, implementing grazing bans, and ecological resettlement of herders has been implemented on the basis of the conclusion that overgrazing was the major cause of grassland degradation and sandstorms. China's grassland policies and projects are normally firstly tested and implemented in Inner Mongolia and gradually extended to other pastoral areas across the nation (Zhang et al. 2007). As a result of the LGDCRS and GEPP, pastoralists from the steppes of Inner Mongolia to the alpine meadows and cold deserts of the Tibetan Plateau across the dry steppe and desert of Xinjiang in western China are facing unprecedented transformations of traditional livestock grazing and grassland management practices. A better understanding of consequences derived from these policies and projects is necessary to assist pastoralists and policymakers to envision new models for promoting sustainable pastoral production and grassland management. Therefore, this case study was conducted to evaluate the feasibility and sustainability of these policies and to explore pastoralists' adaptive capacities to those institutional changes in Inner Mongolia.

4.3.2 Methods

Integrated approaches including literature review and fieldwork visits were used in this case study by the third author. General information about the LGDCRS and GEPP in Inner Mongolia and the whole of China was collected from public reports, government documents, and online libraries. Through literature reviews, data on and information about the implementation of the LGDCRS and GEPP were collected from scientific publications, online documents, and expert opinions. During three fieldwork visits between 2004 and 2007, participatory observation and in-depth interviews were used to collect data from 56 households who have been affected by the LGDCRS, GEPP, or related interventions in three prefectures: namely, Alashan, Xilingol, and Hulunbir. These areas are also sites representative of three types of grasslands, desert steppe, typical steppe, and meadow steppe, in Inner Mongolia (Fig. 4.5). For household interview, the hand-written survey questionnaires included (1) local traditions of pastoral production and grassland management; (2) attitudes of interviewees to the LGDCRS,

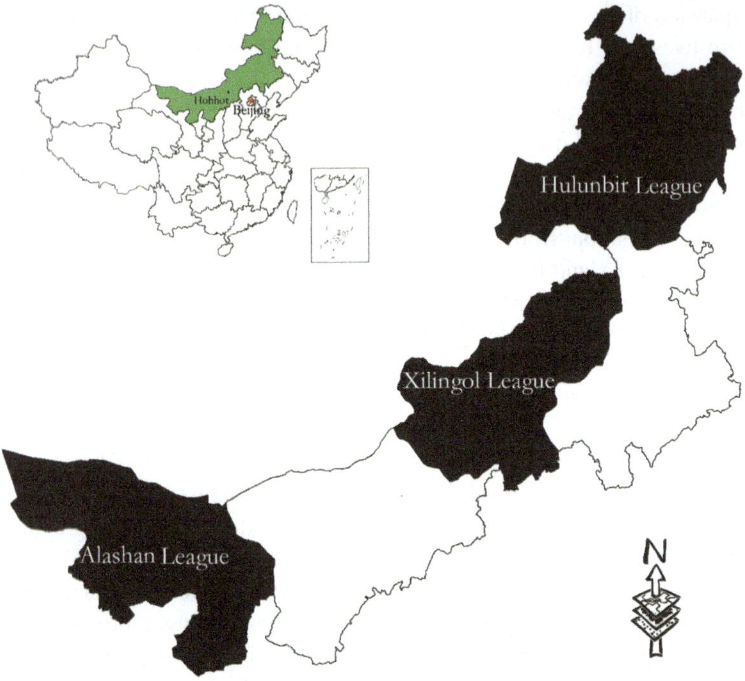

Fig. 4.5 Sampling sites in Inner Mongolia, China

GEPP, and related interventions; (3) interviewees' perception about grassland conditions and understanding of the importance of grassland protection; (4) local adaptive actions to improve pastoral production and grassland condition; and (5) interviewees' suggestions for sustainable pastoralism and grassland management. Additional information about challenges, opportunities, and changes related to implementation of the LGDCRS and GEPP, external support, and internal partnerships was collected and recorded through group discussion and personal communications. The data quality was ensured by careful investigation and cross-checking with different sources. Systematic qualitative techniques recommended by Patton (1990) and Miles and Huberman (1994) were used to analyze all the data.

4.3.3 Ecological Vulnerability of Pastoralism with Institutional Transformation

The survey indicates that the local pastoralists across all sampling sites have historically practiced transhumant grazing on the communal rangelands by moving their livestock from winter pastures to spring–autumn pastures to summer pastures in a collective way called *otor* on the basis of the traditional norms or agreements made among them. By doing so, the local pastoralists can not only avoid overgrazing of the rangelands by adjusting the grazing time and intensity according to plant production, but can also ensure the livestock's feed and water requirements and keep livestock healthy through frequent movement. With the implementation of the LGDCRS in the early 1980s, the livestock were divided by each household according to the numbers in each family, but the rangelands were not contracted to the individual households until the mid-1990s. As a consequence, each household increased its livestock numbers to graze the communal rangelands without any control since the traditional institutions of pastoral collectives were abandoned in the name of household responsibility advocated by the LGDCRS. Some interviewed pastoralists stated that they have doubled or even tripled their livestock population within 10 years from the initiation of the LGDCRS, leading to problems of rangeland overgrazing and eventually rangeland degradation and desertification. One of the interviewed pastoralists in Xilingol said: "I have seen the dramatic declines in grass height and cover of my pastures since I increased my cattle population by two times since the beginning of livestock contract responsibility, some of my grazing pastures have become desert lands due to overgrazing."

In the mid-1990s, the grasslands were divided among individual households on the basis of the contracts between the government and the pastoralists, which maintained that ownership of the pastures was controlled by the government and the use right of pastures was given to the pastoralists. As each pastoral households wanted the pastures close to water resources or their house, the large pastures originally shared among the pastoralist collectives for transhumant grazing use had to be segmented into small pieces. Each household received smaller portions of pasturelands far from their original winter, spring–autumn, and summer grazing areas. Because of shortage of labor, some households had to abandon or transfer the summer pastures or spring–autumn pastures that were too far away to other pastoralists. The pastoralists raised more livestock on the remaining pastures with the expectation of high profit from high-intensity livestock grazing. However, in reality, their expectations were defeated by the degradation or desertification of intensively grazed pastures. Some pastoralists have experienced vicious cycles of "increased grazing livestock number–deteriorated rangeland conditions–declined livestock production–lowered family income." In contrast, the abandoned pastures were either lightly grazed by a quite low number of livestock or overgrazed by a huge number of livestock as the communal properties and the leased pastures were often heavily grazed since the tenant did not care anymore about protection of other people's property. As a consequence, these pastures have been degraded in the form of either shrub encroachment or land desertification. Moreover, the field observation shows that the communal pastures (passages) for seasonal livestock movement have been often overgrazed and degraded with the fencing of individualized rangelands under the LGDCRS (Fig. 4.6).

In the early years of the first decade of this century, the GEPP was initiated in Inner Mongolia to mitigate the dramatic rangeland degradation and desertification. This policy was designed to restore the degraded rangelands mainly through compensation of the pastoral households on the b asis of their pasture sizes to reduce the grazing livestock population, to ban livestock grazing on degraded rangelands, to raise livestock in the stall, and to practice pasture fallow at seasonal or yearly intervals. However, evidence indicates

Fig. 4.6 Degradation of the communal pastures out of the fence (Xilinguole). (Photo by Xueliang Bai, 2012)

that this policy was far from effective. In numerous site visits, it was found that in most areas unpalatable or poisonous grass dominated the plant communities of the banned pastures, although some of the pastures banned for grazing were higher in plant cover than the grazed pastures. During a field visit to Alashan in early May of 2007, an old camel herder said when looking at the fenced pasture: "The plants growing in the fenced pastures turn green very late in spring and look worse than the plants growing in the open pastures. The new branches and leaves of edible shrubs are unable to grow well without browsing by camels when livestock grazing is banned. We know from generations to generations that livestock grazing can promote the growth of foraging plants and suppress the appearance of weeds. However, the role of grazing livestock has been totally overlooked by policymakers. This has resulted in reduced plant production and increased weedy plants." In most cases, the local rangeland monitoring agencies guarded the rangelands during the daytime, but the local pastoralists grazed their livestock on the banned rangelands secretly at night. As a result of illegally heavy "night grazing," some of the rangelands were seriously degraded, especially in the dry years.

4.3.4 Economic Vulnerability of Pastoralism with Institutional Transformation

The LGDCRS was originally designed to improve the production efficiency of pastoral systems and to prevent rangeland degradation on the basis of the belief that the collective system was highly associated with the low production of the pastoral system and uncontrolled livestock population growth. However, as stated already, the LGDCRS did not work well in preventing rangeland degradation. Moreover, the survey showed that the LGDCRS was not effective in promoting production efficiency of pastoral systems. Instead, the LGDCRS led to economic vulnerability of pastoral production systems in most cases, as it lowered the ability of pastoralists to benefit from the rangelands (Li and Huntsinger 2011). The pastoralist interviewees stated that they had practiced the *otor* for free livestock grazing without a cent of investment before the implementation of the LGDCRS, so they could deal with the problems of feed and water deficiency in dry years by moving their livestock from one pasture to another under their collective's coordination. However, now they have to rent other pastoralists' pastures to meet feed and water requirements of livestock in the dry years, as they cannot practice *otor* within their fragmented pieces of pastures. Alternatively, they have to buy supplementary fodder from other pastoralists or outside cultivators to balance the livestock's feed requirement and to dig deep wells to meet the livestock's water requirement in dry years. As a consequence, their investments in pastoral production dramatically increased and the risk of losing profits from livestock production was also greater. For example, a pastoralist in Xilingol said: "In the past [before the LGDCRS], we did not need to pay fees to anybody for practicing *otor*, now we have to rent the *otor* pastures in harsh years by paying a high amount of money, even paying animal pass-by fees and animal watering fees. Often, there are many uncertainties to find the *otor* pastures and there are no guarantees we will make profits by renting *otor* pastures." This is verified by field surveys of pastoralists in the same district (Li and Huntsinger 2011):

> Last year [2006] I [a herder called Bater] went out early in June to try to seek a place to otor, but failed. Quite often you hear of a possible pasture in some place, but you can never believe what you hear. You need to go there and see the real situation. Like in my case, once I heard of a place that would allow otor for a lower price, so I rode a motorcycle to the place to see, and found the price was actually very high for what was there. Due to this delay in finding suitable rangeland, I couldn't practice otor on time last year.
>
> We [another herder called Ale and his lessor] had agreed to a charge of 8 yuan [US$1.1] per sheep per month [in 2006], but later when another herder promised to pay 10 yuan [US$1.4] per sheep per month, the lessor immediately violated our agreement and rented to the herder offering more money. Then I had to search for another pasture.

Moreover, buying the supplementary fodder to meet the livestock's requirement was not an economically sustainable way to maintain the pastoral production in rangeland areas of Inner Mongolia. The interviewed pastoralist in Xilingol also stated: "Some households spend a lot of money to buy the supplementary feeds for their livestock in harsh years, while their gains from selling livestock are often lower than their payments for supplementary feeds. They have to borrow money or make loans for to make their living, making them fall into a poverty trap." This statement is supported by field investigations by Wang and Zhang (2012) in another pastoral district of Inner Mongolia, Chifeng, as follows:

> Zha Lazeng, the former [Gonger] village chief, bought forage for four years. In 2009 he sold livestock for about 40,000 yuan but spent 20,000 on forage. Drought made the livestock production a loss. A few years ago incomes might have been lower than the present, but costs were also relatively low so he was never in debt. Now, after he had paid the forage and other

costs, he could not make a living by just relying on income from livestock. He borrowed 10,000 yuan in 2009.

Another woman named Si Qin married into the village in 2004. Since her marriage, her family borrowed money every year. As the weather became drier, their life became much worse. In 2005, her family rented a piece of rangeland for 800 yuan and harvested 10,000 kg of forage. As the weather became drier it was difficult to rent pasture which they could harvest for forage. They started to buy-in forage at very high prices, especially in a dry year. In 2007, she had spent a few thousand yuan for forage, but in 2009 it rose to about 30,000. To afford the cost of forage Si Qin borrowed a large amount of money. By 2010, the loans totaled 70,000 yuan.

Similarly, the GEPP did not work well in promoting the development of a pastoral economy and even led to economic vulnerability in pastoral societies of Inner Mongolia. Although the GEPP provided some eco-compensation to pastoralist households in the name of "Payment for Ecosystem Service" on the basis of the size of their pastures banned for grazing, the high expenditure of building sedentary houses and livestock sheds, cultivating and harvesting forage, buying and transporting supplementary fodder, and caring for animal health resulted in no benefits from pen-feeding/stall-feeding livestock production associated with the GEPP. Some of the pastoralists in Alashan noted: "We are traditionally camel nomads in desert areas, we have never practiced forage cultivation and stall-feeding, we do not have techniques to process the feedstuffs, to raise the camel in stalls. Mostly importantly, camels are semi-wild animals that need free movement in open pastures to retain their health. Once the camels are fed in the stall, we lose the benefit from the pastoral production." The interviews indicate that because of the high cost of fodder in Alashan and Xilingol districts, families have abandoned livestock production as a livelihood strategy. This phenomenon can also be found in other pastoral districts in Inner Mongolia. For example, in field investigations in the village of Gonger in Chifeng, Wang and Zhang (2012) stated:

All the herders [in Gonger village] paid high costs to buy fodder. According to their calculations, if a sheep was fed solely with purchased fodder, then at least 3 kg were needed each day, which cost about 3–5 yuan. If the period of feeding lasted for six months, then the forage alone would cost 500–700 yuan, whereas the best price for one lamb was 400–600 yuan. As a result, herders' livestock decreased but their loans increased. In 2010, about twenty households, or 25% of all households in Gonger Village, had no livestock. It was evident that feeding animals with forage purchased from the market was unsustainable.

The resettlement strategy connected to the GEPP has increased the economic burdens of pastoralist households, as they have to invest a lot of money in housekeeping and family expenses. The government provided some subsidies for building houses (normally 8000–10,000 yuan) and livestock sheds (normally 4000–5000 yuan) according to the GEPP, but the pastoralists spent more than twice the amount of these subsidies to build a house and a shed (Fig. 4.7). The government also paid eco-compensation (about 5000–10,000 yuan per year per family on the basis of their family size and banned pasture areas) to the pastoralist households for their living expenses, although these payments were far less than their living expenses. One of the interviewed pastoralists stated: "When we lived in the yurt on the rangelands, we did not need to pay for construction materials for the house, electricity, fuel wood,

Fig. 4.7 Newly constructed feed stall thanks to the government subsidies in Xinlinhot, Inner Mongolia. (Photo by Li Yang, 2013)

animal feed, our daily food (milk and meat). Since we moved into resettled buildings in town, we have had to pay for everything, electricity, coal, animal feed, and our daily food, even water. All the countable and uncountable expenses are far beyond the compensation provided by the government. Very often, we have to get loans or borrow money to make our living or transform our livelihood from livestock keepers to something else". The field visit indicated that most of the resettled pastoralist households fall into a poverty trap, although some of the resettled pastoralist households succeeded in livelihood transformation and life improvement.

4.3.5 Social Vulnerability of Pastoralism with Institutional Transformation

According to the survey, the implementations of the LGDCRS and the GEPP led to not only the breakdown of the pastoral collective, the operational unit responsible for customary norms, regulations, and actions in pastoral production systems for centuries, but also the loss of indigenous knowledge, cultural traditions, and pastoral identity. Moreover, social conflicts and disparities emerged among the pastoral communities and between the pastoral households. Social vulnerability of pastoralism was thus increased and accelerated. As stated by one old male herder in Xilingol: "Before the LGDCRS, we followed the traditional mobile routines to herd our livestock between the *otor* pastures which were divided among different pastoral communities according to customary regulations. We can borrow other pastoral communities' *otor* pastures to herd our livestock during severe drought on the basis of oral or written agreement that we gave some gift livestock in return or we lent our *otor* pastures to them for grazing when they faced similar problems. Within our own communities, we negotiated among all the household representatives with the coordination of community head (normally a distinguished elder) to make decisions such as how many livestock that each household should keep, what time and what kind of animals should be grazed on what pastures, how many people and who should be responsible for herding livestock, and how the profits from pastoral production should me distributed throughout the whole community. In such a way, all the pastoral groups can coexist harmoniously. However, we have been losing all of these indigenous institutions with the implementation of the LGDCRS, which may result in frequent conflicts among the pastoral households over water resource sharing, livestock passage utilization, and pasture boundary clarification. Sometimes, there are fights and violence among pastoral households due to communal livestock passage use or unclear pasture boundaries." This was verified by Zhang's (2012) interview with a pastoralist named Baolidao in same district as follows:

> He was once again agitated when he complained about the trampled rangeland. His rangeland is around 16,000 mu and borders his sister (interviewee) Gaowa's rangeland. His sister's herds often move to graze on his rangeland and his rangeland has been destroyed. He complained to his sister once but she did not think it was a problem since it is impossible to control the movement of animals. Afterwards, he turned to the county Grassland Station for a solution. 'What is the purpose of ecological resettlement? The state says that it is for the rehabilitation of the rangeland. I asked the officers if they will regulate or not [the invasion by my sisters' animals].' However, the officer suggested that he had better negotiate with his sister or else should catch the invasion activities in the field and then call them to come. 'How can I get the time to watch in the field every day? Is that not their job?' He had no plan to set up fences because it was rather costly.

In addition, an elderly male herder in Xilingol also stated: "In the past, we collectively grazed the livestock on the rangelands through division of labor among different households in the whole community. Different households took different responsibilities, such as herding the animals, caring for children and elders, collecting fuel wood, harvesting feed, etc. In this way, we could use the human labor efficiently and maintain the pastoral production effectively. With the implementation of the LGDCRS, individual household had to shoulder all the workloads, herding, milking, caring children, collecting fuel woods, and harvesting feed, etc. Because of labor shortage, some households in my community have abandoned some of their pastures or some households have totally abandoned livestock grazing by leasing their pastures to others. As a result, there are disparities between poor pastoral households and rich ones. Moreover, the undesirable things such as criminals, violence and divorces have increased in the society."

The GEPP has promoted the resettlement of pastoralists with the purpose of reducing grazing pressures on the rangelands. Roughly, 8 % of the rangelands in Xilingol were projected for ecological resettlement (Brown et al. 2008) and about 49,000 pastoralists in this district were resettled between 2003 and 2010 (XLDRC 2011). According to the strategy of the resettlement connected with the GEPP, the pastoralists live in the areas where cultivated forage-based livestock stall feeding cannot be performed and should be moved out and resettled near the towns or cities to develop livestock stall

Fig. 4.8 Resettled pastoralist households thanks to the government subsidies in Xinlinhot, Inner Mongolia. (Photo by Li Yang, 2010)

feeding systems or work in the secondary and tertiary sectors. For those resettled households, the government has allocated a detached or semidetached brick house and other facilities such as a livestock shed/stall (Fig. 4.8). Additionally, the government has provided the resettled pastoralists with some ecological compensation for their living expenses, loans for purchasing livestock and feed, and training for alternative livelihoods. However, the interviews indicate that there remain many social problems in the ecological resettlement process. Some respondents claimed that they failed to adapt to the resettled life, since they cannot find an alterative livelihood to livestock grazing as they said: "Herding animals on rangelands is our traditional life, we can't do any other jobs than livestock herding from old generations. Stall feeding (livestock) is hard work, which needs advanced technology and higher input. We can't afford to do it." Moreover, social tensions have appeared among resettled communities. As one elderly female noted: "After we moved to this resettled community, I found more conflicts arose among us. In the past, we lived far from each other in the yurts on the rangelands, and we treated each other in a very friendly manner when we met. But now, we live in a crowded community and the neighbors can easily break friendships because of minor conflicts." Because of discomfort with living in town or urban areas, some resettled pastoralists have moved back to their fenced pastures for herding livestock secretly such as night grazing.

In addition, the survey indicates that the GEPP has led to more conflicts between government officials and pastoralists. The pastoralists often wanted more compensation for living expenses, stall feeding and shed construction costs from the government, whereas the government officials forced them to move into the resettlement buildings without more support. The pastoralists struggled with the government officials for more benefits. Government officials frequently monitored the illegal grazing activities on the banned rangelands and they often fined pastoralists or confiscated their livestock as punishment for the illegal night grazing. As passionately explained by one male camel herder in Alashan: "We play the game of 'cat and mouse' with government officials, we mostly lost the game as 'the mouse'. We know the importance of rangeland conservation, but how can we survive without herding animals? What are the ideal options to mitigate the contradictions between livestock grazing and rangeland conservation? What are the ways to alleviate the conflicts between us and government officials in the process of GEPP implementation?"

4.3.6 Local Adaptations To Institutional Transformations

According to Kreutzmann (2003), "pastoral practices have always adapted to new and threatening challenges and found an outlet to cope with mounting constraints." However, implementation of the LGDCRS and GEPP has caused many difficulties for pastoralists in Inner Mongolia, and the local pastoralists have strived to develop adaptation strategies to mitigate these problems and even to convert the disadvantages into opportunities. Collective action is one of the key strategies which has been successfully adopted in some pastoral societies in Inner Mongolia. The survey in Hulunber indicated that the pastoral groups in one *gacha* (Mongolian term for "village") in Xinbaerhuyou *banner* (Mongolian term for "county") have practiced collective grazing systems without dividing the rangelands into individual pastoralist households from the very beginning of the LGDCRS. Instead, they distributed the communal rangelands to a group of pastoral households and established the collective institutions for livestock production and rangeland management according to the old grazing tradition. They have sustained the *otor* pastures and kept mobile grazing the whole year round, and they have practiced the division of labor and profit sharing among all the pastoral groups on the basis of agreements and regulations made by the collective. In such a way, they can sustain livestock production and maintain the rangeland health, even in adverse weather conditions caused by climate change or climate variability. As stated by one of the interviewed pastoralists: "Although we may not have gained big profit from this production mode (collective grazing), we can get relatively stable and reliable incomes for a good living, even in dry years. The risks of livestock loss in the disasters of drought, snowstorm, and (rangeland) pests have been greatly reduced. Most importantly, we maintain the rangeland conditions very well. There is less rangeland degradation in our *otor* pastures." With the release of the 2002 revision of the Chinese New Grassland Law, which allows pastoralist groups to make contracts for using the rangelands with the government, this production mode (collective grazing) has been promoted as one of the successful models of the pastoral system in China.

Revival of *otor* practice is another adaptive way applied by some pastoralists in Inner Mongolia. From site visits in Xilingol, it has been found that some pastoralists whose contracted pastures are close to each other or who are relatives and friends have joined together to form grazing groups and to reactivate the *otor* grazing system on their allocated winter, summer, and spring–autumn pastures in a collective way based on the oral or written agreement among them. This phenomena has also been observed by other scholars in other pastoral areas of Inner Mongolia. For example, Wang and Zhang (2012) reported the following from a survey in the village of Gonger in Chifeng District :

> Suri the village head overcame these difficulties unbalance winter and summer pasture uses through cooperation. In contrast to other herders, Suri did not stop grazing the winter pasture. Every winter, he coordinated with his brother-in-law.... The two households had worked together to enclose their winter pasture. In winter the two households would take turns to send their labour to care for the livestock grazing there... in the face of continuing drought Suri collaborated with seven other households to form a group to graze cattle on the summer pastures. The village enclosed a piece of summer pasture in 2009. From 2010 the village heads decided to give the pasture to the sub-village to use. Single households could not use it because the labour in any one household was insufficient. However, eight households were able to use the pasture collectively. All of their cattle grazed there. Each week the eight households sent three herders from different households to stay in the summer pasture to care for the animals.

"Company + farmer" is a new production model supported strongly by the government to build cooperation between dairy or beef companies and local pastoralists, especially the resettled ones. This new model encouraged the individual livestock producer to enter into a contract with professional dairy or beef companies such as Yili and Mengniu (two of the biggest dairy companies in Inner Mongolia) as the livestock product (milk, beef) suppliers. The companies have provided the feedstuffs and milk cows for their stall raising. Some of the resettled pastoralists have practiced this production mode as an adaptation. From the cooperation with professional companies, they can earn a considerable income for family, and mitigate conflicts with neighbors and reduce the risks of livestock loss in droughts or snowstorms. However, some negative consequences of this model, such as low milk price for sale to the contracted companies and lack of technical support for improving their skills in livestock rearing, have limited the massive extension of this model among the pastoralists. In addition, some young members in the pastoralist households have changed their livelihood strategies by migrating as laborers to cities, starting small businesses, or becoming tourist guides. As a consequence of livelihood diversification, the pressures of human and livestock populations have been lowered to some degree. However, the livelihood transformation of young generations in pastoral societies may lead to the problem of increased marginalization of pastoralism. As one worried old herder in Xinlingol expressed: "If our next generation moves to the town or city, who will do the herding in the future. We may lose our pastoral traditions one day."

4.3.7 Discussion

Since the 1980s, the LGDCRS has been implemented in Inner Mongolia and expanded to all pastoral regions in China with the aim of mitigating "the tragedy of the commons," described by Hardin (1968), that unclear property rights were associated with the degradation of a common pool resources such as the community pastures. However, in recent decades, there has been lot of debate about Hardin's solution for alleviating the "tragedy of the commons," which is the privatization of communal land. Some scholars insisted that the "the tragedy of responsibility" might be a more accurate term to describe the situation of pastoralism in Inner Mongolia (Li and Huntsinger 2011) and even in the whole of High Asia (Kreutzmann 2003). The clarification of property rights by individualizing the rangelands did not help the pastoralists effectively manage these natural resources in Inner Mongolia (Li and Huntsinger 2011). We argued that the failures of the LGDCRS and the related interventions in addressing the grazing livestock production and rangeland management can be well noted as "the drama of the commons" (Ostrom et al. 2002), implying that land grabbing and expropriation of resources occurred in an environment in which customary rights can easily be breached and community practices do not count. The increased rangeland degradation with the implementation of the LGDCRS pushed the government to implement the GEPP, which includes a grazing ban, grassland fencing and fallow, and pastoralist resettlement. However, evidence from Inner Mongolia shows that the GEPP did not work well in preventing rangeland degradation as expected by the government. We can conclude that the institutional changes associated with the LGDCRS and GEPP have broken the coupled human and natural system of pastoralism, leading to ecological, economic, and social vulnerability of pastoralism there.

To address the ecological, economic, and social issues in the pastoral realm in Inner Mongolia, it is necessary to rebuild the indigenous human ecological relationship of pastoralism. The approach of coupled human and natural systems suggested by Liu et al. (2007) can be used to activate the revival of indigenous knowledge, customary norms, and traditional practices such as *otor* in Inner Mongolia. The coupled human and natural system approach can help pastoral societies find appropriate ways to cope with institutional changes by facilitating effective collaboration among social scientists, biophysical/physical scientists, practitioners, managers, and users. Moreover, the implications of the coupled human and natural system approach are critical to sustain pastoralism in Inner Mongolia in both policy and research dimensions. Human components need to be emphasized and well integrated with scientific objectives and policy priorities to equitably balance local people's needs with national or regional conservation and development policies and strategies. The coupled human and natural system approach can help researchers identify the complexities such as reciprocal effects, the influence of differing scales of biological and social organization, and emergent properties (Liu et al. 2007), which could lead to innovative scientific insights that are essential for the development of effective policies that will promote and maintain the ecological and socioeconomic sustainability of pastoralism (Dong et al. 2010). It can also help policymakers understand the interface between social, economic, physical–biological, and ecological models in promoting sustainable pastoralism, which may result in innovative policy decisions that can balance the needs of society with the best scientific knowledge available. Future programs of institutional changes in pastoral society such as small-town urbanization must include interdisciplinary investigations of socioeconomics, human dimensions of natural resource use, adaptive management processes, information management systems, and syntheses of the state of scientific and indigenous knowledge.

4.4 Role of Ecological and Sociocultural Diversity in the Pamirs of Badakhshan, Afghanistan

4.4.1 Context

The Pamirs, neighboring the Altay Mountains and the Tian Shan, are located between Europe to the west and Asia to the east and between the Middle East and northern Eurasia. This region of Inner Asia has historically sustained extensive nomadism, agropastoralism, and agriculture in its valleys, producing food for subsistence and marketable crops through glacier-fed irrigation (see Fig. 4.9). As the Pamirs were part of the Silk Road, diverse ethnicities engaged in trade which also facilitated exchange of ideas; they were not isolated, as is commonly asserted of mountainous societies (Bliss 2006; Felmy and Kreutzmann 2004; Grotenhuis 2002;

Fig. 4.9 Strategic location of the Pamirs

Kassam 2009a; Kreutzmann 2003; Olimova 2005; Wood 2002). The notion that mountains offer both refuge and isolate human communities is not tenable given the historic evidence of agropastoral activities combined with mining, trade, porterage, smuggling, and even raiding (Kreutzmann 2003). The physical remoteness of the Pamirs has not prevented outside political interference nor limited commercial relations and other exchange within the area. Because of its strategic significance, Inner Asia has been the target of invasions from Arabia, China, Mongolia, and Persia. Most of Inner Asia was under Persian influence until the Arab invasions under the Umayyad and Abbasid dynasties starting in the seventh century. Fatimid religious and cultural ethos also contributed to a flowering of pluralistic Islamic thought, philosophy, and mysticism in Inner Asia (Daftary 1990; Hunsberger 2000).

Since the nineteenth century, the Pamirs have been within the imperial vision of Euro-American interests. Transformation is a continuous and dynamic process in the Pamirs, and the changes from the nineteenth century onward can be viewed as results of the imperialist impulse and are characterized by two phases: (1) the European colonial presence and (2) unfettered globalization. The peoples of the Pamirs have been at the forefront of violations of their autonomy and self-determination in the form of imperial machinations of the British Empire and Russia, and subsequent Cold War alliances between the West and the Eastern Bloc countries. By 1979, the Pamirs had become a major deployment point for the Soviet military poised to invade Afghanistan. Ultimately the Soviet military withdrew amid fierce local opposition with significant financial, military, and logistical support from the USA. In the wake of the Taliban victory, and the subsequent defeat of the Taliban by the US-led alliance after the events of September 11, 2001, a world war manifested as an internal war continues indefinitely with a significant cost to Afghan lives. Now not only are the traditional rivals of the Cold War such as Russia and the USA participating, but China, India, Iran, Pakistan, and Turkey are also exerting their strength as regional powers with global reach. The 36-year global war localized to Afghanistan has left a fragmented state, warlordism, and opium cultivation for global markets, and contributes to regional instability. The consequences are very real and potentially fatal for the people of the Afghanistan as well as those from outside who seek to contribute to their livelihood security and well-being.

Under these conditions, livelihood systems are compromised and the threat of famine is ever present. With sustained political instability, economically the Afghan Pamirs have remained largely ignored by the central government, so local agropastoral knowledge continues to sustain livelihoods of the population and small-scale production prevails. Physical and institutional infrastructure such as roads, health care, education, and electricity have been limited, if not entirely absent (Bliss 2006; Felmy and Kreutzmann 2004; Kassam 2009a). Nonetheless, evidence from the Pamirs of Afghanistan reveals a narrative of pluralism and resilience under conditions of war, dramatic climate change, and potential food crises (Kassam 2009a).

4.4.2 Qualitative Examination of Diversity

As in the case study from the Altay Mountains and the Tian Shan, the livelihoods in the Pamirs illustrate that the inner workings of a system are revealed when it is subjected to systemic stress or perturbations (Kassam 2009b: 233 n5). This is the case of relations between pastoralists and farmers in the Pamirs of Inner Asia. We will qualitatively examine the role of diversity at the level of ecological niche, cultural and religious difference, and ecological professions such as farmers and herders to understand its potential impact on food and livelihood security.

4.4.3 Methods

In 2006, the first author interviewed a group of people from the village of Pul-i-Zirebon. The interviews were based on participatory action research methods (Chambers 1997; Greenwood and Levin 1998; Kassam 2009b). These group interviews were subsequently complemented by individual interviews to obtain greater detail and triangulate information from a variety of sources. Thirty-eight individuals, all male, were interviewed as part of this preliminary research. In the course of the interviews, it became clear that the survival of these people in the face of war and the uncertainties of sociocultural and environmental change depended on mutual support between ethnic groups. Difference seemed to be central to mutual livelihood security for a variety of ethnicities in the region. In 2008, follow-up interviews were conducted to validate the information from 2006 and examine in more qualitative detail the role of sociocultural and ecological difference in providing capacity to adapt to systemic perturbations and stress. To examine this finding, 61 individuals were interviewed, included 45 men and 16 women. In 2009, additional interviews were conducted with Arab Pashtuns (13 women and 7 men) while they were in their encampments near Pul-i-Zirebon, as were more follow-up interviews with the Shugnis in Pul-i-Zirebon (nine women and three men), a total of 32 individuals. The iterative nature of the interviews facilitated exploration of the rather complex interconnections between diversity, ecological zones, and adaptation to sociocultural change. The research was complicated by border crossings from the different regions of the Pamirs of Tajikistan into Afghanistan. In terms of safety and logistics, these were challenging undertakings.

4.4.4 The Role of Difference in Livelihood and Food Security

Although the first author found supporting and complementary evidence that Kyrgyz herders and Wakhi farmers collaborate in the Wakhan region of Afghanistan for mutual food and livelihood security (Kassam 2010), this case study will focus only on the Arab Pashtuns and the Shugnis in Badakhshan, Afghanistan to illustrate this complex and symbiotic relationship.

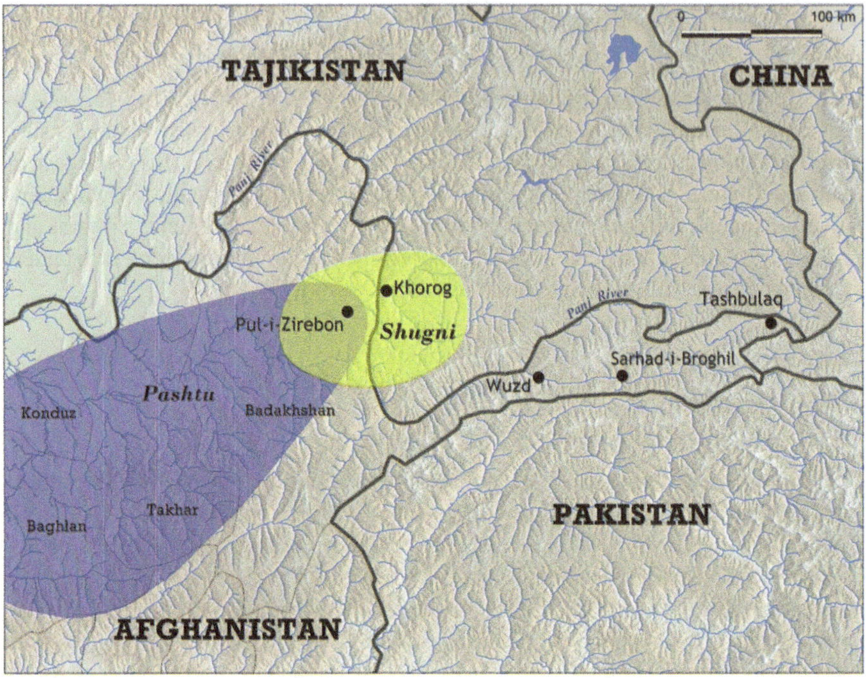

Fig. 4.10 Map of ethnic Pashtun migration from the lowlands into Shugni homelands

The Arab Pashtuns are pastoralists. In the spring, families migrate with their animals from lowlands in the provinces of Baghlan, Konduz, and Takhar to the highlands near Pul-i-Zirebon, in the province of Badakhshan (see Fig. 4.10). Since both humans and livestock depend on salt, villagers from Badakhshan have historically traveled to lowland markets such as Faizabad to purchase it (Barfield 1981). These interactions established trade relationships between the two groups. The Arab Pashtuns are Sunni Muslims and speak Dari, an Indo-European language related to Persian. The Shugnis are highland farmers who live in the region of Pul-i-Zirebon near Lake Shiva, Badakhshan, who also have animals. In the summer, Pashtun encampments and pastures border their villages and pasture lands. The Shugnis are Ismaili Muslims and speak Shugnani (like Wakhi, an Indo-European language of the Pamir group).

The ecological professions of these ethnically and religiously diverse groups are distinct, and seasonally their habitats overlap. Instead of the potential conflict between herders and farmers, it is noteworthy that their interaction is complementary to mutual needs. The Arab Pashtuns arrive in Badkhshan in June and return to the south in September, traveling for 3 weeks to 1 month in each direction (see Fig. 4.11). The Pashtuns consider themselves the wealthier members of the relationship. The measure of their wealth is the number of animals: while the Pashtun nomad is said to have 800–1000 sheep and goats, a Shugni farmer is considered

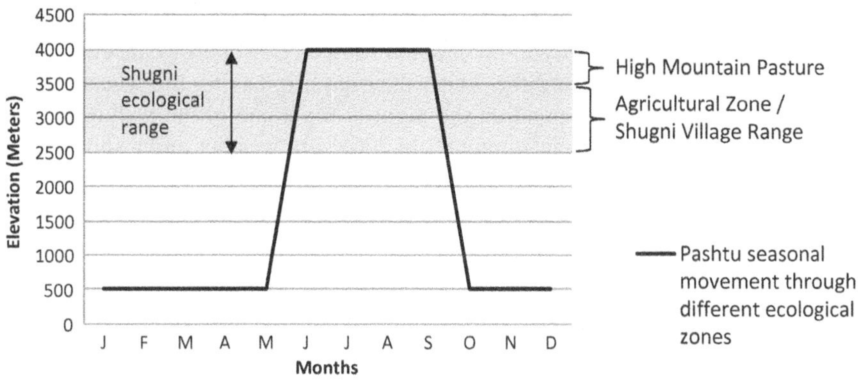

Fig. 4.11 Seasonal overlap of the ecological space of the Pashtuns and Shugnis

wealthy if he has 50 animals. The Pashtuns openly acknowledge the relative poverty of their Shugni neighbors: "We do not fight with them [the Shugni] because they are so poor. Instead, we consider them our brothers." Equally the Shugnis acknowledge the relative wealth and political power of their nomadic trading partners.

The Shugnis grow mainly wheat, barley, and peas and keep livestock such as goats, sheep, and a few cattle, as well as horses and donkeys. When they have surplus crops, the Shugnis are unable to move these commodities to the lucrative southern markets, so they rely on trade with the Pashtuns. Although the Arab Pashtuns carry sufficient rice on their animals in their migration to the highlands of Badakhshan to sustain themselves, they also buy wheat from the Shugnis, as well as dried yogurt while they are in the highlands. The Arab Pashtuns are an important (albeit seasonal) force in Badakhshan as they are the link between the lowlands of the south and the highlands of the north. The difference between these ecological zones works to their advantage in trade and facilitates a symbiotic relationship with the Shugni farmers in Badakhshan. The relationship has been mutually beneficial. The Shugnis obtain tea, salt, oil, ironware, cloth, and kitchenware from the Pashtuns, and sometimes donkeys, cows, sheep, and goats. Mostly, items are exchanged and not purchased with cash. The subsistence agriculture of the Shugnis does not provide the villagers in Badakhshan with sufficient cash to purchase salt, tea, cloth, and ironware from distant markets, and they must make long journeys to Faizabad and Rustaq to obtain necessary goods. Aware of the cash needs of villagers, the Pashtuns bring sufficient cash to the highlands to purchase wheat from the Shugnis (Barfield 1981). This allows the Shugnis to get access to currency for other purchases.

These transactions, which occur between individuals (generally men), are based on relations established between the Shugnis and the Arab Pashtuns over a few of generations. During the interviews, both villagers and nomads reported sustaining relations that were first established by their grandfathers more than 47 years ago. We observed Arab Pashtuns arriving in the village with their camels, horses, or donkeys, having tea at the *Mamon Khana* (guest house), meeting their friend in the village, securing the wheat they require, having the wheat milled into flour, and sometimes spending the night at the home of their Shugni host, and then returning to their encampment. In 2008, dry weather and a shortage of rain resulted in a poor harvest. The cold winter and greater snowfall in 2009 exacerbated the food problem, as supply roads were closed. Conservative estimates indicate that 66 people died of severe malnutrition: 57 children, 7 pregnant women who died while giving birth (newborns, not counted here, did not survive either), a 75-year-old man, and a 60-year-old woman. As the villagers were receiving emergency food aid, the Pashtun nomads arrived to purchase wheat. It is clear that some villagers traded the emergency supplies for cash needed to buy other necessary items. For the Shugnis, there is a delicate balance between survival and famine.

As a result of relations with the Pashtuns, a Shugni villager may ask his nomad friend to bring some items from southern markets, such as cloth and kitchenware, on his next trip north. When the Shugnis go south, the Pashtuns extend similar hospitality. Whereas Shugni women do not visit the homes of the Pashtuns in the lowlands, the Pashtun women do visit the homes of the Shugni women when they are in the highlands. In the villages, the Pashtuns not only have an assured place to sleep, but also experience the stability of long-term hospitable relations.

The Pashtuns also help their Shugni friends to secure seasonal employment in the lowlands, particularly in the winter when agricultural activities are at a minimum. The less wealthy Shugnis seek such employment in the southern lowlands and often live at the homes of their Pashtun friends. Their work tends to involve caring for and feeding livestock, collecting fuel for heating the Pashtun homes, and fetching water. They may also work as agricultural laborers, plowing fields in the lowlands and planting rice. They are paid in cash and payment is mutually decided before they come to the south to work. This type of seasonal employment lasts for 1 or 2 months.

As noted earlier, the Shugni farmers also keep animals, using mountain pastures in the summer. However, as they lack the resources to retain a large group of animals through the harsh winter, the Shugnis trade their goats and sheep with the Arab nomads. The Pashtun nomads can pay in cash or exchange the expensive items they have transported from southern markets for goats and sheep to renew or increase the size of their herds. This trade saves the Shugni farmers from potentially time-consuming and expensive travel to lowlands markets. The wealthier Shugnis, those who have more than 50 animals, will give some of their male goats and sheep to the Pashtuns to tend in their pastures during the summer months. In the autumn, on their journey back, the Pashtuns return the animals to the Shugnis. Similarly, during their stay in the highlands, the Pashtuns will bring their injured animals to the Shugnis to tend in the vicinity of their villages. In the winter season, the Shugnis give their male horses and bulls to the Pashtuns to take south and in the spring they bring them back. The Pashtuns also store their extra supplies such as tea and salt in the homes of the Shugnis.

The Shugnis maintain that conflict with their Pashtun neighbors is rare but may arise when the Pashtun shepherds are careless and let their animals graze in Shugni pastures, on crop land, or on land designated for growing fodder. Although the niches overlap, the presence of spatial boundaries speaks to the old adage that "good fences make good neighbors." The Shugnis also pointed out that conflicts are usually resolved in favor of the side that possesses the most resources to influence decisions made by local government arbiters. This would likely be the Pashtuns.

However, both the Shugnis and the Pashtuns are at the mercy of regional government commanders who are extorting animals from the two communities. These local commanders are particularly vicious to the Pashtuns, who have relatively more wealth to extort. The long war and resulting alliances have exacerbated arbitrary enforcement of law. Use of pastures in Badakhshan is highly competitive, and access to new pastures is acquired through purchase, rental, or theft. Pasture rights were established and reorganized in 1921 by Nadir Khan. The Pashtuns have exclusive rights to pastures in the form of *firmans* (deeds) issued by the government. These rights—which are not tribal or common property, but individual family rights—are guaranteed by the state, and they may be bought, sold, rented, or inherited. Whereas the Arab Pashtuns have individual titles to summer pasture use, the Shugni villages have collective title to their traditional summer pastures. The idea of renting pastures reinforces the notion of private ownership (Barfield 1981). In our interviews, the Pashtuns reported increasing difficulties with local government because their lands are under threat from local commanders. In the highlands, these commanders buy up from the government the pasture land on which the Pashtuns have traditionally grazed their animals. They then rent it back to the herders for 4000–5000 afghanis (US$80–100) per season, a significant capital outlay in this region. In many cases, the Pashtuns have deeds to prove grazing rights from the time of their grandfathers, but the local commanders insist that they pay to use the land. Furthermore, in the spring migration of the Pashtuns to the highlands with their animals from lowland provinces such as Baghlan, Konduz, and Takhar, these local commanders control the trails and demand animals in return for safe passage. When the Pashtun tribesmen refuse, the commanders or their henchmen beat the tribesmen and take their animals by force.

The Pashtuns and Shugnis do not practice intermarriage, thereby retaining their cultural distinctiveness. However, Shugni women recalled that in the past when their families were indebted to the wealthier Pashtun tribesmen and women were given to repay the debt: "In earlier days, our ancestors were very dependent on Pashtuns because they were prosperous, and our ancestors were always in debt, which they could not repay, but they would give away their daughter in return for the debt. Now there are no such cases, and may God prevent their return." The giving of daughters as repayment of debt is no longer practiced. Barfield (1981) reported that sometimes close ties between wealthy Shugni farmers and Pashtun nomads are secured by a one-way marriage relationship between Shugni women and Arab Pashtun males. He maintained that Arab Pashtuns refuse to let their women marry Shugni men. However, the interviews indicate that, in fact, both sides reported no marital connections.

The Pashtuns and Shugnis do not share *Mazars* (sacred places). "We do not say bad things about their [Shugni] *Jamat Khanas* (places of prayer) and we do not visit them, and they do not say bad things about our holy places and they do not visit them. Moses had his religion and Jesus had his religion." The Ismaili Muslims, in their places of worship, pray with both sexes present, men on one side and women on the other. There is no physical barrier, and both sexes are given equal preference of space as both equally occupy the space of the prayer hall from front to back. During the summer, while the Pashtuns were visiting the village of Pul-i-Zirebon, women did not attend the *Jamat Khana* for prayer. The Shugni men explained that they were absent to protect themselves from persecution by followers of the more extreme interpretations of Islam, in other words some of the Pashtun tribesmen. The Pashtun women reported that they feared these extreme elements when they were asked if they would let their photographs to be taken. The first victims of the Pashtun-supported Taliban are the Pashtuns themselves, before their violent and intolerant religious dogma affects others ethnic and religious groups. The perpetrators are the victims of their own ideology.

4.4.5 Discussion

The relationship between the Pashtuns and Shugnis is not a mere narrative of economic comparative advantage. It is not based on a simplistic economic calculus. The ecological context and diversity in ethnicity as well as professions provides a sociocultural mechanism for food and livelihood security under tremendous stress. Table 4.4 summarizes the relationship between the Pashtuns and the Shugnis.

The primary difference begins with the ecological niche and professions of the Pashtuns and Shugnis. This sets the stage for a relationship that includes both ecological context and sociocultural distinctiveness. Between the Pashtuns and the Shugnis there is religious distinctiveness that is most visible in their treatment of women and susceptibility to fanatical interpretations of Islam. Yet there is an attempt at mutual respect under very unstable conditions driven by religious rhetoric. There is a linguistic difference that is driven by cultural heritage, but they have learned to communicate with each other to overcome this boundary. The Sunni Muslim Pashtuns are pastoralists who have agricultural land, whereas the Ismaili Muslim Shugnis are sedentary farmers who keep some animals. There is a difference in ecological professions and yet understanding of the role of each other's ability and expertise because some Shugnis go to the lowlands to work on Pashtun lands and Pashtuns bring weak animals to be tended to by the Shugnis as the Shugnis give their animals to be tended by the Pashtuns. There is an appreciation of the practical knowledge that each ecological profession brings to the complementary relationship.

What insights does this case study reveal? Despite the rhetoric of religions and ethnic conflict in Afghanistan, farmers and herders with different ethnicities are not only able to get along but also ensure each other's mutual well-being. Policymakers concerned about food and livelihood security should take note that multiple professions ensure mutual survival. Instead of a homogenous policy response, taking into account the ecological context and sociocultural differences can produce a complex

Table 4.4 Summary of differences in Pashtun and Shugni relations

	Arab Pashtuns	Shugnis	Comparison
Elevation	500–4000 m	2500–4000 m	The Pashtun pastoralists are at lower elevations, migrating upland to the Shugnis
Ecological niche	Lowlands to highlands: valleys and villages in Baghlan, Kunduz, and Takhar with seasonal use of high mountain pastures	Highlands: valleys and village region of Pul-i-Zirabon with seasonal use of high mountain pastures	Seasonal overlap in ecological niche between the Pashtuns and Shugnis. This overlap is used to retain longer-term relations by pastoralists, who store goods for the next season, or farmers requesting items from the next migration, or through seeking seasonal employment
Religion	Sunni	Shia Ismaili	Demonstrate diversity in religious distinctiveness and attempt to respect each other's faith
Language	Dari	Shugni	Cultural distinctiveness
Profession	Nomadic pastoralists with some agricultural land	Sedentary farmers with some livestock	Sunni Muslims are pastoralists who have agricultural land, whereas Ismaili Muslims are sedentary farmers who keep some animals
Trade items	Livestock, kitchenware, ironware, salt, and other items from southern markets, cash	Wheat, animals, dried yogurt	The Pashtun pastoralists bring items from southern markets to trade them for agricultural items in the highlands with the Shugnis
Employment	Employer	Employee	Pastoralists employ farmers. Farmers also give their animals for care to the pastoralists

and yet sustainable outcome. This case study illustrates that the sociocultural aspects of pastoralism are embedded within the ecological and both need to be taken into account in policy formulation. The war in Afghanistan has been very long and is an effective test because it is a system under constant anthropogenic perturbation from international, regional, and local sources. Nonetheless, the Pashtuns and Shugnis demonstrate agency under these unstable conditions by ensuring that their differences through ecological and sociocultural contexts are a practical asset.

4.5 Conclusion: Reflection on the Three Case Studies

The first case study illustrates an effort by the central government in China to homogenize and transform livelihood strategy by sedentarizing ethnic communities that have historically practiced pastoralism. The net effect is removal of sociocultural and ecological diversity from the system. It is an example of the core trying to assert administrative authority in the name of ecological restoration while pursuing strictly an instrumental agenda of economic extraction of key renewable and nonrenewable resources. After sedentarization of the Kazakh population, a variety of town-based livelihood practices such as crop production, wage income, and small businesses fail in these indigenous societies in comparison with the practice of pastoralism. The Kazakhs lack the knowledge as well as the sociocultural and ecological context to make such livelihood activities a success, which has been the hallmark of Chinese development. Chinese policymakers are trying to "civilize" the Kazakh pastoralists in their own image. The objective of their policy is to eliminate differences in ecological and economic profession, and the net effect is livelihood insecurity concentrated among former pastoralists with Kazakh identity. Chinese government policy is in fact fueling unsustainability of Kazakh pastoralism.

The second case study, from Inner Mongolia, China, reiterates the first case study, where government policy in the form of the LGDCRS and GEPP increased ecological and economic vulnerability through institutional change. Nonetheless, pastoralists are developing adaptive capacity by drawing on their historical human ecological relations to sustain pastoral livelihoods, again illustrating indigenous approaches to common pool resources is key to survival. This case study also illustrates that diversity as presented through the history of indigenous rangeland management and pastoralism is a necessary livelihood strategy in Inner Mongolia.

The final case study illustrates that weak central government is also adding to regional instability in Badakhshan, Afghanistan, but in contrast, its relative weakness is making posible the existence of diversity to contribute to the food and livelihood security of both farmers and herders alike. Here difference in an ecological context, religious interpretation, ethnicity, and ecological profession is clearly an asset for survival. Although the Pashtun pastoralists are wealthier than the Shugni farmers, when their ecological zone overlaps with that of the Shugnis in the highlands, the Pashtuns are dependent on the Shugnis. This recognition and retention of difference facilitates mutual dependence and contributes to survival of both communities.

Choice by different ethnic human societies to engage in pastoralism is a time-tested practice that has historically proven itself under conditions of stress in both China and Afghanistan. An important insight that these three case studies illustrates is that policy intervention without recognition of sociocultural and ecological context can prove to be a source of instability for livelihood and food security of marginalized ethnic populations under the modern nation state.

References

Amitai R (2005) Mongols, Turks, and others: Eurasian nomads and the sedentary world. Brill, Leiden

Barfield T (1981) The Central Asian Arabs of Afghanistan: pastoral nomadism in transition. University of Texas Press, Austin

Becquelin N (2004) Staged development in Xinjiang. China Q 178:358–378, 10.1017/S0305741004000219
Behnke R (1993) Range ecology at disequilibrium: new models of natural variability and pastoral adaptation in African savannas. Overseas Development Institute, London
Benson L, Svanberg I (1998) China's last Nomads: the history and culture of China's Kazaks. ME Sharpe, Armonk
Bliss F (2006) Social and economic change in the Pamirs, Tajikistan. Routledge, New York
Brown C, Waldron S, Longworth J (2008) Sustainable development in western China: managing people, livestock and grasslands in pastoral areas. Edward Elgar, Cheltenham
Cerny A (2010) Going where the grass is greener: China Kazaks and the Oralman immigration policy in Kazakhstan. Pastoralism 1(2):218–247
Chambers R (1997) Whose reality counts? Putting the first last. Intermediate Technology Publication, London
Chilonda P, Otte J (2006) Indicators to monitor trends in livestock production at national, regional and international levels. Livestock Res Rural Dev 18(8):117
Cliff TMJ (2009) Neo oasis: the Xinjiang Bingtuan in the twenty-first century. Asian Stud Rev 33(1):83–106
Daftary F (1990) The Ismailis: their history and doctrines. Cambridge University Press, Cambridge
Dong SK, Gao HW, Xu GC, Hou XY, Long RJ, Kang MY, Lassoie PJ (2007) Farmer and professional attitudes to the large-scale ban on livestock grazing of grasslands in China. Environ Conserv 34(3):246–254
Dong SK, Wen L, Zhu L, Li XY (2010) Implication of coupled natural and human systems in sustainable rangeland ecosystem management in HKH region. Frontiers of Earth Science in China 2010 4(1):42–50
Dong SK, Ren Z (2015) Animal husbandry civilization and grassland health: understanding the coupling system between grassland animal husbandry ecology and human society. J Lanzhou Univ (Soc Sci) 43(4):105–110
Everitt BS, Landau S, Leese M, Stahl D, Shewhart WA, Wilks SS (2011) Cluster analysis, 5th edn. Wiley-Blackwell, Oxford, http://onlinelibrary.wiley.com/book/10.1002/9780470977811
Felmy S, Kreutzmann H (2004) Wakahn Wolusswali in Badakhshan. Observations and reflections from Afghanistan's periphery. Erdkunde 58(2):97–117
Fernandez-Gimenez ME, Le Febre S (2006) Mobility in pastoral systems: dynamic flux or downward trend? Int J Sustain Dev World Ecol 13(5):341–362
Goodman D (1989) China's regional development. Routledge for Royal Institute of International Affairs London, London
Goodman D (2004) China's campaign to "Open up the West": national, provincial, and local perspectives. Cambridge University Press, Cambridge
Greenwood DJ, Levin M (1998) Introduction to action research: social research for social change. Sage Publications, London
Grotenhuis ET (2002) Along the silk road. Smithsonian Institution, Washington
Hardin G (1968) The tragedy of the commons. Science 162:1243–1248
Harris RB (2010) Rangeland degradation on the Qinghai-Tibetan plateau: a review of the evidence of its magnitude and causes. J Arid Environ 74(1):1–12
Hunsberger AC (2000) Nasir Khusraw, The Ruby of Badakhshan. IB Taurus, London
http://www.ecologyandsociety.org/vol15/iss2/art8/
Kassam K-AS (2009a) Viewing change through the prism of indigenous human ecology: findings from the Afghan and Tajik Pamirs. Hum Ecol 37(6):377–390
Kassam K-AS (2009b) Biocultural diversity and indigenous ways of knowing: human ecology in the Arctic. University of Calgary Press, Calgary
Kassam K (2001) North of 60: homeland or frontier. In: A passion for identity: Canadian studies for the 21st century, Scarborough: Nelson Thompson Learning, pp 433–455

Kreutzmann H (2003) Ethnic minorities and marginality in the Pamirian Knot: survival of Wakhi and Kirghiz in a harsh environment and global contexts. Geogr J 169(3):215–235
Li WJ, Zhang Q (2009) Jiedu Caoyuan Kunjing: Dui Ganhan Banganhan Caoyuan Liyong he Guanli Ruogan Wenti de Renshi. [Understanding the Grassland Dilemma: several problems on grassland use and management in Arid and Semi-Arid Areas]. Economic Science Press, Beijing.
Li WJ, Huntsinger L (2011) China's grassland contract policy and its impacts on herder ability to benefit in Inner Mongolia: tragic feedbacks. Ecol Soc 16(2):1, http://www.ecologyandsociety.org/vol16/iss2/art1/
Liao C, Barrett C, Kassam K-A (2015) Does diversification improve livelihoods? Pastoral Households in Xinjiang, China
Liao C, Sullivan PJ, Barrett CB, Kassam K-A (2014a) Socioenvironmental threats to pastoral livelihoods: risk perceptions in the Altay and Tianshan mountains of Xinjiang, China. Risk Anal 34(4):640–655
Liao C, Morreale SJ, Kassam K-AS, Sullivan PJ, Fei D (2014b) Following the green: coupled pastoral migration and vegetation dynamics in the Altay and Tianshan mountains of Xinjiang, China. Appl Geogr 46:61–70. doi:10.1016/j.apgeog.2013.10.010
Li J (1991) Xinjiang climate. Beijing: Meteorology Press
Longworth J (1993) China's pastoral region: sheep and wool, minority nationalities, rangeland degradation and sustainable development. CAB International, Wallingford
Liu JG, Dietz T, Carpenter SR, Folke C, Alberti M, Redman CL, Schneider SH, Ostrom E, Pell AN, Lubchenco J, Taylor WW, Ouyang ZY, Deadman P, Kratz T, Provencher W (2007) Coupled human and natural systems. Ambio 36:639–649
Mi'erzhahan J (2004) Kazak people, 2nd edn. Minzu Press, Beijing
Miller DJ (2000) Tough times for Tibetan Nomads in western China: snowstorms, settling down, fences and the demise of traditional nomadic pastoralism. Nomadic Peoples 4(1):83–109
National Development and Reform Commissions (2011) New policies for returning rangeland to grassland. http://www.sdpc.gov.cn/xwfb/t20110831_431891.htm
Miles MB, Huberman AM (1994) Qualitative data analysis: an expanded sourcebook, 2nd edn. Sage, Thousand Oaks, CA
Olimova S (2005) Impact of external migration on development of mountainous regions: Tajikistan, Kyrgyzstan, Afghanistan, and Pakistan. Strategies for development and food security in mountainous areas of Central Asia. In: International workshop, Dushanbe, Tajikistan, June 6–10, 2005
Ostrom E, Dietz T, Dolšak N, Stern PC, Stonich S, Weber EU (2002) The drama of the commons. National Academy Press, Washington, DC
Patton MQ (1990) Qualitative evaluation and research methods, 2nd edn. Sage, Newbury Park, CA
Sandford S (1983) Management of pastoral development in the third world. Wiley, Chichester
Starr S (2004) Xinjian : China's Muslim borderland. ME Sharpe, Armonk
Toops S (2004) Demographics and development in Xinjiang after 1949
Wang XY, Zhang Q (2012) Climate variability, change of land use and vulnerability in pastoral society: a case from Inner Mongolia. Nomadic Peoples 16(1):68–87
Wood F (2002) The Silk Road. Folio Society, London
Wu ZZ, Du W (2008) Pastoral nomad rights in Inner Mongolia. Nomadic Peoples 12(2):13–33
Xinhua (2007a) China's economic and social development plan. China Daily. Retrieved from http://www.chinadaily.com.cn/china/2007-03/19/content_830762_2.htm
Xinhua (2007b) Hu Jintao proposes scientific outlook on development for tackling China's immediate woes, challenges. http://news.xinhuanet.com/english/2007-10/15/content_6883135.htm
XLDRC (Xilingol League Development and Reform Commission) (2011) Survey report on the implementation of Xilingol League Beijing-Tianjin sand source control program, 9 Dec 2011. http://www.xlgldrc.gov.cn/gzdt/vzhgggzdt/201112/t20111209_721907.html
XUAR Chorography Committee (2010) Xinjiang Yearbook 2010. Xinjiang Statistical Bureau Press, Urumqi, China
Yeh ET (2009) Greening western China: a critical view. Geoforum 40:884–894

Zhang MA, Borjigin E, Zhang H (2007) Mongolian nomadic culture and ecological culture: on the ecological reconstruction in the agro-pastoral mosaic zone in northern China. Ecol Econ 62(1):19–26

Zhang Q (2012) The dilemma of conserving rangeland by means of development: exploring ecological resettlement in a pastoral township of Inner Mongolia. Nomadic People 16(1):88–115

Chapter 5
Building New Human–Natural Systems for Sustainable Pasture Management in South America

Hermes Morales, Fernando Coronato, Soraya Carvalho, Alejandro Saravia, Alejandro Schweitzer, Amaury Burlamaqui, and Jean F. Tourrand

Abstract On the basis of research conducted in three contrasting South American ecoregions (southern Patagonia, Argentina; the central Pampas and Campos, Uruguay; and western Amazonia, Brazil), this chapter aims to better understand the complex sets of reasons that have recently led local societies to adopt more sustainable pasture management in South America. After a brief overview of each of the three biomes, representative social–ecological systems of pastoralism are identified with the objective to describe their respective pasture management history, especially the colonization process and the settlement of the pioneers, and the successive farming systems, mainly the practices related to herd and pasture management, are compared. Finally, the main local and national policies regarding the livestock sector and landownership are analyzed. The evolution of the social–ecological systems of pastoralism in these ecoregions was assessed with a three-dimensional model of vulnerability based on the agroecosystem resilience, livelihood improvement, and institutional capability. The evolution of the mental models about livestock has also been analyzed to better understand the current perceptions of the local people and their scenarios for the future of livestock in their social–ecological systems. The results focused on the dynamics of rangeland management, the vulnerability of the pastoral social–ecological systems,

H. Morales • A. Saravia
IPA, Montevideo, Uruguay

F. Coronato
CENPAT-CONICET, Puerto Madryn, Chubut Province, Argentina

S. Carvalho
UFPA, Belém-PA, Brazil

A. Schweitzer
UNPA-CONICET, Rio Gallegos, Santa Cruz Province, Argentina

A. Burlamaqui
Embrapa, Boa Vista-RR, Brazil

J.F. Tourrand (✉)
Cirad, Montpellier, France
e-mail: tourrand@aol.com

S. Dong et al. (eds.), *Building Resilience of Human-Natural Systems of Pastoralism in the Developing World: Interdisciplinary Perspectives*,
DOI 10.1007/978-3-319-30732-9_5

and the evolution of the mental models of local people about livestock, so as to discuss the different rhythms of transformation and the existence of critical thresholds. Besides diverse land and social issues, it is imperative to intensify livestock production and increase the offer of new technologies, as well as to identify some relevant human dimension factors, such as the demand of local societies for collective livelihood improvement, the recent national and international environmental policies related to global warming and biodiversity, and the stronger awareness of consumers for sustainable development. In the context of global change, the practices of rangeland management will change in a more sustainable way, resulting in better society–environment interaction and human–nature integration.

5.1 An Overview of Pastoralism in South America

In geography and history, South America offers an excellent opportunity to explore the main drivers of change in the interaction between society and environment referred to pastoral production systems. Geography, or indeed paleogeography, explains the fact that the pastures of this continent have evolved void of the big herds of herbivores that conditioned the early relationship between rangelands, herds, and humans in the Old World and North America (Melville 1997; Milchunas et al. 1988). There are large areas of grasslands, steppes, savannas, and shrublands for grazing pastures according to the vegetation map of South America (Fig. 5.1), which were void of big ruminants until the arrival of Europeans in the sixteenth century , which is the point where history contributes to the uniqueness of the opportunity mentioned above. The multiple ecological processes triggered by this dramatic transplantation can be termed "ecological imperialism" (Crosby 2004); within the huge extent of this biogeographic revolution, the concerns regarding pasture-based animal husbandry are but one topic we intend to consider in this chapter.

Since the sixteenth century and for half a millennium, the human–nature relationship in South America has been based on the practices involved in mining natural resources of the soil (water, natural vegetation, and biodiversity) and the subsoil, especially gold and silver in the past, and more recently oil, gas, coal, and numerous minerals. European settlers and their descendants found that the best way to settle new lands and progressively develop the hinterlands from the already colonized areas. In less than five centuries, new societies have been built mixing Amerindian people with migrants from diverse regions of the world, especially from Europe, but also from sub-Saharan African, the southern and eastern Mediterranean, Japan, and China. In these societies, the pioneer spirit is a part of the heritage, as in North America. Furthermore, the advance of democracy since the end of the last century has opened up new opportunities to develop sustainable social–ecological systems.

Arriving in the plains, pioneers early introduced cattle and sheep to the savannas and steppes, profiting from the natural forage of the rangeland to feed the herds. So, cattle breeding quickly developed in several South American areas of rangeland, including the Pampas and Patagonia, the lowlands of the Pantanal and the Llanos, the

Fig. 5.1 Vegetation map of South America. (Adapted from Blanco et al. 2013)

alpine meadows of páramo and puna in the Andean highlands, and some Amazonian islands, such as Marajó. Livestock was of major importance at the onset of rural societies in these regions. On the other hand, in the forest ecosystems, such as the Amazon rainforest and the Atlantic Forest located along the Atlantic tropical coast, the European settlers usually started to slash and burn forest plots to cultivate the land, using ash from natural vegetation to fertilize their annual and perennial crops. Ruminants were farmed to produce food for home consumption, goods to sell, saving, and diverse services. In exchange, the cropping system produced forage and grain to feed the herds. The situation was intermediate in the savanna biomes, such as the Cerrado and Gran Chaco located in the center of South America, and Caatinga in the northeast of Brazil. In the farms in these regions, animals and herds also had essential functions; for example, for tillage, transport, and other farm work. So, in all South America, animal breeding had a major and diversified role in the postcolonization farming systems and livestock was always a key factor in the coupled human–natural systems and their dynamics.

The period of colonization differed according to the location of the ecosystems. Those near the coast where the migrants arrived, in the case of the Pampas and the Atlantic Forest on the Atlantic coast, were usually settled much earlier than the ones located in the center of the continent, such as Amazonia, the Cerrado, and Gran Chaco. Despite its maritime position, Patagonia was colonized quite late, because of the severity of the environment. Obviously the ability to progress was a determinant factor everywhere. So, the steppes and savannas were generally colonized before the forest areas. The resistance of local populations interfered a lot; for instance, in some Andean zones where the local Amerindian societies were strongly organized, or Araucanians in the Pampas of Argentina and the Tobas in Gran Chaco.

The weak sustainability of pasture management has led to a gradual degradation of natural and cultivated grasslands, at a pace depending on the bioclimatic conditions, the ecosystem, and the set of practices, especially overstocking, bad use of burning, or no respect for the vegetative cycle. In the center of Patagonia for example, the degradation process was very fast, in just four decades, between 1880 and 1920, mainly due to too high overgrazing coupled with low rainfall and strong wind erosion. In the case of Amazonia, the grassland degradation resulted from a set of unsustainable practices, starting from forage seeding until grazing. In the Pampas, light degradation of the natural grasslands/rangelands had been noticed for some centuries, especially regarding the native trees of the rangeland used for a long time for construction, cooking, and heating. However, the main impact on Pampa rangeland is the transformation of natural vegetation into cultivated grasslands, and more recently the replacement by tree plantations (eucalyptus and pines) and annual crops (mainly soybean) for exportation.

The Pampas, Patagonia, and Amazonia are three contrasting South American biomes in term of bioclimatic conditions and natural vegetation, but also the period of colonization and the settlement process, current socioeconomic development, the National context, etc. In this comparative research, the reference social–ecological systems are the Pampas in Uruguay, Patagonia in Argentina, and Amazonia in Brazil. The Pampas were colonized from the sixteenth century to the eighteenth century, Patagonia was colonized at the end of the nineteenth century and the beginning of the twentieth century, and Amazonia was colonized in the past half century. Rainforest is the natural vegetation of Amazonia. Arid steppe and prairie are respectively the natural vegetation of Patagonia and the Pampas. The current pastoral societies herded the natural socio-ecosystem, although these have been strongly transformed since the beginning of the colonization. They are currently changing even more because of global changes and their diverse components, especially changes to the global economy and the huge demand for food and feed, for which South America is becoming one of the world's breadbaskets. Moreover, other global changes are impacting the social–ecological systems, such as climate change, whether in Amazonia or in the Pampas and Patagonia, new demands of young people in terms of life and work conditions, social networks, and new information and communication technologies.

In the context of global changes affecting the South American biomes differently, the objective of this chapter is to use case studies to describe the changes in the structure and the functioning, and the current and future trends of the social–ecological systems in the three livestock production ecoregions, mainly the new human–nature partnership in the rangeland/grassland management. Comparisons between the three ecoregions allow us to identify similar and dissimilar factors acting in each case that determine singularities in the process regarding spatial extension and temporality as well as the existence of thresholds and nonreversible situations. The comparisons are based on three different tools: the vulnerability of the social–ecological system, the mental models of local people about livestock, and the identification of different ecological thresholds. Here, the concept of three-dimensional vulnerability (agro-ecosystem, livelihood, and institutional capacity) of pastoralism proposed by Dong et al. (2011) is used to compare the resilience of the social–ecological systems in the three cases. The successive situations concerning the stability of the agroecosystem, the livelihood of farming families, and the institutional capacity are defined assess the resilience and the vulnerability of the social–ecological systems of pastoralism and their trends. The comparison of mental models of local people about livestock activities in the past, now, and in future scenarios is another tool used to better understand the trends of the social–ecological systems in the three cases.

5.2 Dynamics of Social–Ecological Systems in Pasture Management in Campos of the Pampas, Uruguay

The area now occupied by Uruguay was qualified as good for nothing on the arrival of the European conquerors in the sixteenth century. With this qualification, the region was settled very late compared with other regions that offered the beloved metals, gold, and silver, mainly in the Andes, or the natural vegetal and animal resources, principally in forest areas. On the other hand, this low interest in Uruguayan soil and subsoil was compensated for by the easy land access linked to the proximity of the Atlantic coast, the ability to move in the rangeland, and the mildness of the climate. For these reasons, European settlement occurred very early compared with that in the two other regions we are studying in this chapter.

At the beginning of the seventeenth century, some cattle from Paraguay were left in the open rangelands in the south of Uruguay. This was the start of an activity that has evolved since then but remains the main one in terms of area involved and economic importance. At the same time, Jesuits where organizing their famous “missions” and occupied the north of the country and left their cultural traits and the names of rivers (Uruguay, Cuareim, Tacuarembó) as heritage. According to Moraes (2008), the south of Uruguay was organized as private property and was occupied by about 10,000 people by the middle of the eighteenth century, whereas the north, organized as communal lands and ruled by Jesuits, was inhabited by more than a 100,000 people.

The Uruguayan rangelands or prairies are usually called “Campos” (Allen et al. 2011). These grasslands consist mainly of grasses, along with herbs, small shrubs, and occasional trees on an undulating and hilly landscape, with variable soil fertility (Fig. 5.2). Campos differ from the Cerrado by having fewer trees, longer and severer winters, and a relative abundance of native legumes. Campos are found in the northern part of the Pampas biome, located in Uruguay, southern Brazil, and northeastern Argentina. The subtropical climate is humid, warm in summer and mild in winter.

Fig. 5.2 Photographs of the Pampas with cattle and sheep in Uruguay

Fig. 5.3 Gaucho culture based on rural housing (*left*), meat food (*middle*), and collective/cooperative structures (*right*)

The annual grass production is about four metric tons of dry matter per hectare, with a typical seasonal distribution, with a minimum of 14 % of total production in winter and a maximum of 35 % in spring (Guido et al. 2014). This production distinguishes natural grasslands of this region as being among the most productive in the world. The stocking rate of the natural grasslands has not changed very much since the early days—no more than one cow per hectare. The production, its distribution, and its variability are now fully described and understood by remote sensing. The grasslands have certainly changed, but not very much during the past four or five centuries (Morales 2007). It is thought that originally taller grasses and perhaps more trees were present. The presence of big ruminants has an impact on carbon and nitrogen cycles, among others. Piñeiro et al. (2006) and Paruelo et al. (2010) estimated that a fifth of soil carbon has been lost in the last five centuries.

However, the living conditions and livelihoods of farming families have changed a lot like the gaucho society, which has built one of the most famous and productive cattle breeding systems in the open range in the world (Fig. 5.3).

Moreover, the gaucho society became a reference in breeding societies in just a few centuries, when others needed several centuries or millennia; for example the Peulhs in West and Central Africa, the Bedouins in the southern and eastern Mediterranean, and Mongols, Uighurs, and Tibetans in Central Asia. Events strengthened the livestock image of gaucho breeders, especially during the First World War, when the Pampas provided meat and cereals for European countries.

One important characteristic of gaucho social–ecological systems is the ecological uniqueness. The region is subtropical humid. It is not tropical because it has too many frosts, on average 25 per year, and is not temperate because of a too hot summer, with temperatures up to 42 °C; rainfall is about 1200 mm/year in an irregular but on average well-distributed manner.

The first product was meat for local consumption, and the practice of slaughtering a cow to eat only a little portion of it astonished many voyagers. The industry was chiefly launched by the end of the eighteenth century and consisted of hides. Before then, the main product was leather. The meat was a by-product without any value until the beginning of another important activity; salted meat to be consumed by the slaves in the plantation industries; cotton, coffee and sugarcane that flourished in Cuba, Brazil, and North America. This activity was dominant until the end of the nineteenth century, when a typical industrial product—Liebig's Meat Extract—was sent to nearly everywhere in the world to feed the British Army. This industrial serving activity was established with the legalization of land as property, fencing, and genetic improvement to make the original cattle more suited to the new situation. The change was nearly complete by 1910, when new and modern slaughterhouses sent refrigerated meat to Europe, beginning an activity that is still very vigorous.

From an ecological and also economic point of view, the development of the wool industry was also very important. The presence of two species—cattle and sheep—is to some extent complementary, and another important reason was to manage two types of risk: economic and climatic. In times of drought, sheep are better adapted and can to some extent compensate for the loss of cattle, and also wool and meat are slightly correlated to markets, although when one of them has a low price, the other can have a high price, contributing not only to the resilience of the farm enterprises but also that of the whole industry.

Wool and meat accounted for more than 85 % of Uruguayan exports from 1870 to 1970 (Moraes 2008), and the resulting wealth created a type of dependence for the country, which even with great effort failed to develop other successful economic activities, resulting in a big political crisis in 1973 as one of its consequences (Fig. 5.4).

The most important attempt to promote development started with the mechanization after the Second World War by the plantation of cultivated pastures in the natural grassland/rangeland to increase the productivity per hectare. By the end of the 1970s it was clear that the traditional beef and wool production would only change a little with the promoted technology (Jarvis 1981), but it was also clear that milk

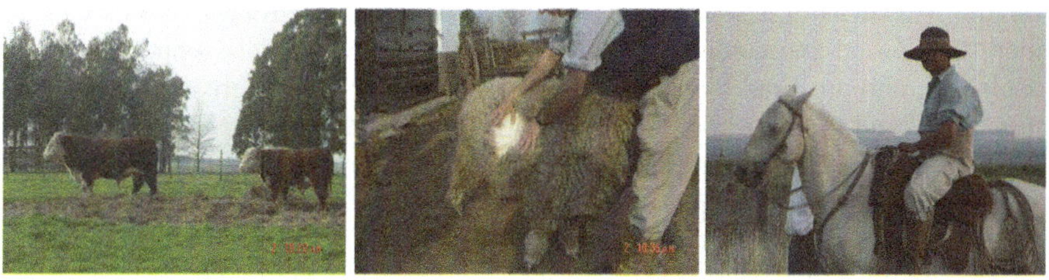

Fig. 5.4 Cattle (*left*), sheep wool (*middle*), and a gaucho on his horse (*right*)

production would benefit from an improved and enlarged pasture management technology. As this become evident, major change continued with the development of agribusiness, tree plantations from the 1990s, and the production of crops for exportation, with an expected result. For the first time, agricultural products were more important as exports in 2010 because of the huge increase in soybean culture that had taken place since 2001 and had boosted grain production. Another important activity that has replaced grasslands is afforestation. Eucalyptus and pines grow very well in Uruguay, and with the support of a law passed in 1987 have become important industries and a pole of attraction for foreign investors. North Americans, Chileans, Spaniards, Finns, and others have bought land, developed plantations, and set up related industries in Uruguay.

The relationship with the open spaces is one of the main features of the cultural dimension of cattle raising in this area, and its main output is the ever presence of horses. Ranches are organized in several manners according to the general strategy of the owner, who sometimes can be a lawyer who comes every 3 months to his property, but the family farm, based on family labor, is the most important type of farm. Farms are divided into several paddocks to facilitate cattle handling but also manage the natural pastures, which are very variable, and one paddock can be more suitable for calves, another for culled cows, and so on. Until now, horses have been used to move cattle inside the properties (Fig. 5.4), and also sometimes among properties, even if trucks are now the main transportation method. Farmers unions are very important, and are present all over the country, being the loudspeakers for farmers and also centers of technical and social exchange.

As we have already explained, young people do not think that this is a fashionable way of living, so it is rather difficult to imagine what the farm organization—or even the industry organization—should be in the next few decades. Even if there are fewer and fewer people working or living in the countryside, traditions are increasingly present in social activities such as commemorations or sports or simply local feast days, where horses and riders are more and more becoming central attractions.

At the same time, some controversies are being examined by Uruguayans and will probably remain unresolved in the near future. The first question addresses national disposal of natural resources. Is it worth promoting nonlocal users of local natural resources? The beneficiaries and affected people are different and are located in different parts of the world. Should Uruguay put at risk its electricity production, which is partly dependent on runoff water, which is diminished by tree plantations? The second question refers to the necessity of introducing norms for the use of natural resources, where two different but related issues can be identified. Good knowledge and accepted mathematical models support land use decisions, and are used by the government to induce limited erosion of agricultural soils, and the Livestock Production on Natural Grassland National Board is deliberating about whether to promote natural grass conservation. Both measures should be adopted by people without their receiving extra money for actions of this type, as it is usual in the USA or Europe (Cattan 2014), and discussions are being held about the possibility of enforcing these norms. How much should a little country engage Pastoralism in South America: itself in global issues such as global warming or biodiversity loss?

5.3 Dynamics of Social–Ecological Systems in Pasture Management in Southern Patagonia, Argentina

Until the late nineteenth century, the dry steppes of Patagonia were void of permanent settlement and supported only a tiny population of nomadic hunter-gatherers. Once the young South American republics of Argentina and Chile had achieved internal organization, they could afford territorial expansion over neighboring Patagonia. By 1880, the "Conquest of the Desert" by the Argentine Army appeared to be not only an internal issue but much fostered by western European countries willing to expand agriculture, especially animal breeding in temperate regions (Fig. 5.5).

This led to the fast occupation of the fertile plains of the Pampas with cattle and crops, whereas the less fertile Patagonia was settled by huge sheep farms for the greater benefit of British, Flemish, or German wool companies (Fig. 5.6).

According to the recorded stocking rates during that early period in most of Patagonia, the early settlers overestimated the carrying capacity of the rangelands and overstocked them. The following figures must be considered in relation to the low primary productivity of the Patagonian steppes (610 kg/ha/year on average for the whole region, according to Paruelo et al. 1998): stocking rates of about two sheep per hectare may sound very modest in productive rangelands elsewhere but they are extremely high in Patagonian terms, and too high for the arid steppe. Nevertheless, those were the initial stocking rates in extensive areas regardless of

Fig. 5.5 The huge Patagonia was settled to develop livestock production using the natural steppe

Fig. 5.6 Wool production at the beginning of twentieth century (*left*), mainly based on the merino breed (*right*)

their actual (and by then ignored) carrying capacity. It is likely that the initial overstocking could have been maintained for a couple of decades because of the biomass reserve the region had accumulated in the past, considering it was void of large herbivore herds (because guanacos cannot be considered as such).

Sheep farms boomed at the turn of the nineteenth century, and for four decades occupied 0.75 million square kilometers. The First World War especially fostered this explosive expansion of sheep. The process continued and sheep numbers peaked in the 1950s, when about 22 million animals were recorded. From then on, overgrazing started to become evident to open-eyed scientists, who anticipated a decreased productive capacity of Patagonian rangelands and, thus, advised there should be lower stocking rates. However, the question of the health and sustainability of pasture did not really start to get the attention of farmers until the 1980s, when the damage became too obvious and costly in terms of reduced production. For wool, this decrease was estimated as 0.5 % per year from 1940 to 1987 in well-managed ranches in northwestern Patagonia (Soriano and Paruelo 1990).

Even if many small farmers (fewer than 1000 sheep) exist in marginal areas, the core of sheep farming is performed on a very extensive basis in plots ranging from 10,000 to 30,000 ha, which support flocks of 2000 to 15,000 sheep. The numbers are much larger for company-owned ranches. Only differing in the larger size of the paddocks, this model was mostly based on the ones that had built the wealth of the Pampas and the Falklands, wetter regions next to Patagonia in which sheep colonization of the latter started.

Decreasing productivity, along with difficulties in wool and meat commercialization, forced the adoption of nonsustainable ecological and economic dynamics in the recent decades (Ares 2006), yet the stocking rate steadily decreased during the last three decades, as evidenced by a regional Patagonian stock as low as ten million sheep in the late 1990s. The current average stocking rate is about 0.3 sheep per hectare but may be as low as 0.1 sheep per hectare (Fig. 5.7). Beyond this threshold it is considered that economic sustainability no longer exists. This explains the closure of many sheep farms in Patagonia, (about one third or two thirds of them, depending on the area), since in most cases reconversion is utterly impossible because of ecological constraints.

Thus, less than a century after the beginning of pastoral colonization, the process has transformed much of the vast steppes of central and eastern Patagonia into desertlike areas. Despite the dramatic decrease in the stocking rates in the last three decades—or even the closure of ranches—pasture recovery (if any) is extremely slow.

Patagonian rangelands have proved to be more resistant than resilient. In fact, they "resisted" heavy stocking during the early period—thanks to the biomass previously accumulated—and maintained a good secondary production for several decades. Once the pasture degradation was evident and stocking rates started to be conscientiously diminished, resilience did not appear as the traditional model predicted, and it became obvious that the "state and transition model" (Westoby et al. 1989) was more accurate for most of the Patagonian rangelands, in which, most likely, a new lower state had been reached.

Fig. 5.7 Harsh conditions—low rainfall, wind erosion, and overgrazing—explained rangeland degradation

Although the lessons had been learned by academics and decision makers, ranchers did not realize that changes in range management should be made to cope with the new (poorer) environmental context. Compelled by the exigencies of everyday life, they insisted on overestimating the carrying capacity of their pastures, and used to blame the climate and the vagaries of the Argentine economy as well as the great variations in the international wool and mutton markets.

Some intensive campaigns in the 1990s and a constant effort to adjust public policies in the first decade of this century eventually led to a widening of the point of view of the traditionally conservative rural milieu. The learning is reflected by new words freshly incorporated into the rural Patagonian lexicon and currently quite often used. Desertification, shrub encroachment, rotational grazing, electric fencing, prelambing shearing, or even feedlot are now meaningful concepts among Patagonian sheep ranchers.

The learning of this agronomic lesson was not a mere top-down process but in many cases a painful personal experience about how to not perish in a degraded environment and an unstable market. In the first few years of the last decade, sheep ranching in Patagonia suffered a deep crisis owing to pasture impoverishment and economic instability added to some episodes of drought and volcanic ash spill. However, this almost generalized collapse served to somehow "reset" sheep farming in the whole region, by causing the abandoning the nonviable ranches and forcing survivors to adopt a more careful rangeland management.

The abandoning of ranches could be interpreted as an ecological victory, because once sheep have been removed, the grasslands/rangelands can be rested and given a (tiny, as seen) chance to regenerate. With both sheep and fences removed, remnant wildlife populations will have access to an unfragmented habitat and the chance to survive and flourish. In such a case, collision with contiguous persistent sheep flocks is unavoidable, either with herbivores (guanacos, rheas) competing for pastures or water, or with predators (foxes, pumas) attacking sheep.

Eventually, these "threats" to sheep farming could be considered as ecologically tolerable. The same cannot be said of the increasing threat of Metal mining and shale hydrocarbon production are fairly well distributed in Patagonia. The discussion about the environmental future of this region is currently open.

It is clear from this case study that at the beginning of the colonization the Patagonian rangelands resisted sheep grazing because of the cold and dry conditions of the natural steppe. The livelihoods of farming families improved thanks to the income from sheep's wool. The local society started to organize itself, focusing on the wool trade and the supply chains for sheep production. However, the permanent overgrazing, the overuse of shrubs for heating and cooking, and with low rainfall and strong wind erosion progressively destroyed the rangelands. At that time, just a few scholars mentioned the challenges to change the practices to sustain the sheep farming system. So, three decades after the beginning of colonization, the decreasing productivity of the sheep farms because of pasture degradation was partially hidden by the high demand for animal products linked to the First World War in Europe, even thought the opening of the Panama Canal significantly reduced the major role of Patagonia in transoceanic trade. Hence, sheep production continued to increase slowly during the first decades of the twentieth century. Then mining exploration, especially oil and gas, and tourism became the pillars of the Patagonian economy. And now, sheep farming is perceived as a heritage, especially by young people and those in urban areas.

5.4 Dynamics of Social–Ecological Systems in Pasture Management in Eastern Amazonia, Brazil

From the middle of the 1960s until the end of the twentieth century, the Amazon rainforest was the northern agricultural frontier of the South American colonization process led by the descendants of European pioneers (Morales et al. 2011). In the two decades before, the Cerrado had been the agricultural frontier (Fig. 5.1). This forest biome was colonized by transformation of the natural forest and savannas of Brazil and Bolivia into crop–livestock landscape.

Currently, the main land use in the Cerrado is maize and soya bean production for human food and animal production, and pastures for cattle ranching. Only 10–15 % of the natural Cerrado biome area has been preserved. The same process was logically applied to the Amazon rainforest until it was progressively stopped from the 1990s with increased environmental awareness among local populations and global governance. Sayago et al. (2010) showed all nine Amazon countries have been affected by the same colonization process. It was more significant in Brazil because of the huge area of Brazilian Amazonia, 65–70 % of the Amazon rainforest.

As decided by the Brazilian government, the colonization of Amazonia aimed to achieve a triple objective: social, economic, and political. First, the Brazilian government wanted "to give land to landless" (*terra sem homem para homens sem terra*). This aimed to attract migrants settle Amazonia but also to face the rural exodus linked to mechanization and land concentration in southern and eastern Brazil. Second, the Brazilian government and national companies wanted to exploit the huge natural resources of soil and subsoil of the Amazon rainforest, to incentivize and directly support the national development of all the country. Third, the military, which ruled the country at that time, wanted to secure the borders and avoid the interest of neighbors, considering that diverse plans had been proposed by foreign companies to exploit the Amazonian resources. One of them was the Rockefeller plan in 1933, which proposed the building of a huge dam at the mouth of the Amazon to facilitate oil exploration and transportation (Veiga et al. 2004).

So, the clear aim of successive Brazilian governments was to use the Amazonian natural resources to support local and national development. Years, decades, and centuries before, there had been diverse valorizations of Amazonian natural resources. Scholars reported that leathers and hides of animal wildlife, drugs and medicines, and fruits were collected by explorers who traded with Indian and *Cabocla*[1] communities from the seventeenth and eighteenth centuries (Vaz 2013; Veiga et al. 2004; Poccard-Chapuis 2004). It was the time of *drogas do sertão*, the word *sertão* meaning "bush," the hinterland of the Northeast Region, far from the coast, where the ports and cities were located. At the end of the nineteenth century, the rubber of *Hevea brasiliensis* became the main resource for exportation. The rubber trade allowed the development of and resulted in the wealth of the main cities, such as Manaus, Belém, and Santarém. The Brazil nut was also a significant Amazonian resource for national consumption and exportation.

Fifty years after colonization, several minerals (especially iron, aluminum, nickel, copper, and zinc) are being extracted from Amazonian subsoil and have significantly contributed to Brazil's development. Hydropower is another Amazonian resource with a significant contribution on a national scale. For example the huge Tucuruí Dam produces energy for local populations but also for mining and transformation (e.g., aluminum). The effect of mining on the social–ecological system is different from that of agriculture. The impact on the ecosystem is usually low because of the subsoil activity and the small land area involved. However, the effect is high on local employment and distributed salaries.

The main impact on the social–ecological system was linked to the development of the agricultural frontier. "The Amazon colonization through the cattle foot" (Santiago 1972) was the leitmotiv during the two first decades, until the middle of the 1980s. Slash and burn was the reference practice to transform forest plots into pasture (Fig. 5.8).

The process was as follows: (1) cutting and burning of the natural vegetation in September to October, at the end of the dry season; (2) sowing an annual crop, in January to February, at the beginning of the rainy season, usually rice, and maize if it is a secondary forest; (3) seeding the forage directly into the cereal, in March to April, 6 or 8 weeks later; (4) harvesting the cereal in May to June; (5) leaving the forage to grow until September to October, before cattle graze it. Every year, every smallholder cultivates 2–4 ha of land by following this process.

In this smallholder farming system, cereal production secured food security: human food, feeding of small animals (pigs and poultry), and sale to provide other goods and cover other expenses. Cattle produce milk and can be used as saving. Livestock multifunctionality, including crop–livestock integration, helped smallholders survive. The same process also allowed cattlemen to expand their ranches, planting year after year some tens of hectares. So, cattle ranching is the main land use after

[1] Community settled by *metis* (cross-breeding) of Amerindians and descendants of Europeans migrants.

Fig. 5.8 Cattle ranching expansion transformed rainforest into large and small ranches

deforestation for several reasons (Veiga et al. 2004). Fifty years after the start of colonization, about 1,000,000 km^2 of the rainforest (i.e., approximately 15–20 % of the Amazon basin), has been transformed into pasture, about 70 % occurring in Brazil.

At the beginning of colonization, some farmers settled on rich soil and planted perennial crops, especially cocoa, coffee, and pepper, in Amazonia of Brazil. However, the crisis regarding perennial crop prices at the end of the 1980s forced many farmers to abandon their plantations and plant forage. More recently, some farmers decided to plant perennial crops again but usually in diversified farming systems. Timber exploitation began with colonization but it became a significant activity only in the 1980s, when foresters started to exploit the public rainforest reserves. Timber exploitation had a strong impact on the ecosystems, mainly in opening tracks in the forest for that the pioneers and landless used to access new land to slash, burn, and cultivate it. Except for some individual cases, the foresters never adopted sustainable timber exploitation. On the contrary, for several decades they preferred to cut and exploit the two or three interesting trees per hectare before to slash and burn the plot and transform in pasture (Fig. 5.9).

Exploitation of soil and natural vegetation by pioneers, cattlemen, and foresters does not explain all the transformation of the natural forest into an agricultural landscape. Landownership access has also been a strong factor in deforestation. Several scholars stated that in different part of the Amazon rainforest, migrants used the colonization process to occupy private and public land and developed farming systems while waiting for the regularization of their landownership by the local government. Treccani (2001), Veiga et al. (2004), and Vaz (2013) mentioned that in the 1950s and at the beginning of the 1960s, some pioneers were occupying the land and registering it so they could sell it to new migrants.

Progressively, the migrants settled in communities. Usually, the community is defined as some families living in the same area and sharing some services and social activities, such as church, school, and soccer team. Mutual work is also done at the community level (Veiga et al. 2004). On the basis of the life trajectories of several Amazon pioneers, Tourrand et al. (2012) showed that migrants are not adventurers. The migration to Amazonia is usually a well-prepared event in the life of future migrants. When they migrate, they know where they are going and who will receive him, often a relative who has already migrated. The migrants know in which community they will live when they arrive, what kind of work they could do, and what their prospects are. Sometimes, they have already visited the region to know better the new land and community.

Fig. 5.9 Perennial crops (cocoa; *left*), timber exploitation (*middle*), and cattle ranching (*right*) in Amazonia in the 1990s

A community is not independent. It depends on the county, which has the administrative rights to the land. Some communities depend on two counties, but this is not frequent. According to the demand at the community level and its elective power, mainly after the return of democracy in 1985, the county has to implement institutional services such as health and school services, maintenance of roads and tracks, electricity, mail services and telecommunications, and extension services. Usually, a community has to elect one member to serve on the county council. That person is responsible for defending the rights of the community in the county.

The communities grow with as new migrants arrive. Some of them will become villages with their own communities. When a certain size is reached, a village and its communities can demand the right to create their own county, especially when the county town is far away or when the context is very different. This was the case for the villages of migrants located along the colonization routes which were linked to the administrative centers located near the river (Fig. 5.10).

During the 50 years of colonization, different social organizations have been created at local and regional levels. At the county level there is the Union for Rural Workers (Sindicato dos Trabalhadores Rurais) and the Rural Union (Sindicato Rural) for smallholders and medium-sized to large agribusinesses respectively. Along the Trans-Amazonian Highway, especially in the state of Pará, there is the Movement for Transamazon Survival (Movimento Pela Sobrevivência da Transamazônica). There is the Council of Rubber Tappers (Conselho Nacional dos Seringueiros) in the state of Acre. Some organizations are at national level, such as the Land Pastoral Committee (Comissão Pastoral pela Terra), linked to Catholic Church.

Until the middle of the 1980s, to quickly settle the Amazonian agricultural frontier, the public policies subsidized the deforestation through special loans from two public organizations: the bank Banco da Amazônia and the Superintendência do Desenvolvimento da Amazônia. Moreover, at least 30 % deforested land was the criterion for a potential owner of this land to receive loans. The public policy changed from the middle of the 1980s. Banco da Amazônia and Superintendência do Desenvolvimento da Amazônia, but also the Bank of Brazil, offered only loans to recuperate degraded land, and not to deforest more land.

At the same time, the control of deforested areas became easier with remote sensing, on both a local scale and a regional scale. Progressively during the first decade of this century, farmers were considered responsible for the deforestation of their land. Furthermore, they had to submit an individual plan aiming to recuperate

Fig. 5.10 Settlement of pioneers (*left*), urbanization (*middle*), and development of a social society (*right*) in Amazonia

a significant part of natural forest on their land. In addition, all the titles of landownership had to be proved, especially for medium-sized and large farms, and have been linked to the recuperation plans for natural forest.

It is clear from this case study that during the 50 years of colonization of Amazonia that the Brazilian government has strongly supported the settlement of the agricultural frontier to profit from the huge natural resources of the rainforest for national development. The Brazilian government is now trying to reduce the impact of deforestation and recuperate part of the natural ecosystems through a set of policies based on subsidies, but also penalties and strong sanctions, including removal of landownership. At the same time, the Brazilian government is aiming to alleviate poverty through special policies focused on the livelihoods and living conditions of smallholders.

5.5 Strategies for Sustainable Pasture Management in Tropical and Subtropical South America

On the basis of interviews with diverse livestock stakeholders (breeders, traders, agroindustry, extension services, local governance, etc.), information in public databases, especially the successive census, and a review of literature, we examined the three cases in tropical and subtropical regions of South America to create a better understanding of the actions developed by societies regarding the livestock sector. We analyzed the interactions between a livestock-raising area (currently or potentially) and a society willing to occupy such territory and practice cattle breeding as the tool for doing so. We describe how the different sets of historical conditions and ecosystem capacities have modified the dynamics of the process in three case study areas, a process lasting five centuries in the Pampas, 150 years in Patagonia and 50 years in Amazonia. One of the essential aspects that we have identified is the nonlinearity of the interactions and the different periods, which characterize the arrival of irreversibility in these contexts. The presence of thresholds, which mark the irreversible transition between two successive states of the social–ecological system, shows how two more and less similar initial situations could diverge a lot in a short time. Consequently, it is not easy to assess and monitor the evolution of the social–ecological system of pasture management and adapt to its dynamics, justifying many surprises.

5.5.1 Similarities Among the Three Case Studies

Although roughly 3000 km separates the core of the Amazon rainforest from the Uruguayan prairies and a further 2000 km separates the latter from the southern Patagonian steppes in Argentina, with the entailing climatic and agronomic differences, the three case study areas share some geohistorical features, allowing a comparative analysis among them. Firstly, the three were marginal areas for the imperial powers that colonized South America from the eighteenth century. In fact, the Portuguese and Spanish empires never achieved permanent settlement either in Amazonia or in Patagonia, and even Uruguay was a frontier (and litigious) area between both imperial powers. None of the three regions had a well-developed Amerindian civilization, such as in the Andes and Central America, which thus became the centers of the Spanish American Empire, nor the mineral wealth supporting the core of the Portuguese Empire in southeastern Brazil (Droulers 2001). So, if the marginality of the fertile prairies of Uruguay can be explained because of their situation at the fringe of two territories, and which has geopolitical origins, that of Amazonia and Patagonia is clearly due to environmental constraints, which repelled permanent economic occupation. In the three cases, regardless of the time at which they happened, the process of permanent economic occupation started with livestock. Since the three case study areas also shared the capitalist way of arranging territory, livestock husbandry was in all three areas the first step to eventually secure landownership, except until the end of the eighteenth century in the Jesuit/Amerindian area in the north of Uruguay. This process occurred quite early in Uruguay, at the turn of nineteenth century in Patagonia, and as late as the end of the twentieth century in Amazonia.

Livestock farming triggered the permanent occupation of these areas previously void of attractive activities. In fact, animal husbandry (cattle in Uruguay and Amazonia and sheep in Patagonia) was the pioneer activity that attracted migrants and initiated ensuing economic activities. Cattle rearing were almost the only economic driver in Uruguay during most of the nineteenth and twentieth centuries. Sheep rearing boomed in Patagonia in the first decades of the twentieth century but collapsed in the last decades. Tropical breeds of cattle are now booming in Amazonia. Other economic drivers followed and currently compete with pastoralism in leading the regional economies. This is forcing adaptive changes of animal husbandry in the three regions.

The three landscapes were part of a continent with a low human pressure. Some 10,000 years ago, humans entered America by the north of the continent, and by the time of the arrival of Europeans some very impressive civilizations existed and others had already disappeared. In the areas that we are studying, the populations were very low, there was only some very primitive agricultural activity, and the societies were not very complex. People organized themselves into somewhat errant tribes, without an important architecture and without any appreciably sophisticated agriculture.

The three regions offered an opportunity for people to earn a living who had suffered difficulties in their original locations, and this was the real reason behind the colonization. The reasons of the immigrants were always similar: they needed a new area where they could develop a better life.

The existence of organized societies that were eager to occupy more space for its functioning was another common point. The eastern part of Uruguay, along the Atlantic coast, was a disputed area between the Portuguese and the Spanish empires. Patagonia was claimed more or less clearly by Argentina, Chile and England. Brazilians were eager not only to give opportunities to poor Brazilians but also wanted to establish firmly their presence in Amazonia.

The original existing wildlife was very rich in the three case study areas, with an important common point being the absence of big herbivores. It has been established that an important loss of big mammals occurred some 10,000 years ago, and this characterized the situation. In grassland/rangeland areas of both Uruguay and Patagonia, little ruminants (deer and guanacos, selvage llamas, etc.) were present but their impact on vegetation was not important.

The drivers that induced pastoral production were the same in the three cases. Initially, local consumption explained its development, but soon after the arrival of Europeans, foreign and distant markets influenced and boosted the process. For Uruguay, initially hides and wool for the European cloth industry and then beef and recently soybean—wool in the Patagonian situation and beef ("the hamburger connection") and also soybean in the Amazonian situation—explained the characteristics of the advance of humans over nature in our case studies, the organization of local societies, and the environmental impacts.

Another important shared characteristic is the difficulties to develop crops. Even though in Uruguay and in Amazonia some areas are dedicated to crops or wood production, in both situations this activity is risky because of fragile or shallow soils, weather variability, market vagaries and the distance to harbors, and also because of problems of adapting temperate crops that originated in Europe or Asia to a different situation. In the case of Patagonia, dryness is so marked that cultures are not possible without irrigation, and water is not abundant.

5.5.2 *Differences Among the Three Case Studies*

The three case study areas are, from south to north, southern Patagonia in Argentina, the Campos in Uruguay, and eastern Amazonia in Brazil. Several sharp differences exist among these three case study areas, and more generally between the three biomes.

First, as already stated, the climate is dry and cold in southern Patagonia, temperate in Uruguay, and tropical in Amazonia. This climatic difference has some important consequences for the human population. The agricultural potentials for European settlers were very different. European agriculture did not have the knowledge, tools, and genetic resources to develop a tropical agriculture, so it is easily understood why this region was the last one that was incorporated by the colonialists.

Clearly, because of these conditions, Uruguay was best suited for European herbivores: cattle, horses, and sheep. The Campos are very good for cattle, and their excellent availability of surface water promoted the herds. The Campos are not always well suited for temperate crops (wheat, oats, and maize), but they can be cultivated for local consumption, and also in some circumstances, as nowadays, can produce grain for export. This is not the case for Patagonia, where it does not rain enough, there are not good wells, and there are only a few rivers. Some cultures are present in the valleys, but their importance is low.

When we compare pastoral potential, there are huge differences. The Patagonian steppe, with harsh winters and scarce rainfall, allows annual grass growth of less than 500 kg of dry matter per hectare, compared with 4000 kg of dry matter per hectare in Uruguay and from 3000 to 5000 kg of dry matter per hectare according to the soil conditions for the grass patches in Amazonia. Good grass management norms allowed commercial production while maintaining the resources, which was easy in Uruguay, but extremely difficult in Patagonian conditions, where the price paid for learning has been very high in some cases. It consisted in destroying the resource. The Pampas case study in Uruguay is very different for the reasons previously mentioned. In contrast to the Patagonian case study, at the beginning of colonization in the sixteenth century, pasture management started with a very low stocking rate regarding the carrying capacity, both in the south and in the north of the country, which were settled by European migrants and Jesuit communities respectively.

The situation is contrasted among the Amazonian social–ecological systems. The natural prairies located on several islands of the Amazon, especially the eastern part of Marajó, and along some riverbanks have productivity near to that of the natural prairie of Uruguay, with a sustainable stocking rate of about one cow per hectare. In contrast, in the Amazon rainforest there were no grass management norms because the grass can grow only after being slashed and burned. A few tropical forages come from South America, but the main part is imported from Africa, especially diverse *Panicum* and *Brachiaria* species. These cultivated pastures opened up a whole new range of opportunities, making possible beef and milk production, and transforming Amazonia to first-class world player for these products. The livestock breeds are also different. In Patagonia and in Uruguay, the Europeans breeds proved to be well adapted, but this was not the case in Amazonia, where the grazing systems are the result of human-induced synthesis of initially not connected living beings. Cattle breeds came from India, buffaloes came from India and the Mediterranean, and grass species came mainly from West Africa to this South American location with a human population dominated mainly by European descendants.

Moreover, even though the three case study areas were colonized by Europeans settlers and their descendants, their histories differ, especially the colonization and the origin of the migrants. From the sixteenth century in Uruguay, European settlers from Spanish Kingdom in the south and Jesuit communities, including Europeans migrants and Indian native families, in the north developed livestock farming systems based on the grazing of the grassland/rangelands. From the end of the nineteenth century, the Argentine Army seized Patagonia from Native Americans to allow colonization by shepherds from the Falklands and the Pampas, and by European and Mediterranean migrants. From the 1960s, the Brazilian government decided to colonize the Amazon rainforest with the three objectives of securing Brazilian landownership, exploiting the natural resources from the soil and the subsoil, and giving land to the landless from other regions.

The contrast regards the land-use change and the herd management practices. Large areas of the Patagonian natural steppe have been seriously degraded by inadequate pasture management, mainly overgrazing associated with low rainfall and strong wind erosion. At present, many farms have become unviable or have even been abandoned because of desertification. The natural vegetation has disappeared and the steppe has been degraded to desert. Large areas of the Amazon rainforest have also been changed by pioneers using slash and burn practices to plant pastures for their cattle herds. However, the bioclimatic conditions allow the progressive regeneration of the rainforest after several years. In Uruguay, the natural grasslands had been softly grazed by cattle and sheep herds for almost five centuries before they were recently partially destroyed to plant cultivated pastures from the 1970s and tree plantations and grain crop systems from the 1990s.

Finally, global environmental stakes probably make the most important difference. Turning the Uruguayan campos into tree plantations or crops, adjusting grazing management in Patagonia to maintain sheep farming, and transforming the Amazon rainforest into crop–livestock land have an impact on global issues such as climate change and biodiversity loss, but also human dimension issues. The impact of transforming the Amazon rainforest into pastures is suspected of changing human life on Earth.

5.5.3 Three-Dimensional "Vulnerability" Coordination Framework in Assessing Vulnerability

The three social–ecological systems of pastoralism in Amazonia, the Campos, and Patagonia have been projected into Dong et al.'s model (2011), which is based on the three-dimensional vulnerability/resilience. We consider that before colonization, the three social–ecological systems of pastoralism were in similar positions regarding vulnerability. The of agroecosystem dimension (vertical axis) was abundant with limited livelihoods (transverse axis) and low institutional capacity (horizontal axis). The red star gives this position, located in octant 7 of the model (Fig. 5.11).

Assessing the present positions of the three social–ecological systems of pastoralism gives the following result. The Amazonian system of pastoralism is located at junction 1–2. The few counties located at the border of the Amazon basin were the first to be colonized. They were almost totally deforested, mainly along the roads where were settled by the migrants. The other counties, colonized more recently, usually still have a significant area of rainforest, especially far from the roads. Mostly, the forest is not exactly the natural rainforest, because the best trees have already been exploited. Despite the differences between the counties, a significant part of the ecosystem has to be considered as degraded, about 20–25 % of the area. This justifies the evaluation of the agroecosystem as neutral, not robust, and not fragile. In contrast, the livelihoods on the Amazon frontier have been strongly improved compared with the beginning of colonization, mainly in the past two decades with specific public policies. The improvement concerns all the public services, including health, education, road maintenance, loans, rural extension, energy, and communications. The same can be said for the institutional capacity, particularly after the return of democracy in the 1980s. Directly linked to the current social and environmental policies implemented by the Brazilian government, the resilience of the agroecosystem should be improved, as should the livelihoods and the institutional capacity (Fig. 5.11).

Nowadays, the Campos social–ecological system of pastoralism is located near the one for Amazonia one in the model in Fig. 5.11. The livelihoods and the institutional capacity have been improved greatly since the beginning of colonization, especially in the last few decades, and for the same reasons as in Amazonia. With regard to the agroecosystem, the natural grasslands/rangelands did not change much from the sixteenth century to the seventieth century until the middle of the twentieth century, when the development of mechanization allowed the transformation of natural rangeland into cultivated grassland with the objective of increasing the productivity per hectare. However, the no change concerned only a small part of the area, especially in northern Uruguay. In contrast, the recent development of tree plantations and soya bean for exploitation is significantly changing the context. Of the 16 million hectares of natural or cultivated grassland, about two million has been transformed into tree plantations or soya bean fields in less than 10 years. Even though the future of the Campos social–ecological systems of pastoralism will greatly depend on the choice of social and environmental policies, it is assumed that the core of the basaltic area—three million hectares—located in the north of Uruguay and unfit for cropping or forest production, will remain in state ownership.

The current position of the social-ecological system in Patagonia is different because of the degradation of the natural ecosystem and the low natural recuperation process. Even after 15–20 years without grazing, there is no significant improvement of the land cover. Moreover, the livelihoods are not good because of the degradation of the common ties that stems from the loss of the rural population (whose density may be as low as 0.01 inhabitant per square kilometer) and the distance between the farms, usually some tens of kilometers. These reasons also explain the weak institutional capacity, even if the institutions are present in the urban centers and extension staff is quite active. So, the Patagonia system has been located as straddling octants 6 and 8 in Fig. 5.11. The outlook for the future is pessimistic because of the low natural recuperation process of the Patagonian steppe and the high cost to replant it. Nevertheless, if resilience is seen not as recovery but as transformation, it must be said that Patagonian society is looking for different ways of transformation to cope with a henceforth degraded ecosystem. The choices range from opencast mining to national parks and reserves, with intermediate alternatives such as agrotourism or durable methods of sheep farming (Coronato et al. 2011), but once again, the final outcome will depend on competition between a broad set of actors, from local to international (Schmink and Wood 1992).

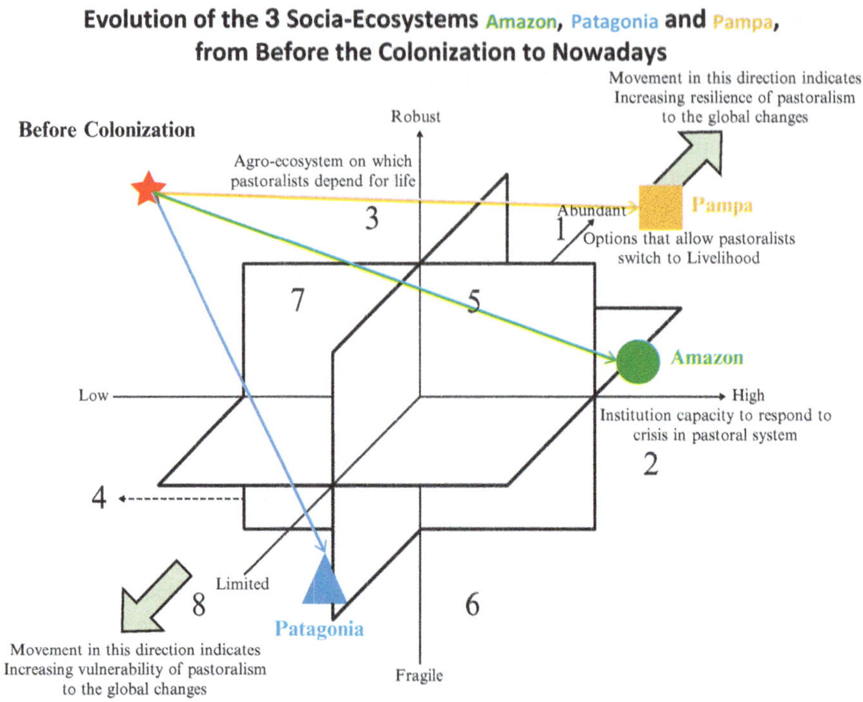

Fig. 5.11 Evolution of social–ecological systems of pastoralism in Amazonia, Patagonia, and the Pampas from before colonization to the present time

5.5.4 *Irreversibility Thresholds and Learning Adaptation Possibility*

As analyzed in Chap. 2, resilience can be seen as stability (buffer capacity), recovery (bouncing back), and transformation (creativity). These three processes mark important differences among our three case studies. Moreover, as stated by March (2010), it is not always possible to learn from experience, especially if the effects and consequences of actions are distant in time or space. Furthermore, the ability to learn from experience is central in adaptive management (Walker and Salt 2006), which consists in taking practice as an experiment, monitoring, reflecting, and proposing changes to increase adaptation, and can be seen as the equivalent of creativity as already discussed. But even if this definition is broadly accepted, successful examples of adaptive management of natural resources are scarce in the international literature (Williams and Brown 2014). Our case studies give us the opportunity to analyze this important subject: can social–ecological systems of pastoralism adapt to different dynamic environments? And what are the learning results from the comparative analysis of our three contrasting case studies?

The usual answer is: it depends. Initially we discuss these characteristics in the two grassland/rangeland regions: Patagonia and Uruguay. Then we include the Amazonia case study.

After five centuries of grazing, the Campos ecosystem appears to be mostly in a very good situation, even though it has been stated that a fifth of its initial content of carbon has been lost because of the presence of European grazers during this time (Piñeiro et al. 2006). It is difficult to establish if this trend will continue in the next five centuries. But for the moment there is no evidence of production capacity loss. The pasture management practices did not significantly degrade the natural ecosystem. In other words, the natural ecosystem is not vulnerable to grazing applied by the ranchers. So, the Campos social–ecological system is resilient, and in this case, resilience means stability (buffer capacity).

Furthermore, the recovery (bouncing back) appears to be very present in the Uruguayan grassland/rangeland conditions. It is well documented that grass growth can completely recover after a drought, even a very severe one as that experienced in 1989.

In contrast, in the Patagonian case, the type of social–ecological system caused large areas to collapse in less than 50 years from the end of the nineteenth century to the first decades of the past century; in less than a half century, large areas of the central Patagonian arid steppe were destroyed by inadequate management practices. The social–ecological system collapsed and some of these lands have been abandoned because they are no longer productive. So, there was no stability because thresholds have been surpassed (buffer capacity). It is interesting to notice that, at this time, the shepherds have adopted these practices, thinking they were in a ecosystem similar to that in the Pampas or Campos where many of them had previously settled.

In the past and even recently, some ranchers tried to restore their rangeland by reducing the stocking rate or adopting pasture rotation until they stopped grazing for many years. But when the steppe is strongly degraded, the natural regeneration (if any) is very slow and reduced to some small plots, because a too large part of the land is uncovered, the soil having disappeared with wind erosion. Artificial restoration is possible, but the very high cost limits this to very small plots, such as those destroyed by mining or oil exploration. Moreover, the low price of wool for many decades was not attractive to possible investors. So, the central Patagonian social–ecological systems would not be resilient because they too degraded according to their restoring capacity. In other words, the resilience of these systems would reach an irreversibility threshold.

The same management practices were adopted in the southern lands of Patagonia, and near the Andes. In these zones, the climatic conditions are better—namely, more rainfall and less dry and strong wind, which means the climatic conditions are quite similar to those of the driest zones of the Campos (even if much colder). Consequently, the degradation of natural rangeland was lower, and sometimes there is no significant degradation of the natural steppe, as it was noticed in the Campos. Most obviously, there is almost no abandoned land in these areas. So, the vulnerability and the resilience of the Patagonian steppe appears strongly linked to the type of pasture management, mainly the balance between the grazing impact and the regenerative capacity of the local rangeland.

The context is different in the Amazon rainforest, where the natural vegetation was destroyed to plant pasture. Many scholars have reported the strong degradation of the pasture in a few years, leading to the temporary abandoning of the degraded plots and consequently the deforestation of new plots to feed the herds. Veiga et al. (2004) linked this degradation to inadequate pasture management. Planted just after the burning of natural vegetation or directly in the first food crop, the pasture profits from the ash to grow, quickly and cover the soil and compete with weeds, including the regrowth of natural vegetation. But after some years of grazing, the soil fertility decreased, areas without forage cover appeared, and natural vegetation grew and competed with forage, until it dominated. However, the natural regeneration allows the progressive restoration of secondary forest in the degraded pasture in time, which depends on the level of degradation and the type of soil. So, after some years, usually 4–5 years, ranchers slash and burn the secondary vegetation to start a new cycle. It was the process most commonly applied during the three decades of colonization.

According to Veiga et al. (2004), sustainable pasture management needs (1) a stocking rate adapted to the forage biomass, (2) rotational grazing management to respect the cycles of pasture, (3) control of weeds, and (4) use of some fertilizers to replace the soil nutrients. Since the end of the 1980s, many degraded pastures have been recuperated on the basis of this set of practices. The process occurred firstly in large and medium-sized farms, which can obtain the funds to finance it, and secondly in small farms.

On the basis of this set of practices, the resilience of the new social–ecological system in Amazonia would be high (buffer capacity) with real restoration potential (bouncing back), even it results from the destruction of the natural forest ecosystem. In this case, the ranchers have transformed the natural ecosystem into a resilient social–ecological system (creativity) where the natural ecosystem is still present in protected forest reserves in specific areas such as near the sources, along the riverbanks, and on the strong to avoid soil erosion and maintain a minimum of biodiversity. In this case, resilience as appears as the capacity of radical transformation of the environment, from forest to grasslands, and eventually cultures, is the objective to allow people to earn their living. Moreover, this ability is directly connected to the ability to interact with distant regions. Without the interaction with African rangeland scientists, who created forage varieties, Indian farmers, who created resistant zebu breeds, and the skill of Brazilian scientists to build efficient farming systems and the pioneers who had no alternative to survive, the grasslands/rangelands in the Amazon rainforest would have never existed.

In this context, in the face of the dynamical complexity of social–ecological systems, stability can be present for centuries as in the core Campos region, the basaltic zone, it can last only some decades, or it can adopt a truly new configuration as in Amazonia. Adaptation uncertainty is linked to the presence of not perceived thresholds, as illustrated by the Patagonian farmers.

5.5.5 Mental Models of the Human–Nature Relationship in South American Pasture Management

On the basis of data collected by the Livestock Farming and Local Development (LiFLoD) network in 13 livestock social–ecological systems located in contrasting biomes on the five continents, Tourrand et al. (2014) have identified similarities between groups of stakeholders regarding their mental models of livestock. The main groups are ranchers (from smallholders to large-scale cattlemen), traders and input providers, managers of agroindustries, extension services, funding agencies, scientists, policymakers, union representatives, association leaders, and consumers.

Tourrand et al. (2014) defined a mental model as a collective representation system—livestock in our case—shared by a group of people. Built over the long term, the mental model is relatively stable and, consequently, does not change much in a short period. Any representation of livestock can be defined as complex set of eight livestock mental models (Fig. 5.12), which were defined as:

- *Lifestyle*: Livestock is the basis of the rancher's life, mainly represented in pastoral societies.
- *Security*: Livestock provides goods, income, and saving. This is particularly dominant in the communities of smallholders.
- *Business*: Livestock is a business on farm, local, and global scales. This is very frequent in the discourse of people working in the supply and market chains, and a part of extension services.
- *Investment*: The livestock sector is a good investment option. Always in the past but more frequent nowadays.

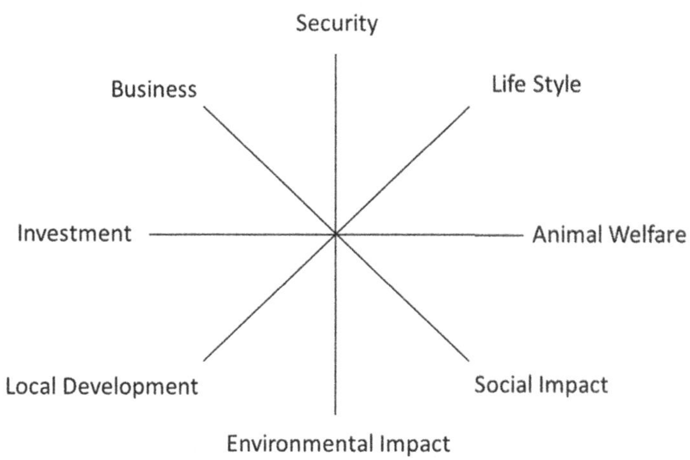

Fig. 5.12 The eight elementary mental methods

- *Local development*: Livestock has a significant role in local development. Discourse mainly in local governance and policymaking, development agencies, universities and research centers, etc.
- *Environmental impact*: Especially focused on greenhouse gas emissions, but also the other impacts. This mental model is more frequent nowadays with the "Green Wave" and the stronger awareness of environmental issues liked to global warming.
- *Social impact*: Livestock leads to social differentiation. This mental model frequently occurs in the discourse in social science community.
- *Animal welfare*: An animal is first of all a living creature very close to humans and which has to be respected, and not eaten.

When we apply these to the three social–ecological systems in the case studies, we find different mental models are interacting according to the period. For example, if we adopt the leitmotiv "the colonization of Amazonia by the feet of cattle," the mental model of the Brazilian policymakers should be a complex combination of mental models including *investment*, mainly national to make a profit with natural soil resources; *security*, especially food security for the survival of migrants; *business* because livestock is business at the frontier; and *local development* to settle and integrate the social–ecological systems into the national territory. At that time, nobody talked about the environmental impacts of livestock, with the exception of some scientists and local NGOs, which were aware of the negative social impacts of livestock (Veiga et al. 2004).

In the same way, Coronato (2010) stated that the colonization of Patagonia at the end of the nineteenth century was also the result of set of factors defined by different and complex mental models. At that time, the Argentine Army aimed to secure the region and to incentivize local development. The British wool companies wanted new lands for investment and to develop their business. Migrants were mainly interested in their food security, some of them using their skill in sheep farming and their lifestyle with livestock. At that time, nobody yet talked about the environmental impact of livestock. In contrast, the size and the location of the land were doing Significant differences among between the settlers.

According to Moraes (2008), in the eighteenth century, at least two main mental models coexisted during the settlement of the Pampas. South of the Rio Negro, the Spanish Kingdom settled the area by allocation of small and large plots of land to European migrants and their descendants to develop cattle and sheep breeding. They did not have an alternative because of the dominance of natural grasslands/rangelands. In the north, the Jesuit communities aimed to join Amerindians and European migrants or descendants in common management of the land. Directly linked to the importance of exportation of animal products, the livestock business always had a significant place in the Uruguayan mental models. The gaucho society has been progressively built during the past three centuries. It is one of the results of the interaction of livestock and natural rangeland (Litre 2010; Morales et al. 2011).

The situation is now changing, especially in Brazilian Amazonia. For different reasons, the Brazilian government recently decided to implement more constrained environmental policies regarding development in Amazonia. Those measures have strongly motivated the stakeholders with mental models dominated by *environmental impact*. At the same time, the same Brazilian government has launched efficient programs to develop small farm agriculture. These programs incentivize and directly support the stakeholders with mental models more oriented on *social impact* and *local development*. So, facing the new context focused on sustainable development, all other stakeholders have to adopt a friendly discourse regarding environmental issues. For some of them, this is easy because of their natural skills. For others, it is more difficult, and they have to force themselves to present more sustainable strategies. Even so, the discourses have changed; the mental models have not radically changed until now, because they are strongly resilient, as already mentioned. They need more time to change, frequently some years to one generation.

In Patagonia, the situation of stakeholders is different from that in Amazonia because of the lack of direct responsibility of current breeders for the pasture degradation that happened several decades ago. Just a few stakeholders recognize the significant contribution of overgrazing to pasture degradation. Some others consider the low annual rainfall and the frequent drought periods to be the main factors for this pasture degradation. Moreover, several stakeholders do not assess the enormous decrease of grassland/rangeland productivity compared with the beginning of colonization, at the end of the nineteenth century. For them, the weakness of the sheep and wool market and the lack of efficient farming policies have led many breeders to stop breeding sheep and then to abandon their farms. Despite this, the mental models are changing. Linked to the development of tourism, some farmers and policymakers are aware of the role of sheep farming in local development. Coronato (2010) showed that sheep are one of the most frequent images of Patagonia for tourists, along with the beautiful landscapes and wildlife. Furthermore, for almost all stakeholders, sheep farming is the only activity capable of maintaining settlement in the region. In addition, in the long term strong international demand will encourage sheep farming and incentivize investors in this sector and in the region.

The context is also different in Uruguay regarding both the market and the environmental impact. Until a few years ago, all stakeholders considered livestock farming in Uruguay as a natural activity with no significant environmental impact (Morales 2007). Livestock was a very relevant component of the natural landscape in Uruguay. So, the publication in 2006 of the FAO's report "Livestock's Long Shadow," which described in detail the strong environmental impacts of animal production, was a shock in the Uruguayan livestock sector. Despite the credibility of the FAO in that country, many people thought it an aggressive strategy of environmental lobbies against livestock activity on a global scale. Nowadays, some stakeholders are aware of the significant greenhouse gas emissions of ruminants, especially in the extensive livestock production as in the Pampas ecosystems. Other stakeholders know of the problem but do not accept it. Finally, others do not believe in the environmental impacts of their livestock farming systems. And they continue to think about the strategy of environmental lobbies. In contrast, many stakeholders are aware of the impact of soya expansion on soil erosion or the impact of pine and eucalyptus plantations on soil acidification and biodiversity loss (Arbeletche et al. 2011). Consequently, policymakers are currently launching a set of environmental measures aimed at controlling soil erosion through the crop rotations to avoid more serious environmental impacts in the future. However, the environmental policy is not yet very restrictive for animal production and tree plantations. In addition, livestock mental models in Uruguay have been built on the change in the international meat market, especially the frequent and sudden variations of the meat price with the alternation of long periods of low prices and short periods of high prices. Linked to strong international demand, especially from Asia, the high price of meat since 2008 has affected the entire Uruguayan livestock sector. Some stakeholders think that it is a permanent change in the international meat market, whereas others consider the process is just a temporary artifact.

It is clear from this analysis that the livestock mental models are engaged in a process of change, mainly due to the new environmental policy from the global to the local scale, but also due to the strong international market for animal products and social development issues in rural areas. The relevancy of these factors is building a new context with consequences for the livestock mental models. Nowadays, all the mental models have to integrate an environmental component, at a minimum to know that any question about environmental impact is relevant. External speeches have radically changed both those supporting environmental policies and those against them. In contrast, internal talks are moving slowly because of the resilience of the mental models. The changes due to the market and social issues are a little different because all the stakeholders recognize these changes. The difference is in the permanence or the temporality of this change. Furthermore, another relevant change in the consumers on other continents, especially in China, Southeast Asia, and Europe, is the higher importance of food safety and food quality. Directly linked to recent crises concerning animal products (e.g., melanin in milk powder, horse meat mixed with beef meat, and avian flu), this change could durably impact the consumption and international demand in terms of meat origin and production systems.

5.6 Conclusion

In the past, new agricultural frontiers in South America were available for European migrants and their descendants, associated or not associated with Amerindian communities. They developed diverse social–ecological systems focusing on grassland/rangeland-based animal production. Some of these social–ecological systems are sustainable, as in the case of the gaucho society in the north of Uruguay, cattle ranching in the natural grasslands/rangelands of the Amazon basin, and sheep farming in a few areas of Patagonia. In other places, nonsustainable livestock social–ecological systems and pasture management practices have strongly damaged or destroyed the natural ecosystems. The process was slow, in two or three centuries, or very fast—in a few decades—as in central Patagonia. Sometimes, it was based on the destruction of the ecosystem, as in the case of the Amazon rainforest. Moreover, linked to the strong international demand for food, the process is continuing; for example, in the biomes of Gran Chaco, the Pampas, and the Cerrado, where there was a huge expansion of exportation agribusinesses, mainly soya bean, cellulose, and biofuels, in past years. Also, the process is still continuing in the Amazon rainforest of the Andean countries for several sociopolitical and economic reasons. In contrast, the process has strongly been reduced in Brazilian Amazonia because of severe environmental policies.

However, presently global ecological challenges make it very controversial and probably impossible in the short term to continue the expansion of agricultural lands, transforming natural grasslands/rangelands and especially forests into croplands. As mentioned for Brazilian Amazonia, the environmental law require ranchers to preserve or restore and maintain the natural forest in at least half of their lands, above all specific areas, such as sources, riverbanks, and strong slopes. Moreover, the law compels them to adopt sustainable pasture management practices (e.g., no burning). In Uruguay, new environmental policy is forcing the exportation agribusiness to integrate sustainable practices, especially crop–livestock integration, to reduce the soil erosion.

Progressively, all the agricultural sectors have to adopt ecological and also social sustainable practices because this is becoming a real exigency of the global economy directly linked to the growing awareness of consumers and the demand of some key actors who have a very important and strategic role and use new information and communication technologies. One of the most relevant examples is that of the Brazilian beef agribusiness, which is forcing the Amazon rainforest producers and partners to reduce their environmental impacts so it can keep its new dominant position in the international beef market.

To monitor these changes, the model of three-dimensional vulnerability allows us to share with breeders and local stakeholders the trends of the resilience of social–ecological systems, and define the possible scenarios on the basis of local knowledge, scientific results, and possible policies or changes. Because it its easy use, and can be applied in both interviews and participative workshops with focus groups or a large population, this model can be used to define the hopes and fears of people, debate scenarios, and above all produce relevant results for policymaking. Moreover, it can be applied on farm, local, and regional scales.

One of the main challenges of participative actions using this model is also to share and discuss the changes (stability, restoration, and creativity) and some irreversibility thresholds along the three axes— agroecosystems, livelihoods, and institutional capacity—which define the vulnerability and the resilience of the social–ecological systems.

However, as already mentioned, the mental models are stable and change slowly. Facing the new exigency in terms of environmental policies, the individual and institutional agents generally choose between resistance with the risk of being marginalized and progressively disappearing, or adapting their discourses and practices to the new environmental contexts. This decision does not mean they are changing or have changed their mental models. They have just accepted the idea to start an adaptation process. The case studies from both Amazonia and the Pampas showed the slowness of these processes, usually at least one generation. The set of environmental policies aiming to build sustainable social–ecological systems started about 25 years ago in the Amazon region and have achieved some significant results, especially in terms of stopping the deforestation. They regret the period before the environmental constraints. In the Pampas, the process is different because of the progressive change from grassland/rangeland to pastureland, then to cropland, and it not being necessary to destroy the natural ecosystem to develop animal production. However, even with the awareness of agribusiness to the growing environmental demand of the market, the implementation of a more sustainable policy is and will be a great challenge for many years, mainly because of the lack of an accepted mechanism to enforce the respect of norms in the absence of financial incentives. The challenge to develop sustainable social–ecological systems seems huge in central Patagonia because of both the degraded agroecosystem and weak livelihoods, which are not attractive for young farmers and investors. With regard to other sheep farming systems, including environmental issues and services, ecotourism development and significant subsidies could be the right way. In all the cases, the scenarios for sustainable management need efficient long-term policies and strong monitoring based on participative methods involving all the local people, from breeders to local governance.

Sustainable development as resilience can have many meanings. Livestock production seems to be perfectly able to adapt to ecological, social, or economic constraints and the very intricate mix of drivers of different types and different scales that operate in the grassland/rangeland ecosystems. Farms could face many more difficulties, and probably different production organizations will appear soon, as has already happened with crop production.

Acknowledgments Acknowledge to the following research projects: MOUVE (ANR, France), PIC (Uruguay) and G2IES (Argentina).

References

Allen VG, Batello C, Berretta EJ, Hodgson J, Kothmann M, Li X, McIvor J, Milne J, Morris C, Peeters A, Sanderson M, The Forage and Grazing Terminology Committee (2011) An international terminology for grazing lands and grazing animals. Grass Forage Sci 66:2–28. doi:10.1111/j.1365-2494.2010.00780.x

Arbeletche P, Perrugorría A, Saravia A, Correa P, Torres(de) M-F, Morales H, Franco R, Gédouin M, Poccard-Chapuis R, Bonaudo T, Capdevila L, Valarié P, Tourrand J-F (2011) El norte del Uruguay: ¿cómo analizan los actores la dinámica de la ganadería? II jornadas interdisciplinarias de estudios agrarios y agroindustriales, Buenos Aires, Argentina, 1–4 November 2011

Ares J (2006) Systems valuing of natural capital and investment in extensive pastoral systems: lessons from the Patagonian case. Ecol Econ 62(1):162–173

Blanco PD et al (2013) A land cover map of Latin America and the Caribbean in the framework of the SERENA project. Remote Sensing of Environ 132:13–31

Cattan A (2014) La préservation des prairies dans la PAC: Les raisons d'une illusion. Le courier de l'environnement de l'INRA no. 94. pp 91–103

Coronato FR (2010) Le rôle de l'élevage bovin dans La construction du territoire de la Patagonie. Doctoral thesis, ABIES/AgroParisTech, Paris

Coronato FR, Fasioli E, Schweitzer A, Tourrand JF (2011) Rethinking the role of sheep in local development of Patagonia, Argentina. In: 9th international rangeland congress, workshop: livestock farming embedded in local development, Rosario, 02/03-04-11, pp 19–20

Crosby A (2004) Ecological imperialism: the biological expansion of Europe, 900–1900. Cambridge University Press, Cambridge, 390pp

Dong S, Wen L, Liu S, Zhang X, Lassoie JP, Yi S, Li X, Li JY (2011) Vulnerability of worldwide pastoralism to global changes and interdisciplinary strategies for sustainable pastoralism. Ecol Soc 16(2):10, http://www.ecologyandsociety.org/vol16/iss2/art10/

Droulers M (2001) Brésil, une géohistoire. Presses Universitaires de France

Guido A, Díaz V, Baldassini P, Paruelo J (2014) Spatial and temporal variability in aboveground net primary production of Uruguayan Grasslands. Rangeland Ecol Manage 67(1):30–38. doi:10.2111/REM-D-12-00125.1

Jarvis LS (1981) Predicting the diffusion of improved pastures in Uruguay. Am J Agric Econ 63(3):495–502

Litre G (2010) GaúchosGlobais. As percepções e estratégias de adaptação dos pecuaristas familiares gaúchos da Argentina, Brasil e Uruguai num Pampa em transformação. Tese de Doutorado, Centro do Desenvolvimento Sustentável—Universidade de Brasília (CDS-UnB), Brasília-DF, Brasil/IHEAL La Sorbonne Nouvelle Paris 3, Paris

March J (2010) The ambiguities of experience. Cornell University Press, Ithaca, 152p

Melville EGK (1997) A plague of sheep. Environmental consequences of the conquest of Mexico. Cambridge University Press, Cambridge, 220p

Milchunas D, Sala O, Lauenroth W (1988) A generalized model of the effects of grazing by large herbivores on grassland community structure. Am Nat 132:87–106

Moraes MI (2008) La pradera perdida. Historia y economía del agro uruguayo: una visión de largo plazo, 1760–1970. In: Linardi, Risso (eds), Montevideo (https://books.google.com.br/books/about/La_pradera_perdida.html?id=RGEVAQAAMAAJ&redir_esc=y)

Morales H (2007) Évaluation des conséquences de décisions stratégiques en élevage à l'herbe en Uruguay. Une approche par les systèmes multi-agents (SMA). Doctoral thesis, ABIES/AgroParisTech, Paris

Morales H, Homem V, Champredonde M, Tourrand JF (2011) South American livestock farming expansion. the long way to sustainability. In: Kammili T, Hubert B, Tourrand JF (eds) A paradigm shift in livestock management: from resource sufficiency to functional integrity. Cardère , Lirac, pp 73–85, ISBN: 978-2-914053-57-0

Paruelo J, Jobbágy E, Sala O (1998) Biozones of Patagonia (Argentina). Ecol Aust 8:145–153

Paruelo JM, Piñeiro G, Baldi G, Baeza S, Lezama F, Altesor A, Oesterheld M (2010) Carbon stocks and fluxes in rangelands of the Rio de la Plata Basin. Rangel Ecol Manage 63:94–108

Piñeiro G, Paruelo JM, Oesterheld M (2006) Potential long-term impacts of livestock introduction on carbon and nitrogen cycling in grasslands of southern South America. Glob Chang Biol 12:1267–1284

Poccard-Chapuis R (2004). Les réseaux de la conquête. Élevage bovin et structuration de l'espace en Amazonie brésilienne. Doctoral thesis, Université de Paris X

Santiago AA (1972) O zebu na India, no Brasil e no mundo. Inst. Camp. Ens. Agric., Campinas, 744p

Sayago D, Tourrand JF, Bursztyn M, Drummond JA (eds) (2010) L'Amazonie, un demi-siècle après la colonisation. QUAE, Versailles, 271p

Schmink M, Wood C (1992) Contested frontiers in Amazonia. Colombia Press University, New York, 387p

Soriano A, Paruelo J (1990) El pastoreo ovino: principios ecológicos para el manejo de los campos. Ciencia Hoy 2(7):44–53

Tourrand JF, Vaz V, Veiga JB, Quanz D, Poccard-Chapuis R (2012) Pionnier d'Amazonie: une vision particulière de l'incertitude. In: Agir en situation d'incertitude. QUAE, Paris

Tourrand JF, Barnaud C, Dobremez L, Lasseur J, Capdevila, Valarié P, Cournut S, Ickowicz A, Moulin CH, Dedieu B (2014). Mental model about Livestock at global scale in a global perspective. Animal

Treccani GD (2001) Violência & grilagem: instrumentos de aquisição da propriedade da terra no Pará. UFPA, ITERPA, Belém, 600p

Vaz V (2013) A formação dos latifúndios no sul do estado do Pará: terra, pecuária e desflorestamento. Tese de doutorado. CDS/ UnB. Brasília

Veiga JB, Tourrand JF, Piketty MG, Poccard-Chapuis R, Alves AM, Thales MC (2004) Expansão e trajetórias da pecuária na Amazônia: estado do Pará, vol 1. Editora da Universidade de Brasilia, Brasília

Walker B, Salt D (2006) Resilience thinking: sustaining ecosystems and people in a changing world. Island, Washington, 192p

Westoby M, Walker B, Noy-Meir I (1989) Opportunistic management for rangelands not at equilibrium. J Range Manage 42:266–274

Williams BK, Brown ED (2014) Adaptive management: from more talk to real action. Environ Manage 53:465–479. doi:10.1007/s00267-013-0205-7

Chapter 6
Adaptation and Resilience in Pastoral Management of the Mediterranean Bedouin Social–Ecological System in the Northwestern Coastal Zone of Egypt

Ibrahim Daoud, Mona Abd-El-Zaher Oman, Veronique Alary, Naeem Moselhy, Ehab Salal, Adel Aboul Naga, Omar Salama, Laura G. Duarte, and Jean François Tourrand

Abstract On the basis of the results of the ELVULMED, MOUVE, and CLIMED research projects, this chapter presents a long-term analysis of the Bedouin society in the northwestern coastal zone (NWCZ) of Egypt, especially the resilience of the Bedouin social–ecological system facing global change. Located along the Mediterranean coast, the NWCZ is bordered by Libya to west, the hinterland of the Nile Valley to the east, and the northeastern Sahara to the south. Settled by Bedouin tribes, the NWCZ is a typical case study of the North African pastoral area. Global change in this arid region is characterized by frequent droughts and water scarcity, structural deficit in food security, strong demographic growth, rural exodus, new social demands, especially from the youth, and serious social challenges currently linked to the Arab Spring.

The first part of this chapter presents some elements relevant to the history of the NWCZ, from the Roman period until the beginning of the twentieth century, to give a better understanding of the context of the establishment of the Bedouin society.

I. Daoud (✉)
DRC/Desert Research Center, Matruh, Egypt
e-mail: tourrand@aol.com

M. Abd-El-Zaher Oman • E. Salal • A.A. Naga • O. Salama
ARC-APRI/Agricultural Research Center–Animal Production Research Institute, Cairo, Egypt

V. Alary
Cirad and ICARDA, Rabat, Morocco

N. Moselhy
DRC/Desert research Center, Cairo, Egypt

L.G. Duarte
Center for Sustainable Development, University of Brasília, Brasília-DF, Brazil

J.F. Tourrand
Cirad & MAAF, Montpellier, France

S. Dong et al. (eds.), *Building Resilience of Human-Natural Systems of Pastoralism in the Developing World: Interdisciplinary Perspectives*,
DOI 10.1007/978-3-319-30732-9_6

Many of these elements are not specific to the NWCZ, and they concern the entire North African and western Asian region. The second part describes the major events that have occurred during the twentieth century in the NWCZ, with the objectives to better define the main phases of the Bedouin social–ecological system and to understand the drivers of long-term change and, consequently, define possible key factors for sustainable development in the face of the new context of global change, including rangeland recuperation and management. The third part considers the Bedouin strategies in the face of global change, especially the 15-year drought from 1995 to 2010, and the changes in the socioeconomic context linked to the building of infrastructures for water supply, tourism development, information and communication technologies, the new demands of the local society, mainly the youth, and more recently the Arab Spring. Maybe the main change is the weak role of the rangeland in the new Bedouin society. In the fourth part, three scenarios for the future are described and which have been drafted with the perceptions of different local stakeholders. The fifth and final part presents some concluding remarks, focused specifically on the future of the rangeland, its management, and the role of the Bedouin society, and more generally the future of the pastoral society in the new North African changing context.

The results presented in this chapter come from data collected during three successive research projects conducted from 2011 to 2013. The first project was a survey based on a large questionnaire completed by 180 breeders in six villages in the NWCZ. The questionnaire included data about the history and structure of the family, the tribe, the land, the crop and livestock farming system, the family and local networks, the perception of change, and the family's projects for the future, etc. The second project was a more detailed survey, based on the same questionnaire with 50 families in the Nagamish wadi located in the central part of the NWCZ, near the city of Marsa Matruh. This project used several tools from diverse disciplines, including crop and livestock farming systems, geography, economics, water management, and policy science. The third project involved a set of interviews with 25 local stakeholders in the NWCZ focusing on the following topics: their mental models about livestock, their perceptions regarding diverse topics such as regional trends, the main drivers of change, past, current, and future functions of livestock on a local scale, their hopes and fears for the local society, current challenges and conflicts, especially those linked to drought, water supply, and rangeland management, and their scenarios for the future.

6.1 Introduction

The northwestern coastal zone (NWCZ) of Egypt is located in the southeastern Mediterranean, between the Nile delta and the Libyan border. The NWCZ extends from the Mediterranean coast in the north toward the Western Desert, part of the Sahara, in the south. Established by breeders several thousand years ago, the NWCZ is a good case study of the pastoral social–ecological system in the North African area. The NWCZ is settled by Bedouin tribes who have developed an efficient social–ecological system based on complex land access and animal husbandry, mainly flocks of small ruminants and some camels grazing on the rangeland. Probably the efficiency of the Bedouin pastoral system was based firstly on the large amount of pastoral resources in relation to the needs of the flocks and secondly on the local control by the tribes. Remains from history show that the NWCZ was in the past an important place of irrigation, especially aqueducts, cisterns, and reservoirs built during the Roman period. This attests that the water availability was already a critical issue. A significant change in land access occurred at the beginning of the twentieth century with land allocation at the tribal level. Free access to the rangeland did not exist anymore; it was limited at the tribal level or with the consent of tribe leaders. Competition and probably conflicts between breeders over rangeland resources were among the major reasons for land allocation. Some decades later, in the middle of the twentieth century, the breeder families started to cultivate barley in small fields around the villages for human consumption, and more recently to feed their flocks during the dry season and to fatten their lambs and rams before sale. Breeders also grew some fig and olive trees in the bed of the wadis. From 1995 to 2010, the NWCZ experienced a severe drought; the annual rainfall did not exceed 140 mm for 15 years. The Bedouin farming system is able to face up to 3–5 dry years, depending on animal sales to maintain the livelihood of the family. Beyond this period, and because of the weak productivity of the rangeland and barley fields linked to the low rainfall, breeders had to move to better areas or to purchase by-products and concentrates to feed their herds. For that, they overused their flocks by selling female lambs and ewes, after having sold the male lambs. Consequently, the size of the herds started to decrease dramatically after about 5–6 years of drought. Hence, breeders were obliged to find other sources of income, especially through economic migration to Libya or the Arab Gulf countries. Besides, during the last 30 years, the NWCZ has experienced other significant changes, especially the strong development of tourism along the coast and the building of water infrastructures to expand wadi agriculture. So, rather than migrating, some breeders have invested in wadi agriculture, which is less dependent on rainfall than rain-fed agriculture, whereas others preferred to take off-farm jobs, occasional jobs in tourism or, if they were educated, permanent jobs in the private or public sector. However, only a few alternatives have emerged for the rangeland, and because of the lack of family labor, because of family members preferring to leave agriculture and partake in other activities, and an insufficient budget to employ shepherds, many breeders have reduced or stopped long seasonal transhumance. Nowadays, flocks stay all year round on rangelands near the villages, inducing a rapid degradation of the most productive rangelands. Facing the conjuncture of factors such as low rainfall for 15 years, strong wind erosion and overgrazing, the rangeland progressively disappeared, especially the shrubs and perennial crops, and was followed by severe erosion of the soil at the end of the process, resulting in wide areas of stony desert, particularly in the inhabited area. In 2014, after 3 years of good rainfall, the rangeland productivity was still weak. On the basis of data collected from 182 farms, more detailed data obtained from 50 farmers in the Nagamish wadi, and 25 interviews with local stakeholders focusing on the factors changing Bedouin society, the adaptations of the rural families to the new farming context, and the degradation of the rangeland, the authors drafted in partnership with local stakeholders the scenarios for the future, especially the successive steps and the needs in terms of public policies. The first scenario is "wait and see or continuing the current trend", the second one is the "strong development of irrigated agriculture," eventually using desalinized water and other water supplies, and the third one is "sustainable rangeland management." Evidence of these changes is the poor efficiency of traditional rules of Bedouin society in managing the rangeland nowadays in the global change context, although the interviewed stakeholders recognized the permanent relevance of tribal scale to define new rules of efficient rangeland management adapted to the new context. The tribes remain the main interacting body in the social–ecological system, especially between the breeder families and the local government. So, if restoring the rangeland productivity requires new technologies from an agroecological point of view, strategic decisions should be shared among the breeders to clearly define the rules and norms of future sustainable rangeland management. The process of making decisions and their implementation should be coordinated at the tribal level. However, the Arab Spring has changed the context both in Egypt and in its neighbors, especially Libya. Today, the main objective of the Egyptian government in the NWCZ seems to be to control the land and make a profit from the resources of the soil and subsoil, and a second priority is the control of the sustainability of the rangeland and consequently the social–ecological system.

The human–nature interaction is complex in the Mediterranean region because of the high diversity of the social–ecological systems linked to its location straddling three continents and being the melting pot of several cultures which have marked the history of humanity. The Mediterranean is at the confluence of the three continents of the Old World: Africa to the south, Asia to the east, and Europe to the north. This geographical specificity constituted over time a land of migration for plants, animals, and humans, which partly explains the richness and diversity of the region's social–ecological systems. Moreover, the Mediterranean has always been a zone of convergence and not a barrier between living beings, as shown by the names that were given to it. For example, *Mare Nostrum*, in the Roman period, and Mediterranean respectively mean "our sea" and "surrounded by land," this land which is often a mountain. Indeed the Mediterranean is surrounded by several mountains and mountain ranges, such as the Atlas Mountains and Pyrenees in the west, the Alps and Carpathians in the north, and the Caucasus Mountains, Mount Lebanon, and the mountain of the Sinai Peninsula in the east, some of these mountains being higher than 3000–4000 m. All these mountains are rangelands that are potentially used by the herds or ruminants, especially sheep and goats, but also cattle and camels. Over millennia, several societies dominated a part of or the entire basin during a particular period: The Phoenicians sailed in the Mediterranean at the time of Ancient Egypt, 3000–4000 years B.C., trading and connecting with different societies. Later, the Greeks followed, then the Romans, who continued to trade and build partnerships between the north, south, east, and west coasts. More recently, the Arabs, the Turks, and the Europeans successively implanted their cultures on large parts of the Mediterranean. All these societies were based on sea resources and trade. However, they strongly depended on land and agriculture. The main poles of settlement were the cities or harbors, the coastlines and the plains in the deltas, and along the rivers where agriculture was developed, often on the basis of irrigation. In contrast, the hinterlands, mainly mountains surrounding the basin, as already mentioned, and the desert areas, especially in the south and east, have been less inhabited because of their harsh weather conditions during part of or the entire year. Indeed, another specificity of the Mediterranean is its climate, characterized by a dry and long summer (from May to September), a relatively cold winter (from November to February) especially in the mountains, and two intermediary rainy seasons, spring (March and April) and early autumn (October). Droughts are also frequent in the Mediterranean, particularly on the south coast and the east coast. This has led to seasonal migrations of societies so they can use the rangeland areas when pastoral resources are available, and to stock feed resources for the harsh seasons, winter in the north and summer in the south. Furthermore, the seasonality of pastoral resources, and consequently that of animal production, has forced breeders to process food into products with long shelf lives so they are consumable all year round, such as cheeses and salted and preserved meats. In parallel, rural societies have implanted in the plains diverse intensive crop–livestock farming systems for food and forage production, often using irrigation. Hence, there is a strong complementarity between the rangeland with pastoral use, including limited rain-fed agriculture, and the plains dedicated to intensive production and where the large part of the population is settled. Among the wide diversity of social–ecological systems and human–nature interactions in the rural Mediterranean, we have chosen to present the NWCZ of Egypt for three reasons. First, the NWCZ is a typical Bedouin area, one of the many in the south and east of the Mediterranean. Second, the social–ecological system of the NWCZ has changed over the past few decades, but is still characterized by traditional human and tribal rules and typical breeding practices. Third, the NWCZ is directly facing global change, especially a very harsh drought, strong social conflicts, and solid demand of young people for new perspectives. Moreover, research projects conducted in previous years have resulted in reliable databases on the Bedouin farming systems and the strategies of the breeders.

6.2 The History of the Northwestern Coastal Zone

The NWCZ of Egypt is a 400-km band of arid lands bordered by the Mediterranean coast to the north, the Western Desert to the south, the Nile delta to the east, and Libya to the west. Figure 6.1 shows the location of the NWCZ in the Mediterranean basin and its main cities.

The average annual rainfall between 1944 and 1995 was 140 mm. From 1996 to 2011, the area suffered from a strong 15-year drought which seriously impacted rural activities. Precipitation did not exceed 40 mm in some years, and it reached 150 mm in only 3 years (Fig. 6.2). Moreover, the low rainfall and its erratic distribution during successive years deeply disturbed the vegetation cycle, and affected the cropping systems of the breeder families more than the rangeland.

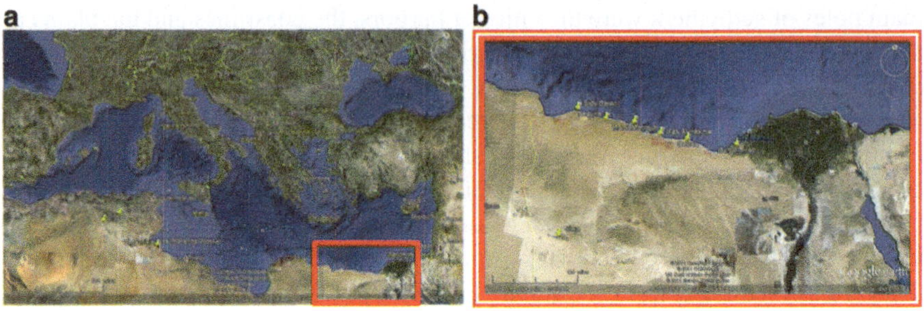

Fig. 6.1 **a** The Mediterranean and **b** the northwestern coastal zone, Egypt

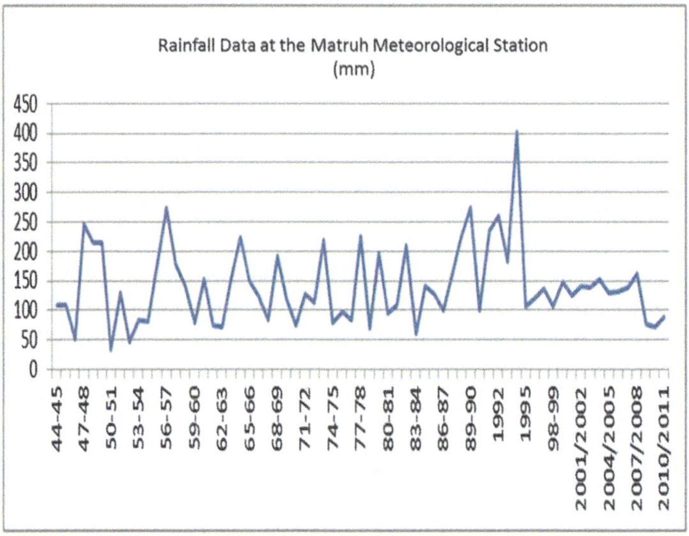

Fig. 6.2 Rainfall at the Matruh Meteorological Station between 1944 and 2011

The NWCZ is divided into three parts in accordance with the availability of water resources (Fig. 6.3). The first part extends from the Libyan border to the village of Fuka (located between the towns of Marsa Matruh and Ras El Hakma) along 285 km, with a depth of 70 km to the south. This area highly depends on rainfall as the only water resource for agricultural activities, rain-fed and wadi agriculture; it includes 218 wadis, or valleys (Fig. 6.4). The second part extends from the village of Fuka to the east to the city of El Hamman near Alexandria. The Nile water reaches this area through the El Hamman canal. The third part is the Siwa oasis, located about 300 km to the south, where a huge aquifer constitutes the main water source for irrigation.

From a social point of view, the NWCZ is inhabited by a Bedouin society. The map in Fig. 6.5 shows the distribution of the NWCZ land among the different Bedouin tribes who live there.

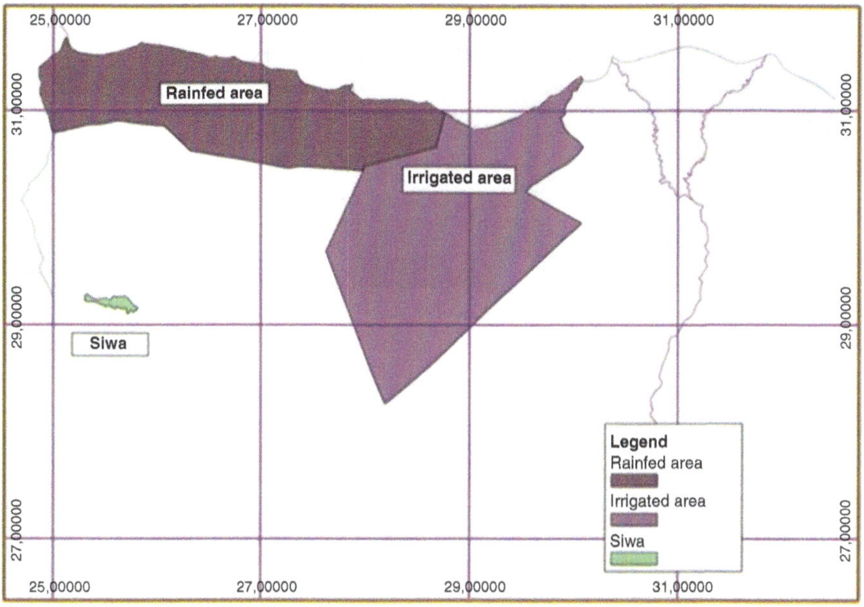

Fig. 6.3 The three areas in the northwestern coastal zone. (From Daoud 2015)

Fig. 6.4 The main wadis and watersheds in the northwestern coastal zone. (From Daoud 2015)

The five main tribes are the Awlad Ali Abiad, Awlad Ali Ahmar, Qutan, Gomiat, and Snena, with a few or many subtribes for each one. The map in Fig. 6.5 gives an idea of the land integration and the complexity of land control, especially with pastoral farming systems based on movement of herds according to the available pasture. Fortunately, in any tribe's area, the farmers, breeders, and their shepherds know the land borders of their tribes and subtribes. Some different marks in the landscape, such as those in Fig. 6.6, define the land borders. Moreover, efficient local committees are responsible for solving land conflicts and even avoiding conflicts.

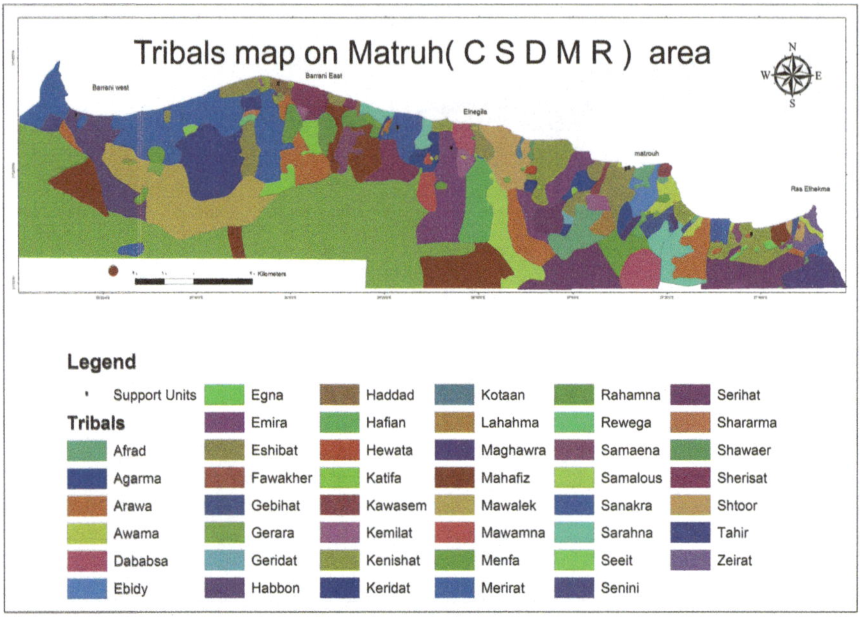

Fig. 6.5 Tribal map of the northwestern coastal zone (Matruh Governorate, Egypt)

Fig. 6.6 Milestone indicating the border between two subtribes in Neguila

However, land issue is a serious problem and challenge in the NWCZ for the Bedouin tribes and their relationship with the Egyptian central government. The borders of the tribes' land were clearly defined at the beginning of the twentieth century, as explained later. These borders are usually recognized by all tribes; however, local conflicts may occur from time to time. During the past few decades, the central government has allocated large rural land lots to people and institutions, especially the Egyptian army, land allocations with which the Bedouins did not agree.

Because of the harsh dry conditions, the Bedouin breeders have developed complex pastoral livestock farming systems based on animal husbandry; these systems include sheep, goat, and camel farming, seasonal migrations, some barley fields, and some wadi agriculture depending on access to wadi lands. The region is also impacted by the development of beach tourism attracting local and regional tourists especially from Arab Gulf countries, and offering seasonal and permanent urban jobs for the local population, including farmers.

6.2.1 A Long Time Ago, the Roman Period

History tells us that the NWCZ was one of the main grain baskets of the Roman Empire, and even before that time an important farmland of ancient Egypt and Greece. In the old literature, the NWCZ used to be named Mareotis, Mariout, or Marmarica, with strong relevance to the Greco-Roman times. Nowadays, from the current landscape, it is difficult to picture this huge farmland, once used by famous foregone civilizations, as the main source of cereals for their empires. With the exception of a narrow coastal strip of 2–3 km with some fields of barley and fig and olive trees, the NWCZ is a desert of sand and stones with dispersed dunes and rocks. Could this change of landscape use be the result of climate change or an effect of irrational use of agriculture? Or could it be the result of the interaction of both climate change and unreasonable farming?

In his review of the literature focusing on the rise and fall of Mareotis, Kassas (1972) mentioned the water management policies and practices. The topographic features of the Mareotis region comprise a series of about ten parallel limestone ridges (bars) running along an east–west direction to form a Pleistocene sequence of shoreline bars as reported by Ball (1939) and Said (1962). These ridges are an effective means of natural redistribution of rainwater. Farms were usually located at the feet of these ridges, where runoff water can be manipulated to accumulate. These remnants prove the existence of water supply and water management strategies at this time.

As reported by Kassas (1972), efficient utilization of water resources necessitated the establishment of several infrastructures; he referred to Hume and Hughes (1921) and de Cosson (1935) to describe one of these infrastructures: the *karm*. Structurally, the *karm* is an artificial hillock, which could be an old house, destroyed and covered by sand. The *karm* is usually less than 3–4 m high and 40 m^2 at the base, built on the coast at a very low altitude, such as the testimony of an old village.

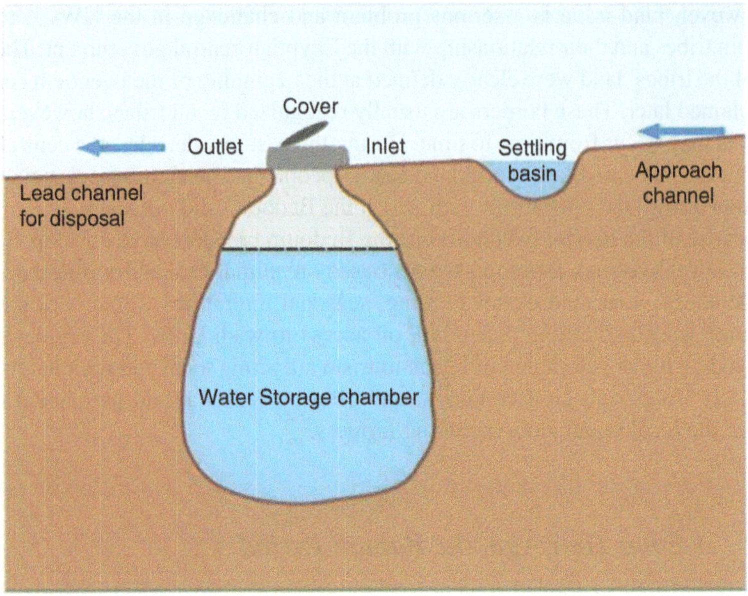

Fig. 6.7 Scheme of a Roman cistern in the northwestern coastal zone. (From Daoud 2015)

The flanks of the *karm* allow runoff water from rainfall to flow to the inside, and consequently to be collected and concentrated in limited areas where plants are grown. The *karm* can also serve as a cistern for water storage; small circular Roman cisterns are located near groups of karms to supply the old villages with water, and large rectangular cisterns are placed in the field for agricultural use. Shata (1953) and El Miniawy et al. (1990) described Roman cisterns, which are sometime named "Roman wells," as large underground chambers excavated in a limestone ridge, serving as a cover, and where rainwater converges (Fig. 6.7). The chamber or reservoir is plastered from the inside to prevent leakage. The capacity of the chamber ranges from 200 to 500 m^3, and the sites of these chambers are carefully chosen to collect runoff water from large land areas. This water is channeled through subsidiary canals to an orifice leading to the chamber.

Shata (1971) estimated that between Alexandria and El-Salloum there are 3000 cisterns with a total capacity of two million cubic meters of good-quality water. Most of these cisterns are still in use today, and were until the beginning of the last century the only efficient water source in the area.

Moreover, after Walpole (1932), Kassas (1972) described a subterranean very special aqueduct discovered near Marsa Matruh in 1931 that was used to harvest water in one place to irrigate land in another place a few kilometers away. This aqueduct consists of a main channel and numerous side galleries that collect and store water. The system is about 854.5 m long, with an average height of 2.1 m and average width of 1.1 m, and contains 25 manholes. The site of the facility was not randomly chosen; it was dug through a limestone ridge associated with a considerable body of sand dunes on its seaward side and an extensive catchment area of rocky plateau dissected by an intricate system of wadis on the inland side. Without doubt, all these water supplies and infrastructures helped Mareotis to be a significant farmland region in ancient times.

Some other water-harvesting techniques and practices that existed in the time of Maerotis and are still known and used today in the NWCZ and in other regions. Kassas (1972) reported after Weedon (1912) that the first record of the decline of Mareotis dated to A.D. 950. By the tenth century, the area had gradually declined, and the vineyards had been replaced by desert. Only a small town remained to the west of Alexandria by 1400. Five centuries later, the district was described as covered with ruins of towns and villages with cisterns dating mostly from Greek and Roman times.

These remnants of water-harvesting techniques in the Roman period, and probably the Greek and ancient Egyptian times more than 2000 years ago, show the strong adaptation of the successive NWCZ societies to drought conditions, allowing them to develop farming systems based on crop irrigation and rangeland management. However, if the local society of the NWCZ was able through irrigation to produce food for its subsistence, and maybe export a little, it is difficult to imagine the zone as a major grain basket of these ancient civilizations. The relatively low rainfall of less than 250 mm would not be able to secure grain production even if the climate had been better in several periods during the past two or three millennia.

Furthermore, Renouard (2006) considers that unlike northern Libya, the NWCZ was not included in the Mediterranean land that was directly controlled by the Romans, since it depended on the kingdom of Ptolemy in the Nile valley and to which the NWCZ was only a hinterland. Hence, the NWCZ was not necessarily a major grain basket for the Roman Empire as were the coastal plains of Libya, Algeria, and Morocco. For the same author, the Sinai to the east of the Nile valley also depended on the kingdom of Ptolemy (Renouard 2006), and was not directly controlled by the Romans.

In conclusion, in contrast to the idyllic image of an ancient rain-fed cropland, the NWCZ would have been an excellent rangeland for the herds of small ruminants and camels, but also for cattle. Rangelands with a low population density and separated by croplands along the rivers with a high population density were a very common feature around the Sahara. An annual rainfall about 250–300 mm is usually sufficient to maintain a natural rangeland in these arid areas, especially with low-density human and animal populations, which was the case of the NWCZ at the time.

However, the lack of abundant freshwater, for consumption by both families and their herds, must have been a strong constraint for the breeders during the dry season. Thus, the livestock systems probably depended on seasonal migrations: wide distribution of herds on the rangeland during the rainy season (from November to April in the north and from June to October in the south), and gradual regrouping around water points, near the rivers and wells, during the dry season (from May to October in the north and from November to May in the south). The herds grazed in the NWCZ during winter and spring, and moved to the Siwa oasis or the Libyan Oases at the beginning of the dry season, in April–May.

6.2.2 The Second Millennium: Nomads and Pastoral Bedouin Systems

De Cosson (1935) added that time, wind, rain, and sand contributed to and completed the fall of Mareotis, leaving only the foundations of countless buildings to tell the tale of this once prosperous land. Some thought that this decline was caused by climate change and the drier conditions. De Cosson (1935) referred to some earlier writers who supported this hypothesis on the basis of the extinction of ostriches from the nearby Libyan Desert. Another school argued that the intellectual activity of Greek times in Egypt, when Alexandria was the center of arts and sciences, would have necessitated a temperate climate, more suitable than that which prevails in Egypt today.

However, Weedon (1912) presented evidence to indicate that the climate of the region had not changed during the previous 2000 years. His most convincing evidence is the presence of some ancient plaster constructions. He argued that the existence of these plaster walls is therefore presumptive that the climate in A.D. 400 was not markedly wetter than it is today. This plaster was made with yellow soil, seawater, and charcoal.

De Cosson (1935) recollected that the area was the theater of extended wars from A.D. 913 to A.D 969, after which the Fatimids were able to successfully complete their migration from Tunisia to Egypt. During the battles, the people of Mareotis took refuge in Alexandria, where many of them settled and stayed. The lack of maintenance of the water supplies and main infrastructure contributed to the decrease of the agrarian production system. The end of this period occurred in the eleventh century with the destructive invasion of Beni Hilal and Beni Soleim nomads, when these tribes were pushed out of Egypt toward the west and settled in the NWCZ and eastern Libya, where they implanted their pastoral system and imposed their nomadic life style. However, during the next few centuries, the NWCZ was crossed by several tribes and social groups who were migrating between the Libyan and Egyptian lands that are known today as being affected by the wars and conflicts in the region.

The available data about the Bedouin pastoral system show that the entire land was common and families could freely move from one place to another in search of pastures. Sheep, goat, and camel farming was the main activity of the Bedouin tribes. During this period, the borders between the tribes had not yet been defined. The nomadic nature of the Bedouins has not imposed fixed boundaries between tribes. However, the sources of drinking water for humans and animals were very limited, and the only water supply system was the Roman cisterns and belowground saline water for animal use only. Water resources and supplies were also common among tribes, who lived in tents because of their continuous need to be mobile in search of grazing lands.

Daoud (2015) reported the story of the settlement of the Aiit Sebak family in 1840 in the south of Naghamish wadi near the city of Marsa Matruh. The Aiit Sebak family is a branch of the Menfa tribe. Shahin, the head of the family, came to the area from Libya in 1840 with his son Hamed, who had studied at Gagbob University in Libya. At this time the land was still common, and each family had the right to freely move in search of water and rangeland.

Hamed was married with 11 sons. The family spent winter and spring in the south of Naghamish wadi, where Shahin settled with his family when they first arrived. They moved between the south of Naghamish wadi and the Tamira oasis, located between the city of Matrouh and the Siwa oasis, about 150 km to the south, where water and grass were available for them and their flocks during summer and autumn.

At that time, the Aiit Sebak family was still living in tents, until 1984 when the Egyptian government in partnership with the World Food Programme (1970) started to settle the Bedouin society by constructing houses for them. The last tent in the area of the Sebak family disappeared in 1989.

In 1920 the land tenure appeared in the Bedouin area; it is called Al Houz. The Sebak family got their land after a hard effort with their neighbors from other tribes. In the south, the rangeland is still common land because it is a rocky land and no oil has been found there.

In conclusion, this brief historicalaccount has given the impression that the NWCZ should have been a transition zone between the Nile valley and Libya. The low freshwater resource has certainly impeded the development of a strong agriculture and, at the same time, the development of permanent housing for the breeders. Furthermore, because of its location on the Mediterranean coast and between two strategic zones (the Nile valley and Libya), the NWCZ has been crossed over the centuries by diverse groups of people who have settled, developed, and used the lands differently across time. All these factors resulted in the development of a breeder society using the natural rangeland resource to feed their herds of small ruminants and camels and with livestock systems based on seasonal migrations to the neighboring zones, especially Libya and the Nile valley on the west and the east respectively.

6.2.3 The Twentieth Century: From Open Land to Collective and "Private"

With the twentieth century, the NWCZ had already contributed for at least five millennia to the most famous Mediterranean civilizations in human history: ancient Egypt and Greece, the Roman Empire, and more recently the Arab period. During all this time, the successive social–ecological systems have been based on the use of the natural resources of the rangeland, including soil, water, and vegetation, especially to feed the herds and also to produce food for the local populations. From the twentieth century, global change has strongly modified. The agro-ecological contextt because of the decreasing importance of rangelands in the social–ecological system, especially during the last two to three decades, linked to the 15-year drought from 1995 to 2010. Progressively, rangeland had become a factor among the complex set of components driving global change, in the same way as it has occurred in North Africa and western Asia.

6.2.4 From the 1920s to the 1960s: From Open Land to Collective Land

As mentioned before, the border between the tribes began to appear around 1920. The question that occurs is why did the division occur and why at that particular time. It seems that the main reason was the establishment of the border line between Egypt and Libya by Italy, which was occupying Libya at that time. The objective of this policy was to control free movement between the two countries, stabilize the tribes in their land, and perhaps to better control transboundary migrations. Besides this political issue, minimizing high concentration of herds on the best pastures would have been a reason to draw the borders as land conflicts among groups of breeders had occurred.

So starting from the 1920s, the land was shared at the level of the tribe only, and grazing permission was needed from the head of the tribe. Perhaps this could be considered the first failure of the nomadic Bedouin system, because according to a Gnashat tribe leader: "We were depending completely on the rangeland; we were seeking grass everywhere; and for that reason we were nomads, moving from one place to another, searching for grass; and also for that same reason the land was common for a long time."

According to the current chief (*omda*) of the Gnashat tribe, his grandfather, also chief of the same tribe, rented in 1925 the northern area of Naghamish wadi from Egypt when Khdiwy Abas was the king of Egypt. The wadi was cultivated with fig trees, barley, and wheat; there were also pigeons, and the paid rent was a quarter of the wadi production. The wadi land was divided between the three tribes of Gnashat, Mawalek, and Gbihat because og the conflict between the Gnashat tribe and their neighbors.

On the basis of interviews with the stakeholders, the best grazing period during the good years was 3–4 months and ranged from the end of December to March, and sometimes the beginning of April. After April, herds grazed for about 2 months on dry grass. During the other 6–7 months of the year, herders had two choices: either to move to the Siwa oasis, or to another oasis in the desert in the south, or to move to Beheira Governorate, in the Nile delta, to make use of crop residues in the irrigated fields. So, the traditional Bedouin social–ecological system in the NWCZ used the local rangeland for 6 months of the year, from mid-autumn to mid-spring, and then migrated during the other half of the year.

So, one of the most important events of this period was the migration of all tribes to Beheira Governorate, near the Nile valley, at the outbreak of the Second World War. This migration was decided by the British Army so as to the protect the Bedouins living in the area from potential battles. This migration resulted in big changes in the flocks' mobility, before the tribes came back to the NWCZ when the war was over. Another important event in this period was the special public policy deciding to clean the 2000-year-old Roman cisterns.

Herding sheep, goats, and camels was always the main activity of the Bedouin breeders. Herd sizes ranged from a couple of hundred to thousands of sheep, and the herd size was probably related to the family size or even the number of families who owned the flock.

Barley cultivation started to appear progressively; some Bedouin families planted barley on their lands before the Second World War, but most planted barley after the war. Agriculture was a great change in the Bedouin society of the NWCZ, which relied mainly on the pastoral system. This was followed later by the onset of mechanization and tractors, and supported by the national policy at the end of this period. However flocks did not graze on barley fields, and barley grain was only used for making flour and bread. Lambing was just once per year. Animal diseases were rare despite the migration of flocks, maybe due to the arid weather. However, for many Bedouins diseases started to be a more serious problems in the 1980s with the use of concentrates to feed the herds. Some informants consider that concentrates appeared even earlier as a source for feeding animals.

Regarding the 1950s and the first years of the revolution, some informants mentioned the beginning of specific Egyptian policies aiming to attract Bedouin families to the Egyptian zone and to enhance their settlement there. For example, Bedouins going from Marsa Matruh to Alexandria did not need to present their documents anymore after the NWCZ had bene integrated as a regular area of the national Egyptian territory. Supporting the development of mechanization was another example of public policy aiming to strengthen the settlement of Bedouin breeders in the NWCZ.

The postwar years were marked by some significant initiatives for the region. In 1947, for example, one of the first research and development projects began in the area of Fuka; it used groundwater to cultivate a new plant for the Bedouins, the *kataf* also called *bersim hegazy* by the researchers (*Trifolium alexandrinium*). One year later, in 1948, people from the Al Shtor tribe started to cultivate fig trees in the Al Dakhla area. In 1952, the king of Lybia, Idris, started to build some dykes in the Naghamish wadi. The main program started only about 25–30 years later. Many other initiatives appeared with policies that were implemented after the revolution, linked to the efficiency of the extension services during the Nasser period.

The increasing importance of the city of Marsa Matruh was another significant factor during this period. With a strategic place during the Second World War and a major location for the Libyan leaders, the regional capital of the NWCZ progressively became the place where regional policies and agreements between the Egyptian government and the western Bedouin society were debated, and arrangements with Libyan leaders were made taking into consideration the same origin of the tribes of the two countries. The old photographs in Figs. 6.8–6.10 show the city of Marsa Matruh during the Second World War (Fig. 6.8), a local Bedouin leader welcoming the Princess of Egypt (Fig. 6.9), and Nasser leaving Marsa Matruh by train (Fig. 6.10). More recently, Gadaffi, former president of Libya, visited Marsa Matruh twice to show the strong links between the tribes of the NWCZ and their neighbors in Libya.

In conclusion, the progressive permanent settlement of the breeder families is the main feature of the beginning of the twentieth century. It was strongly linked to the implementation of public policies aiming to strengthen this permanent settlement. Land allocation at the tribal level was one of these policies; others were support of rain-fed agriculture, development of infrastructure, recognition of Bedouin needs in

Fig. 6.8 Marsa Matruh during the Second World War

Fig. 6.9 Local leader with Egyptian princess and prince in Marsa Matruh

the Egyptian society, and promotion of local governance based on the city of Marsa Matruh as a strategic place located between the Nile valley and the Libyan border. All these drivers have led to the progressive transition from the pure pastoral system based on seasonal migrations to an agropastoral system integrating rain-fed agriculture, wadi cultivation, and off-farm activities.

Fig. 6.10 Nasser leaving Marsa Matruh by train

6.2.5 From the 1960s to the 1990s: National Policy Implementation for Local Development

The Egyptian national investment aiming at strengthening the Bedouin breeders' settlement in the NWCZ continued in the second part of the twentieth century, in partnership with the FAO and the World Food Programme through a set of specific public policies. The reasons of this strong investment were mainly linked to the national security of the country boarders, because of the Mediterranean coast and the Libyan border, and also to securing the control of potential natural resources of the subsoil, especially oil and gas. But, at the same time, the idea was to maintain a good relationship with Libya at the national level and not put the tribes in a complex position because of their binational identity, Egyptian and Libyan. The closing of the border for 10 years starting in 1979 caused by a war between the two nations and enhanced the settlement policies of the Egyptian government.

Different policies were applied in different sectors. During past decades, special effort had been made to improve transportation through the building of the railway line and the main road crossing the NWCZ, from Alexandria in the east until the Libyan border in the west. The railway line was used a lot during the Second World War. The new investment focused on maintaining these infrastructures and building others, especially the secondary roads. At the same time, new schools and public health centers were established not only in the city of Marsa Matruh but also in the central small towns and villages where Bedouins had recently settled, especially along the road going to Libya.

The cooperation programs of the FAO and the World Food Programme 2270 were launched in the 1970s and the 1980s respectively. Through the second program, houses have been financed for the Bedouin breeders through long-term loans and interest-free banking. At the same time, established policies encouraged the Bedouin farmers to cultivate their land with barley and fig and olive trees. The Bedouins received seeds and fertilizers and plowed their lands with tractors; however, harvesting was always done manually.

Besides the adoption of cropping activities, crop development resulted in the initiation of the land division at the family level (*beat*). This mainly concerned the cropland near the villages, which were often recently built. Each family wanted to have its cropland near its house. Consequently land allocation was fragmented in rural areas. Nowadays, the main part of the rangeland is still common land at the tribal level and has not yet been divided between the families of the tribes.

In the same period, several policies also focused on animal husbandry, given the important contribution of this activity to the income and the lifestyle of Bedouin breeders, especially to incentivize the breeders' cooperative and associations. According to the stakeholders, the policies applied by the cooperatives were centralized but the decision was local in each village. These associations and cooperatives provided the breeders with animal husbandry inputs at low cost, mainly drugs but also animal feed, such as concentrate and hay. One the interviewed key informant estimated the feed subsidizes about 40 % of the production cost. For example, a breeder needs about EGP 250 (USD 30) to produce a lamb. He would receive a subsidy of about EGP 100 (USD 12). The same key informant estimated the total subsidies were about EGP 750 million (USD 90 million) in 1990, equivalent to about EGP 4 billion to EGP 8 billion today.

At the end of this period, special policies encouraged exportation. Sheep export initiatives were coordinated by public agencies, hence enabling the breeders of the NWCZ to export animals to the Gulf countries. Through a particular partnership between the cooperatives, the research centers, and academic institutions, the breeders received technical assistance and financial support to improve the genetics of their flocks.

Moreover, and during the same period, especially during Sadat's government, the tourism sector started to develop in the NWCZ, from the eastern part near the Nile delta (three or four special places located 25 km from Alexandria) and in the city of Marsa Matruh. It progressively expanded in the deltas of the wadis, where it settled on the most fertile agricultural lands planted with fig and olive trees. Breeders began to sell pieces of their land to private businesses, which had not been not possible before becaus no official land tenure had existed in the Nasser period.

Land competition was still a great challenge in the NWCZ, because of the existence of only three types of legal landownership systems recognized by Egyptian law: private ownership, cooperative ownership, and state ownership (Arabic Republic, 1989). Tribal or common land ownership did not exist. The rangeland, called "desert land," was classified as *aradi bur*, which means fallow or undeveloped lands, and was owned by the Egyptian State (*malkiya lil-dawla*). which can use, lease, and even sell these lands. Law 143 (1981) stipulated that the Ministry of Defense has the right to use all the desert lands for strategic purposes and that the Ministry of Land Reclamation can develop these lands with permission from the Ministry of Defense.

As a result of this legislation, the New Reclaimed Lands program appeared in the eastern part of the NWCZ, near Alexandria and the western delta first, and along the El Hamman canal later, in order to develop these lands by means of irrigation. Through the development of new irrigated lands, this ambitious program aimed to reduce the pressure on landownership in the traditional irrigated land of the Nile delta and Nile valley, and allowed the of giving land to young people who did not have access to it. Besides securing the livelihoods of the new small farmers, the leitmotiv of the New Reclaimed Lands program was to develop the Egyptian export capacity of the big farms, but also that of small farmers associated with big farms.

In conclusion, until the middle of the twentieth century, the Bedouin pastoral system relied on open access to natural resources of the rangeland and on the mobility of the herds and families. But a new context started to develop, especially with the settlement of the breeders affected by Egyptian national policy and consequently land allocation. This period strengthened the role of the tribes and initiated privatization of natural resources. However, the situation was more complex because of the lack of laws clearly defining the tribes' rights and the control of natural resources. Many leaders of tribes consider that these policies snatched their landownership and their rights in controlling their lands and the natural resources of the soil and subsoil, especially by allocating several rural and urban lots to the Egyptian institutions, mainly the army.

6.3 From the 1990s, or the Strategies of the Bedouin Society Facing Global Change

According to several stakeholders, two main factors of change have acted since the 1990s in the NWCZ. The first is the change in national policies, compared with past decades. The government stopped much direct support and many special actions that focused on Bedouin society. The second is the 15-year drought from 1995 that was partially compensated for by the implementation of the Matruh Resource Management Project[1] (MRMP) financed by the government in partnership with international development agencies, especially the World Bank. Nevertheless, at least three other factors of global change have strongly impacted the NWCZ during this period: first, the accentuated increase in rural population caused by the high Egyptian demographic growth; second, new information and communication technologies, especially the cell

[1] http://www.worldbank.org/projects/P005153/matruh-resource-management-project?lang=en&tab=overview

phone, satellite TV, and more recently the Internet, which have deeply modified the habits of the breeders and some animal husbandry practices; third, the new demands of the society, especially the youth, which are directly contributing to the building of a new context and the development of new challenges for Bedouin society.

6.3.1 The Drought, a Common Event but Also a Major Factor of Change

Droughts of 2–3 years, and even up to 5 years, are common in this arid area. The resilience of the social–ecological system, coupling the natural resilience of the rangeland and the adaptive practices of the breeders, allows these latter to face droughts. Facing lower productivity of the rangeland, especially because of the weak development of annual forage affected by the annual rainfall, the herds graze more on perennial forages and shrubs and hence succeed in maintaining the base of pastoral farming, including the reproduction process and lamb production. Because of the lack of forage used to feed fattening animals, families resort to selling lambs before fattening and at lower prices than well-fattened lambs. They also sell the oldest ewes. This overselling reduces the size of the herds but maintains the income level and consequently the livelihoods of the families. The size of the herds increases again during the subsequent years when good weather conditions prevail. Moreover, some breeders decide to move to zones with better rainfall, reducing both the number of animals in the area and the pressure on the rangeland. Hence, it can be concluded that during short droughts, the resilience of the pastoral social–ecological system relies first of all on the reduction of the herd size at the family level through overselling, and at the local level on the migration of some herds. The rangeland and herd size recover when adequate rainfall returns. The "rangeland productivity" and "herd size" couple follows Hollings's (2001) notion of resilience in which the social–ecological system can absorb the disturbances without shifting to an alternativee system. Furthermore, the NWCZ example of short droughts, as narrated by the local stakeholders, illustrates the empirical approach recommended by Frazer (2007) to better understand the environment–society interaction.

During the 15 years of drought, from 1995 to 2010, annual rainfall did not exceed 150 mm, as mentioned in Fig. 6.2. The reports of the Aiit Sebak family give an idea of the impacts on the life of Bedouin breeders and their hardness. They summarize the opinion shared by all the breeders and stakeholders of the NWCZ. Until 1995, family herds ranged from several hundred heads to some thousand heads. The main job of the family members was breeding sheep, goats, and camels, in addition to cultivating barley in the beds of the wadis, especially for family consumption, and sometimes for sheep fattening. After 1995, drought hit the area, followed by desertification and rangeland degradation. It completely stopped raining, and the rangeland was deeply affected by the southern dry wind. The vegetation cover decreased progressively year after year, until it completely disappeared, thus exposing the soil to erosion. This harmful process continued for the next 15 years, resulting in a

Fig. 6.11 Natural rangeland in Negila with regular annual rainfall

soilless land, a land covered with sand and bare rocks only. Figure 6.11 shows the rangeland in spring under regular weather conditions, before the drought. Figures 6.12 and 6.13 give the situation of the rangeland near the coast and in the south respectively after the 15 years of drought.

In conclusion, a set of factors, mainly the previous 15 years of drought, have seriously destroyed the pastoral system that was based on rangeland grazing. Nowadays, it is very difficult to survive in the NWCZ with only the traditional Bedouin pastoral system, especially with the lack of specific policies regarding animal production, as previously explained. The situation is similar in different rangeland areas in North Africa and western Asia because of the higher frequency of droughts and the weak management of rangeland, especially the lack of control of the stocking rate in regard to the productivity. Huge herds still exist, however they are owned by large traders or industrial managers who have invested in animal production and have financial resources to face climatic events such as droughts. The old traditional pastoral system that existed decades ago does not exist anymore.

6.3.2 Alternatives to Drought

The 15-year drought resulted in the destruction of the social–ecological system. Breeders depended on livestock to survive, and they sold animals to purchase feed for the remaining herd. The current feeding systems are based on hay and

Fig. 6.12 Sheep grazing on degraded rangeland near the coastline in Nagamish wadi

Fig. 6.13 Seriously degraded rangeland in the desert in the southern Nagamish wadi

concentrates purchased on the market in 7–8 months in good years, and during a longer period in worse years, especially in summer and autumn.

Fifteen years later, the largest flocks of thousands of small ruminants decreased to 100 or 200 small ruminants. Many breeders, particularly the smallest, had no alternative but to seek other sources of income. Most of them became traders to maintain their farming activities; others migrated to the urban centers to look for jobs, or to Libya and the Gulf countries to work as shepherds.

Facing rangeland degradation, all breeders expanded their barley areas, and grazed their flocks in the fields when the weather conditions were not suitable for grain production. The photographs in Figs. 6.14–6.16 show hay selling at the market in Negila (Fig. 6.14), a sheep flock being fed with concentrates (Fig. 6.15), and another sheep flock grazing in a barley field which does not receive enough rain to produce grain (Fig. 6.16).

Nowadays, wide areas of rangeland around the villages no longer exist. The land is now planted with barley and is no longer available for grazing neither in winter nor in spring, which has resulted in some conflicts linked to this new land use. These conflicts are usually solved at the tribal level, thus enhancing the importance of the tribe in the local dynamics. Nevertheless, other breeders decided to invest their labor in wadi agriculture, planting fig and olive trees, barley, and vegetables. However, the available area is not sufficient to satisfy all the demands of all breeders for wadi land, despite the huge investment by the MRMP in water infrastructure in the wadis.

Fig. 6.14 Feedstuff for sheep production at a Negila market

Fig. 6.15 Herd of sheep fed with feedstuff (hay and concentrates) purchased at the local market

Some breeders decided to invest in intensive systems and uncommon livestock activities in the NWCZ, such as indoor sheep fattening and intensive poultry production. However, these two new livestock farming systems need capital and/or good contacts, especially for them to become integrated in the corresponding complex market and supply chains. The photographs in Figs. 6.17–6.19 show a sheep fattening farm (Fig. 6.17), an intensive broiler barn (Fig. 6.18), and an intensive turkey farm (Fig. 6.19).

The willingness to maintain a physical presence on the land through a rural activity is a common point between all alternatives. The strong attachment to their homeland is one major aspect, but so is the need of Bedouins to prove their interest in their homeland to local and national governance. Even when the breeder chooses the alternative of a long-distance migration to the Gulf countries, for example, the farm activity is maintained and the breeder frequently comes back to his farm, at least once a year. In this case, the land, the house, the fields. and the herds are entrusted to a family member, the father, a brother, or an uncle. Moreover, local stakeholders mentioned the systematic return of migrants to the NWCZ after a few or several years.

These migrants usually come back with some investment funds and new ideas and knowledge acquired during their stay away from their homeland. The Nile delta is an excellent place to identify and learn new production methods; the financial capital and new knowledge in agriculture partially explains the diversity of the

Fig. 6.16 Herd of sheep grazing in a nonproductive barley field

choices made by the Bedouins in the NWCZ. Homeland attachment is another reason that ensures the Bedouin develops activities on his land.

6.3.3 Development of Wadi Agriculture

At the beginning, the MRMP was an integrated development project, launched in 1993 and financed by the World Bank with the objective to improve the living conditions in the rural area of the NWCZ. Because of the arid weather conditions in the NWCZ, the MRMP focused on water harvesting, infrastructure, and supply of potable and agriculture water. The MRMP developed several actions in research, research and development, and extension in diverse sectors: wadi and rain-fed agriculture, rangeland, valorization of products, education, public health, etc. The first phase terminated in 2002, and at the end of the second phase, 10 years later, the MRMP became a special program of the Desert Research Center, in partnership with international funding agencies. The actions of the MRMP focused especially on the western part of the NWCZ, from Fuka to the Libyan border, where the wadis are located.

The two main actions of the MRMP were building cisterns in the villages (Fig. 6.20) and developing wadi agriculture through the construction of dykes and

Fig. 6.17 No-grazing sheep breeding and fattening in the hinterland of Marsa Matruh

dams in the beds of the wadis (Fig. 6.21). Through these actions, the MRMP contributed to keeping the Bedouin farmers in the rural area, although the latter did not have as much access to the wadi land as they needed.

Another important feature of the last 20 years is the strong development of tourism, especially along the eastern coast of the NWCZ, 80–100 km from Alexandria, and around the city of Marsa Matruh (Fig. 6.22). The development of tourism and the urban expansion directly impacted the farming land use near the infrastructures; however, it also offered job opportunities in the service sector for qualified people, especially young people, and in construction for those who were not sufficiently qualified or other jobs. It also introduced a new lifestyle, particularly to the families who decided to live in the urban area and those who decided to raise their children in urban settings.

Because of the drought, the MRMP focused its activities on water supply, water harvesting, and the development of wadi agriculture reduce the vulnerability of breeder families. The MRMP invested in rangeland areas, especially in management improvement reduce the degradation of rangeland and recuperate its productivity. The lack of clearly defined alternatives to the rangeland justified this low investment.

To summarize the three last decades, the context of the NWCZ has strongly changed. Even though the permanent settlement of Bedouins started some decades ago, the coastline is now a succession of towns separated by villages

Fig. 6.18 Intensive poultry production in Nagamish wadi

and hamlets linked by highways, railways, and secondary roads. Seasonal tourism has a strong presence in all the urban areas, especially in the buildings receiving tourists during holidays. The landscape near the coastline is dominated by barley fields and wadi agriculture, mainly fig and olive trees. Between the cultivated lands, the rangeland is usually highly degraded because of overgrazing. Far from the coastline, only some wadis are inhabited and equipped, usually by the MRMP, to be cultivated. The rangeland between the wadis is also degraded, mainly because of the effects of the previous drought and overgrazing caused by the increasing animal population and the abandoning of seasonal migrations. Toward the south, there are no wadis anymore and the rangelands are degraded for the same reasons.

Fig. 6.19 Intensive production of turkey for exportation in Nagamish wadi

Fig. 6.20 Cistern built by the Matruh Resource Management Project for a Bedouin family near Nagamish wadi

Fig. 6.21 Water infrastructure in the wadi bed for agriculture development (figs in this case)

6.4 Future Scenarios

On the basis of the interviews with local stakeholders, three main scenarios have been drafted on the future of the NWCZ. The first scenario is "continuing the current trend," defined by the expansion of tourism along the coast with no specific local development policies. The second relies on the strong "development of irrigated agriculture" financed by national and international policy mechanisms, and the third one integrates specific policies focused on "sustainable rangeland management."

6.4.1 Continuing the Current Trend

The current trend in the NWCZ is based on the four following components: the Bedouin pastoral system, which is vanishing with the lack of specific policies; the development of wadi agriculture managed by the MRMP; the mining activities in the hinterland controlled by national companies; and tourism along the coast.

The tourism sector is expected to strongly develop because of the new demand of the high and middle Egyptian classes for specific places, such as the

Fig. 6.22 Tourism development on the Mediterranean coast near Marsa Matruh

Mediterranean coast, at which to spend their summer vacations and short holidays. This population usually works and lives in the big cities, especially in Cairo and Alexandria, with about 20 million and six million inhabitants respectively. During the past decade, the Egyptian population took advantage of the growth of the national economy, and Egypt was considered as one of the pre-emerging countries. According to the local governance of Matruh Province, the city of Marsa Matruh receives approximately half a million tourist in summer, attracted by diverse famous tourism spots (Figs. 6.23, 6.24, and 6.25). At least two million or three million Egyptians spend their vacations each year on the Mediterranean coast, mainly in the several resorts located in the eastern part of the NWCZ, on the coastline near Alexandria.

Facing the urban lifestyle in the Egyptian cities, especially Cairo and Alexandria, people, especially retired ones with a high pension, started to have permanent housing on the Mediterranean coast. Access to the coast is easy through the double-lane highway, flights from Cairo during summer, and the old railway from Alexandria, which could be rehabilitated. The tourism sector and the newly settled population have high demands for urban services, especially quality food, health care, and access to drinking water, efficient energy, and phone networks. Progressively, this demand may lead to urban planning and improvement of urban equipment and services in the Mediterranean coastal towns such as in the case of Marsa Matruh, where a station for water desalinization has been established and a pipe bringing Nile

Fig. 6.23 Aguiba Beach located near Marsa Matruh, in summer

Fig. 6.24 The corniche in Marsa Matruh

Fig. 6.25 Cleopatra Beach in Marsa Matruh

freshwater directly to the city has been constructed, and in the case of the Siwa oasis, which has become one of the main providers of bottled drinking water in the national market.

Furthermore, the tourism sector has so far focused on the beach; however, it is likely to expand in the future to explore the archeological potential of the NWCZ, and include ecotourism and trekking in the hinterland, and further develop the seaside and bathing tourism as well as beach sports.

The development of tourism affects the NWCZ in several ways. Firstly, buildings, hotels, and resorts are constructed for tourists along the coast and near the beach; they are mainly concentrated on the periphery of cities and in the deltas of the wadis, and to a lesser extent in the towns. Deltas are the best agricultural lands; they are planted with trees, olive and fig mainly for the local and national market, and with annual crops, especially barley, for home consumption, and with diverse cash crops for the local market. Nevertheless, farmers are selling land plots to tourism companies especially because the price they are proposing can be very high. Hence, even if the land sold to the tourism sector is small on a local scale, the impact is highly significant because it would be better used for agriculture. This tendency is expected to increase in the future with the expanding development of tourism.

Secondly, the growth of tourism has a great impact on employment in the NWCZ. Many job opportunities are created in the construction sector and the various related services, especially in urban areas, for the local population and mainly

Fig. 6.26 Small wadi near Marsa Matruh managed for vegetable production in summer

young people. Many breeders have become animal traders because of their skills in this domain, but they have also invested in other sectors. And although some started working in construction, a job that does not require a high-level qualifications, others have become taxi drivers or work in security. Many breeders are interested in having nonpermanent jobs, which allows them to continue managing their farm at the same time (Alary et al. 2014a). The jobs requiring qualifications are frequently taken by migrants from the Nile valley or by local graduates. It is anticipated that employment opportunities will continue to grow, particularly with the launching of new services and enhancing of existing ones.

Thirdly, the urban growth and the development of tourism in the NWCZ have created a significant food market. The main share of the tourists' food comes directly from the Nile delta to meet the high demands during short periods. However, there are several opportunities for the local production, such as production of lambs, figs, olives, dates, vegetables, and fruits such as tomatoes and watermelons. Several farmers are developing and focusing their production on this market, adapting their farming systems to this specific new demand. For example, the vegetable production calendar in the wadi bed is set in a way to make sure harvest time meets the period when local demand is high, causing an increase of prices (Fig. 6.26). The case is similar for intensive poultry production (Fig. 6.27) and, more recently, for sheep fattening. Moreover, the current trend opens niche markets for some specific

Fig. 6.27 Bedouin breeder selling his intensive poultry production products at the local Marsa Matruh market

local products, such as lambs, dates, figs, and olives. Stakeholders mentioned the opportunity to better valorize the organic nature of the local production (Fig. 6.28).

Two interviewed traders—butchers of small ruminants located in the city of Marsa Matruh—declared that for some years a significant part of their income derived from barbecues for parties organized by local leaders and politicians for tourists during vacations, especially during summer. They also mentioned the importance of this increasing local market to local animal production. For them, there is a real opportunity in the future for some Bedouin breeders to focus their sheep fattening toward this local market.

In conclusion and as clearly seen, the current trend in the NWCZ based on the progressive development of tourism will continue to offer good opportunities to the rural Bedouin society, through the strong local market demand for farm products, and the side off-farm jobs for the farmers who want to maintain their farming activity. The sustainable use of the rangeland and the development of the Bedouin pastoral system, at least its maintenance, is not a major challenge in this scenario. The main challenge, however, would be water availability. Indeed, the NWCZ has wide lands—even if the best agricultural lands have already been turned into construction sites or will be built on later—enough sun, and an efficient number of farmers and breeders would be needed to sustain a high farming production. Nevertheless, despite the numerous infrastructures and water supply facilities that were built

Fig. 6.28 Shop for traditional herbal medicine at the Marsa Matruh market

during the last 20 years, the water resource remains insufficient. Hence, a specific regional policy to develop the irrigated agriculture in the NWCZ would be the base of the second scenario.

6.4.2 Development of Irrigated Agriculture

As already mentioned, if the NWCZ was really the grain basket of the Roman and Greek empires and, before that, of Ancient Egypt, the production of crops progressively declined during the first millennium, to cover only local consumption. The reasons for this decline were probably the incessant conflicts in the area which forced the farmers to leave, but also climate change and conflicts with other zones, especially in the northern Mediterranean.

The new development of rain-fed agriculture started immediately after the Second World War in the NWCZ. This was directly linked to the settlement of Bedouin families who ceased their seasonal migrations. Weather conditions allowed only barley cropping during the winter and spring, and harvest was often impossible because of low precipitation or bad distribution during the cropping cycles, resulting in barley being grazed by the flocks. Moreover, fig and olive trees were planted in the beds of the wadis to maximize use of rainfall.

From the early 1990s, the MRMP started to build water infrastructures, mainly dams and dikes in the beds of the wadis to stop water flow and improve its infiltration. The water infiltrated the soil (500–600 mm) and made fig, olive, and barley production possible in the beds of the wadis, despite the low rainfall of only 100–150 mm (Alary et al. 2012). The main part of the 218 wadis was partially equipped to better use the low rainfall. Many breeders invested in wadi agriculture during the 15 years of drought from 1995 to 2010. The access to wadi land is normally decided at the tribal level. However, the wadi land is limited, and consequently not all families have access to it. Above all, the first families usually cultivate the best land, that located downstream, and they receive more water. Currently, many families do not have access to wadi agriculture, but they would like which is an answer to the frequent demand of the rural population and local stakeholders.

Moreover, building water infrastructure is very expensive and needs specific policies, especially when the landownership is not clear between the tribes and the local governance. Recently, a new program of constructing water infrastructure has started with the objective of equipping about 65 km of wadi beds. This program is financed by the European Union and is implemented by the MRMP in partnership with Italian institutions. The rainfall collected will be used locally, and only a very small amount will go to the sea.

Furthermore, some breeders who did have access to the wadi lands dug wells to use groundwater to water their flocks, irrigate their gardens, and eventually produce vegetables and fruits to be sold at the local market. Unfortunately, the groundwater is usually salted in the NWCZ. So, some stakeholders proposed new policies to support the purchasing of equipment to desalinate groundwater on both a family scale and a community scale. Because of the high energy cost of the desalination process, the high potential of the use of green energy, especially solar and wind energy, but also tidal energy in the NWCZ, is suggested. The NWCZ is also a gas producer.

Stakeholders also proposed some elements of an ambitious policy to develop irrigation in the NWCZ based on the available water resource on local, national, and regional scales. They are aware that the NWCZ has no priority for using Nile water, except in the eastern part, where the El Hamman canal already brings Nile water. For the central and western part of the NWCZ, stakeholders suggest the use of the groundwater located around the Siwa oasis and on the other side of the Libyan border, even if the recent social problems make this option impossible. Another alternative is to desalinate groundwater and seawater, as already mentioned, with small equipment on family and community scales. Because of the high energy cost and the requirement to desalinate water, a regional project involving the building of several nuclear power plants has been planned.

The development of irrigated agriculture in the NWCZ has two objectives: the first is to ensure food supply to meet the increasing demand of the local market, following the expansion of tourism; the second is to produce enough food for the market of the big Egyptian cities such as Cairo and Alexandria, and also the export market. The first objective is easily understandable because of the already mentioned factors, even if the competition will be strong with the production in the Nile delta. Regarding the second objective, much of the local production is already consumed in the cities of the Nile valley, such as chicken from intensive poultry farms, dates, figs, and olive oil. Some products such as sheep are exported to the Gulf countries and others such as turkey are exported to Europe, especially for the Christmas period. Moreover, some local products from the NWCZ are bought by European import companies for sale on the European market, especially olive oil.

In conclusion, the development of the irrigated agriculture in the NWCZ is an ambitious development policy on national and international scales, especially for the Mediterranean. This great challenge should be seriously analyzed before its implementation, especially assessment of the impact on natural resources—namely, water and soil. Furthermore, this scenario could be developed regardless of the rangeland; which means that the herds grazing on the rangeland, which was the basis of the pastoral Bedouin society of the NCWZ for several centuries, could become marginalized in the local development. Nonetheless, animal breeding should not disappear, at least not quickly, since many families will continue to raise livestock, while progressively adopting practices of agropastoral systems. Some Bedouins will certainly continue with significant animal production on rangeland, especially to supply the local market; however, this will not be considered significant on a local scale. Abandoning and neglecting rangelands may have negative repercussions on local development, particularly water supply. Indeed, the continuing degradation of rangelands will lead to stronger soil erosion, and therefore a faster silting of infrastructures in the wadis. Moreover, the concentration of construction along the coastline and the wadis will leave vast areas without genuine local control. Moreover, the subsoil is rich in natural mining resources. So, a relevant alternative would be to include a specific plan for sustainable development of the rangeland integrated in the set of policies on a regional scale.

6.4.3 Sustainable Rangeland Management

Rangeland is the natural ecosystem of the NWCZ. It was the main feed resource of the Bedouin pastoral system. In the first few kilometers along the Mediterranean coast, the rangeland progressively disappeared during recent decades as a result of urbanization, tourism expansion, and the development of rain-fed agriculture around the villages and in the wadi beds. Nowadays, there is practically no more pasture for the livestock in this zone, except the crop residues of barley fields. More to the south, many rangeland areas are severely degraded from the combined effects of low rainfall, wind erosion, and overgrazing, as already mentioned. During the last 20 years, shrubs have disappeared and the upper soil layer has gone, including the seeds of annual vegetation, leaving today small deserts of stones. Many breeders and stakeholders do not even make an effort to explore the low-producing pastures; hence, many of them stopped their seasonal migrations and are investing in other activities, as mentioned earlier.

Despite the severe degradation of the rangeland, the local stakeholders, particularly the leaders of tribes, still consider that pasture use has always been and will always be a priority for Bedouin society for at least three reasons. First, the landownership of the pasture areas is not clearly defined, and the leaders of tribes do not admit nor accept the national Egyptian law, which does not recognize their traditional rights to these lands. This is certainly the most complex conflict between the Bedouin society and the national governance, especially after the Egyptian Army reclaimed part of these lands. The second reason is also linked to land tenure and concerns the natural resources of the subsoil. Indeed, while oil and gas companies are exploring the southern part of the NWCZ, other mining companies are also interested in the potential of the zone, which makes the leaders of the tribes want to control these resources, eventually in partnership with the national institutions and private groups. The third reason is that rangelands are still a main feeding source for the flocks, and their productivity is expected to increase under good weather conditions. In addition, sheep fattening is developing in the southern lands, even if itis based on feedstuffs purchased in the Nile delta.

In the context of the NWCZ, the implementation of a sustainable management program in a determined rangeland area should take into consideration the social, economic, and environmental issues simultaneously. The social issue regards the rules to be shared and decided on between the families and the tribes concerned in the rangeland area. The economic issue deals with controlling the natural resources of the rangeland subsoil, whereas the environmental issue focuses more on the ecological processes needed to recover the rangeland and then manage it.

According to the farmers and stakeholders, the first step would be an agreement at local scale to better organize the use of the flock tracks, especially to avoid that some tracks were closed (for many reasons) and perturbed the daily and seasonal migrations. The implementation of such action requires awareness among all the herder families in the NWCZ, and a strong outreach by the leaders of the tribes. Consequently, a detailed map would be needed to build a shared and common view of the current tracks and the alternative ones. Finally, a decision process could be launched, respecting the interests and constraints of the different families.

Another action would be the recuperation of the strongly degraded rangeland areas adjacent to the villages. The objective is to launch a social–ecological process aiming to improve the vegetation cover of shrubs and annual plants, which are the basis of the pastoral feed resource. This requires a strict and medium- to long-term control of both the stocking rate and the grazing periods. Formal or informal contracts have to be made between the different families with respect to the medium and long term. At the same time, some practices have been found to be efficient in recuperating the vegetation cover.

In the south, at a distance farther than the coast and the villages, the most urgent action would be to implement sustainable rangeland management to avoid greater degradation of the vegetation cover and to maintain or improve the productivity of the pastures. Again, formal or informal contracts have to be made between the different parties interacting in this zone.

This third scenario is more of a set of recommendations synthesized from the data collected at the level of the local farmers and stakeholders about the rangeland, its management, it degradation, and the means to recover it. The technical issue is not the main challenge; however, the implementation of this third scenario requires first of all collective decisions followed by the application of adequate measures, which are not easy to achieve. Collective decisions might be difficult to reach especially because they require the approbation of all farmers involved; for example, defining the tracks for the herds between barley fields or choosing the dates to start or stop grazing on recuperated rangeland. The measures resulting from the shared decisions will be more complex to apply because tools and control methods will also have to be defined, including penalties, follow-up on decisions, and assessment processes. Maybe 5 years ago, the leaders of the tribes had the authority and the confidence of the farmers to coordinate the implementation of this third scenario, with the institutional and financial support of the local and national governments. Nowadays, even if the local stakeholders always efficiently face this challenge, the Arab Spring has strongly changed the context: first with the local strengthening of the Egyptian Army, second with the lack of confidence between the army and the local population of the NWCZ, and third the chaos in Libya, which has brought insecurity to all the Sahara region.

Furthermore, the third scenario is an alternative to neither the first scenario nor the second one. It is more of a complement regarding specifically the rangeland area, a complement which could be implemented in both cases.

In conclusion, whereas the first scenario is based on the reasonable use of natural resources and only near the coast, the second one requires high use of these resources, especially water, associated with a strong artificialization of the context, including the land in the far south. The third scenario is completely different because it aims to improve the resilience of the social–ecological system, focusing both on the social and the environmental components through sustainable rangeland management. However, even if the third scenario had a small chance of being implemented and coordinated by the leaders of the ribes, the Arab Spring has greatly changed the context, particularly through the new tension between the different groups, including the national and local institutions, the tribes and their leadership, the financial agencies, and the private sector. Moreover, the local stakeholders think that confidence will not return for several years or maybe some decades, and the time to implement rangeland sustainable management in the NWCZ has passed.

6.5 Concluding Remarks About the Resilience of the Bedouin Social–Ecological System

6.5.1 Land Issue: The Main Challenge for a Long Time

Land issue is a big challenge in the NWCZ. Since Ancient Egypt and until today, the history is punctuated with conflicts over land and control of natural resources. More recently, the land issue has been linked to the strategic location of the NWCZ at the Libyan border and along the Mediterranean coast, the expansion of tourism, and the development of agriculture. Hence, land tenure and landownership appeared logically as two key factors in the three scenarios built with the local stakeholders, especially since the national law is not clear regarding the traditional land rights of the tribes. National and international policymakers have to integrate these data into the development policy for the NWCZ. Land issue and landownership rights of the breeders are frequently mentioned in North Africa and western Asia as an essential factor for local development.

6.5.2 *The Tribe, Always a Pillar of Bedouin Society*

The tribe operates as the Bedouin network, acting in all components of the society, on both social and cultural issues in economics and policies, in good times and in difficult times. For example, during the last drought, the leaders of the tribes interacted with almost all the adaptation strategies of the Bedouin breeders, such as supporting a new activity, finding a job or migration for a needy family, and giving access to wadi land to a herdless breeder. So, the contribution of the tribes seems essential in policymaking and in the implementation of these policies. Even if the concept of the tribe is usually considered as a heritage, especially in urban areas, young and capacitated tribal leaders frequently become relevant local stakeholders in rural areas, supported by their tribes and using their tribes to invest in political arenas. Furthermore, the tribe has a major place in the advance of democracy, another challenge on a regional scale. Indeed, until now every vote is in favor of the representative of the tribe, which justifies the interest of young people in the politics of the tribe.

6.5.3 *High Skills of Bedouins in the Adaptation Process*

Bedouin society has demonstrated a strong ability to adapt to severe drought, and more generally to global change (Alary et al. 2014b). This strong resilience relies on several factors, such as (1) the skills of the Bedouin breeder to adapt his herd to harsh animal husbandry conditions, (2) the mobility of the herds, but also that of breeders, especially to migrate and find additional income, (3) solidarity in the tribal system, as previously mentioned, and (4) some efficient national and international public policies (e.g., the MRMP). This strong ability of adaptation, especially the integration of new technologies, is a positive factor of Bedouin society for the future and should be taken into consideration in research and development projects and policymaking.

6.5.4 *Low Attractiveness of Rangelands*

However, the capacity of adaptation and innovation of Bedouin society concerns the diverse production and sectors of activity, but not rangeland management. Some farmers have tried to implement new systems of rangeland management to recover and maintain the productivity of the rangeland, but the results were usually unsatisfactory. Moreover, between the three scenarios suggested by the local stakeholders, only the third one focused on rangeland management. This shows the role of rangeland given by the local population, especially the young people. The rangeland was the basis of the livestock farming system and the main pillar of Bedouin society lifestyle, but now it has disappeared. However, the rangeland has always had significant functions, especially in water supply through the watersheds, the resources of the subsoil, and also the biomass to feed the herds and maintain biodiversity.

6.5.5 The Arab Spring Has Changed the Context

The current sociopolitical situation resulting from the Arab Spring, in Egypt and Libya, has built a special context with no clear perspective. Hence, the technical constraints linked to irrigated agriculture and rangeland management appear as secondary issues on a regional scale and strongly depend on the sociopolitical future and the funds invested by national and international agencies. In the past, the social conflicts, usually over land control, were both the milestones and the main drivers of change in the NWCZ. The Arab Spring is creating a new situation: rangeland management will not be a major challenge anymore. So, maybe nothing will change during the next few years and decades regarding the rangeland, except the ongoing degradation process.

Acknowledgments We acknowledge the ELVULMED, MOUVE, and CLIMED research projects financed especially by the French National Agency for Research and Egyptian and international institutions.

References

Alary V, Daoud I, Abdelzaher M, Salama O, Aboul-Naga A, Merveille N, Tourrand JF (2012) Adaptation des societies bédouines de la Côte Nord-Est de l'Egypte au changement global. Autrepart (62), pp 183–200

Alary V, Hassan F, Daoud I, Aboul-Naga A, Abd El Zaher M, Bastanelli D, Lescoat P, Moselhy N, Tourrand JF (2014a) Bedouin adaptation to last drought in the North Coastal Zone of Egypt: continuity or rupture? World Dev 62:125–137 (http://dx.doi.org/10.1016/j.worlddev.2014.05.004)

Alary V, Messad M, Osman MA, Aboulnaga A, Daoud I, Bonnet P, Juanes X, Tourrand JF (2014b) Livelihood strategies and the role of livestock in the processes of adaptation to drought in the coastal zone of Western Desert (Egypt). Agric Syst 128:44–54. doi:10.1016/j.agsy.2014.03.008

Ball J (1939) Contribution to the geography of Egypt. Survey Department Publications, Cairo

de Cosson A (1935) Mareotis. Notes on the coast road between Alexandria and Mersa Metruh. BSAA, pp 48–61

Daoud I (2015) Bedouin society strategies facing drought in north west coastal zone of egypt: A case study of wadi naghamish. Thèse de Doctorate, AgroParisTech/ABIES, Paris, France

El Miniawy H, Mark F, Tobah S (1990) Qars rural development project. Development plan-phase II draft report development strategy. The Egyptian Environmental Affairs Agency, Cairo

Fraser E (2007) Travelling in antique lands: Studying past famines to understand present vulnerabilities to climate change. Climate Change 83:495–514

Hume WF, Hughes F (1921) The soils and water supply of the Maryut District. West of Alexandria, vol 37. Government Press, Cairo

Holling CS (2001) Understanding the complexity of economic, ecological, and social systems. Ecosystems 4(5):390–405

Kassas (1972) Cited by Bruins HJ, Evenari M, Nessler U (eds) (1986) Rainwater-harvesting agriculture for food production in arid zones: the challenge of the African famine. Appl Geogr 6(1):13–32

Renouard M (2006) Histoire et civilisations de la Méditerranée. Ouest-France, Rennes, 127p

Said R (1962) Geology of Egypt, 1st edn. Elsevier, Cairo

Shata A (1953) New light on the structural development of the Western Desert Egypt. Bull Inst Desert Egypt 3(1):101–106

Shata A (1971) The geomorphology, pedology and hydrogeology of the Mediterranean coastal desert of U.A.R. Symposium Geology of Libya. In: Faculty of Science. University of Libya, Tripoli, pp. 431–446

Walpole GF (1932) An ancient subterranean aqueduct West of Matruh. Ministry of Finance, survey of Egypt. Paper no 42. Ministry of Finance, Cairo

Weedon (1912) Cited by Bruins HJ, Evenari M, Nessler U (eds) (1986) Rainwater-harvesting agriculture for food production in arid zones: the challenge of the African famine. Appl Geogr 6(1):13–32

Chapter 7
Modeling Coupled Human–Natural Systems of Pastoralism in East Africa

Randall B. Boone and Carolyn K. Lesorogol

Abstract Sustainability science studies the drivers of change that alter natural and human systems now and into the future. Systems with natural and human components (coupled systems) may be resilient or nonresilient to these changes. Our focus is on rangelands of East Africa, where drivers of change include rapid population growth, poverty, degraded rangelands, declining wildlife populations, land fragmentation, reduced mobility, and more frequent drought associated with climate change. Computational simulation is a widely used approach to understand coupled systems. Modeling allows us to simplify the representation of systems and to control system elements. Coupled systems applications often include one or more models simulating ecosystem responses, joined to an agent-based model representing individuals or households. With those applications, ecosystem attributes can affect human decision making, and decisions people make affect ecosystems. Internal drivers of change are incorporated into the processes and rules of the models. External drivers are assessed through scenario analyses. We applied a coupled systems approach to Samburu, Kenya, where residents raise cattle, goats, and sheep. We quantified the effects of immigration of herds during drought and changing wildlife populations on local household well-being. Immigration of outside animals caused a 0.5 decline in the number of animals per person and an increase in the amount of supplemental foods needed. Increasing wildlife populations caused declines in livestock numbers, as expected, but payments of the magnitude that may be expected did not offset losses. The coupled systems approach allows more direct inference of changes, and for changes in the ecosystem and human components to influence each other. Challenges include the complexity of the endeavor and the issues addressed, difficulties in model validation, the rates of change, and the

R.B. Boone (✉)
Natural Resource Ecology Laboratory and the Department of Ecosystem Science and Sustainability, Colorado State University, Fort Collins, CO, USA
e-mail: randall.boone@colostate.edu

C.K. Lesorogol
George Warren Brown School of Social Work, Washington University, St Louis, MO, USA

S. Dong et al. (eds.), *Building Resilience of Human-Natural Systems of Pastoralism in the Developing World: Interdisciplinary Perspectives*,
DOI 10.1007/978-3-319-30732-9_7

political will that may override what scientists view as sustainable solutions. Modeling will improve as psychological research refines decision making rules, the temporal and spatial grain of the simulations increases, and the ability to model many thousands of agents expands.

7.1 Introduction

Short-term shocks and long-term drivers of change are taxing the sustainability of ecosystems and the people who rely on their services around the world. Today, questions of sustainability are so broad in scope and so important to the well-being and longevity of societies that scientific fields have been invented to study these systems. This century has brought us sustainability science, a field that seeks to understand the interactions between ecosystems and societies and the ways drivers of change may alter natural and human systems now and into the future. Sustainability science differs from traditional fields because of its explicit focus on problems, the need to span spatial scales, topics that can be ingrained but require urgent solutions, the complexity of problems, and the need to put potential solutions in place to guide development while research continues (Kates et al. 2001). A resilience framework has been developed (Walker et al. 2002, 2004) where the dynamic nature of systems is embraced. A resilient system is one that can absorb perturbations and return to its current state, versus a nonresilient system that is at risk of a permanent shift in state (Benson and Garmestani 2011). Ultimately of most interest is how people respond to change, providing a linkage between the resilience of a system and its capacity to adapt to fast, fine-scale drivers while constrained by slow, broadscale drivers, such as climate change (Gunderson and Holling 2002; Hobbs et al. 2008).

Drivers of change are particularly pressing in East Africa, where multiple stressors make questions of sustainability complex. Economic progress in the region (i.e., Burundi, Rwanda, Uganda, Kenya, and Tanzania) has been rapid (e.g., 6 % growth per annum), but human population densities are high and many people remain poor, with Kenya's average per capita income less than $2 per day, and Burundi's less than $1 per day (Eyakuze and Salim 2013). Agriculture is a major contributor to the regions' economy, and millions of people make their living from the land, many as pastoralists (i.e., those who rely on the livestock they raise) and agropastoralists ((i.e., those who rely on the livestock they raise and cultivated products). This puts both the residents and the economies of East Africa at risk from changes in their environment, such as more frequent droughts anticipated with climate change (Williams and Funk 2011). Degraded rangelands and forests magnify the risk from those changes (e.g., Pricope et al. 2013). Pastoralists and agropastoralists coexist with diverse wildlife species that remain commoner in East Africa than in other parts of Africa (though declines in numbers are occurring), which further complicates the addressing of questions of sustainability (Galvin et al. 2008; Reid 2012).

An outcome of sustainability science has been the adoption of a main pathway of understanding and projecting change, computer simulation. Specifically we will focus on a method of simulation used in this research called "discrete-event simulations." Simplified representations of reality and its processes are coded into programs, where events may occur at scheduled periods as the orderly progression of time is simulated. Simulation has been used in ecological research for many years (Huston et al. 1988). In the social sciences, its use is more recent, but is transforming social sciences by allowing researchers to add in silico experimentation to their toolbox (Brenner 1999; Kohler and Gumerman 1999).

Here we review the nature of research that adopts a coupled human–natural systems approach. We then introduce East African rangelands and their inhabitants. The purpose and nature of simulation to address questions in coupled systems is discussed. We then summarize a case study from Samburu County, Kenya, and conclude by discussing some aspect of the future of coupled systems modeling.

7.2 Coupled Human–Natural Systems

Before the end of the last century, natural and human systems were studied independently. Researchers sought to understand ecosystem dynamics (typically extrapolating from very small study sites; Kareiva and Andersen 1988), either ignoring the roles humans played in ecosystems or treating the effects of humans on systems as troublesome factors to be controlled for. Anthropological studies made strides understanding the ways in which humans lived and their dynamics (e.g., Baker and Little 1976). Cultural ecology, founded by Julian Steward (e.g., Steward 1955), considered the ways in which ecological conditions modified human cultures. Others followed, building a rich theoretical basis for the evolution of human cultures (Bateson 1972; Orlove 1980; Boyd and Richerson 2005) and its own critiques. In general, though, ecosystem science and anthropology moved along parallel pathways, rarely sharing information, and sometimes with interactions viewed with hostility.

The first large research project to consider humans as an integral part of an ecosystem was in East Africa, and was the South Turkana Ecosystem Project, supported mostly by the US National Science Foundation and led by Colorado State University. Through the 1980s, dozens of researchers and students sought to understand how the Turkana people used their landscapes to survive, and how changes in the ecosystem influenced the people (Coughenour et al. 1985; Ellis and Swift 1988; McCabe 2004). In the years since then, members from that team and others have been considering humans as components of ecosystems (e.g., Galvin et al. 2008), an approach now termed "researching coupled human and natural systems." As an example of the shifting emphasis in research, there is now a program in the US National Science Foundation (2014) expected to support more than $16.5 million in research effort using such approaches in 2016.

Four general components define a study adopting a coupled human and natural systems approach (National Science Foundation 2014): The dynamics of a natural system are studied, and the dynamics of an associated human system are studied. Then the key distinction, the means in which the natural system influences the human system, is studied, and the means in which the human system in turn influences the natural system is studied. For example, a study may explore the ways an ecosystem responds to grazing and how the human system responds to changing policy, *plus* that study would explore the ways in which choices people make about where to graze their animals may alter the ecosystem, and how those changes in the ecosystem alter human well-being.

7.3 East African Coupled Systems

There are many types of systems in East Africa and people relying on them, from coastal fisheries, to high-elevation tea plantations, to the urban setting of Nairobi with its more than three million residents. Here, we focus on the rangelands of East Africa. Those rangelands have been used centuries by pastoralists (Reid 2012). The Maasai, Samburu, Turkana, Borana, Pokot, and many other groups raise some combination of cattle, goats, sheep, and camels, relying mostly on milk and meat as livestock products. Livestock join a diversity of wildlife in these lands, animals that must compete with the livestock for forage, with their interactions yielding densities of wildlife and livestock that vary across space in complex ways (Ogutu et al. 2010). The lands these herbivores inhabit are heterogeneous because of differences in climate, soils, and disturbance histories (e.g., fire, fuelwood use, and grazing). These semiarid rangelands have highly variable climates, which leads to landscape patches that vary through time in the forage that they provide for animals. Wild and domestic animals and their herders access these forage patches through mobility. Seasonal movements allow animals to access forage of sufficient quantity and quality, and to acquire the water, minerals, and shelter they may need.

The climatic phenomenon known as the intertropical convergence zone moves annually north and south from about 15°N to 5°S, with rainfall peaking near the leading edge of the zone (Ellis and Galvin 1994). This brings to much of East Africa a bimodal rainfall pattern. For example, in Kenya, what are known as the short rains extend from about October into December, and the long rains extend from March to May. This expands the area that is used for pastoralism rather than rain-fed cultivation; whereas areas with annual precipitation as low as 400 mm may be expected to regularly produce crops in western Africa, in East Africa, annual precipitation must be about 700 mm or higher to regularly produce crops (Ellis and Galvin 1994). People still often cultivate land in drier areas in the region. Indeed, many people who would self-identify as pastoralists are now agropastoralists, tending to small cultivated plots (less than 1 ha). But they may expect a crop from those plots once every 4 or 5 years, and even less frequently with projected changes in climate (Case 2006; Hoffman and Vogel 2008).

Change in part defines pastoral areas of East Africa, but today the pace of change is increasing. The number of livestock that can be supported on the landscape cannot be changed significantly, but human population growth is rapid. That is leading to more people being supported by the same number of livestock, resulting in a reduction in the number of livestock per person. There is also evidence of stratification in wealth and income among pastoralist households, creating the possibility of persistent differences in well-being within communities (Lesorogol 2008). The number of families that can survive from their animals alone is declining, and the need to diversify livelihoods is high. People now often participate in small business activities, livestock trade, handicraft production, or wage labor. Lands are now fragmented by infrastructure such as roads and fences, buildings, businesses, and conservation areas or other areas where grazing is not allowed. These developments render mobility a more difficult adaptive pathway than in the past. In addition, government rules and changes in land tenure make moving livestock more difficult. In some areas, such as those discussed later, land that used to be available for grazing through reciprocal sharing undergirded by social norms is more often rented out, effectively making it inaccessible to many poorer pastoralists. Some pastoralists favor being less mobile. They value individual ownership of rights to a parcel of land, rather than communal ownership and the uncertainty that brings. Families appreciate being more sedentary to be close to schools and hospitals, and to be able to tend to crops. As a result, many pastoralists now maintain a more or less fixed home base from which livestock migrate during dry seasons and drought while other members of the household remain at the home base relatively permanently. The degree of mobility often depends on aridity and the availability of resource-rich ecological patches with more mobility associated with greater aridity.

7.4 Modeling Purpose and Approach

The purpose of modeling is often taken to be prediction (Epstein 2008), but that is rarely the case in coupled human–natural systems modeling. There are far too many changes in any real system to be able to predict specific outcomes a decade on. Instead, simulation approaches allow us to hold most ecosystem attributes constant and vary only selected attributes in scenarios. We may then describe the magnitude and direction of change that may be expected if the manipulation that is the focus of the scenario were performed, *all else being controlled*. The utility of modeling extends beyond addressing scenarios, however. Modeling can deepen an analyst's understanding of relationships, as what may have been mental models or casual understanding of relationships must be quantified and coded, which requires that explicit hypotheses be formed. What is to be included in a model must be distinguished from what may be excluded. For example, a researcher studying pastoral livelihoods may decide to include livestock dynamics in a model but to exclude rare livestock species—that exercise is instructive. The processes and rules governing a system must be decided on and encoded, interactions between elements must be described, and parameter values must be assigned.

The scientific specialties involved in coupled human–natural systems modeling are too numerous for an individual to possess, and team-based research is the norm. Teams may include ecologists, anthropologists, economists, hydrologists, and simulation specialists, for example. For the team to reach some mutual understanding of the ways in which system elements couple requires good communication, and for the processes and rules to be coded correctly in a simulation is a separate challenge. The value of this exercise in forcing researchers to make implicit understanding explicit and to quantify relationships is difficult to overstate; coupled human–natural systems research is an excellent team-building exercise (Axelrod 2006).

Other benefits of modeling include our ability to generalize results to novel areas or into the future, to conduct analyses that would be viewed as too expensive, impractical, or immoral, and to integrate a wealth of data and relationships with a reduced risk of some relationship being forgotten. Simulation can guide data collection, highlighting parameters that require better estimation, can identify novel questions, and can bound the values of parameters to be within reasonable ranges (Epstein 2008). Modeling may be undertaken in real-world settings, or may use hypothetical landscapes and test theory without being encumbered by the circumstances of a specific setting (Griffin 2007). A use often overlooked is for the simulations to form a common starting point for stakeholders in resource management to leverage discussions (Boone and Galvin 2014); people may disagree on the modeling outcomes, but everyone begins from a similar place.

The approach used to study coupled human–natural systems typically includes an ecosystem model joined with another model representing humans and their decision making. Ecosystem models include mathematical descriptions of processes such as water flow, nutrient cycling, and the ways in which plants of different types grow, reproduce, and die. The models may be nonspatial or point based, where a simulation represents a point within a landscape that is considered homogeneous. In that case, conducting scenario analyses on regions involves summarizing many individual simulations, one per homogeneous landscape unit. Other simulations are spatially explicit, and within a single simulation may represent many landscape units, or for a gridded landscape, where the landscape is divided into many square cells and ecological processes within each of those cells are simulated. Discrete time models such as these use climate data read in at regular intervals (e.g., hourly, daily, monthly) to make the simulations more closely emulate observed dynamics.

Joined to the ecosystem model is an agent-based model. Agent-based models are well suited to represent systems derived from the bottom-up organization of elements (Grimm 1999; Railsback and Grimm 2011), such as social systems. Simulations include elements, or agents, that interact with other agents and their environments according to rules. The behaviors of populations of agents are summarized and reported in ways analogous to those in other analytical approaches, with the added flexibility of being able to make summaries across hierarchical levels from a single set of results. In simulations, interacting agents may exhibit emergent behavior, which is an aggregate response that is not part of the constituent agents. In coupled human–natural systems, the agents are often individuals or households; here, for brevity, we speak of households. Values are provided to the model that describe the attributes of each household, such as the age and sex of the household head and household members, initial assets and debts, income sources, and expenses. Other values stored within the simulation track attributes of households, such as monetary holdings, as they change through time. Rules that use thresholds and logical bifurcations are included that allow agents to emulate real-world behaviors, to some degree. For example, a rule may be included where, at each time step, a household inspects its monetary resources and decides whether to sell assets to meet household needs. The individual households represented in the agent-based approach allow for people to have responses that are spatially explicit, meaning that households have specific locations and interact with specific parts of ecosystems. Their decisions are influenced by the nature of their local environments, and the actions they take may alter their environments.

After the application of a coupled human–natural systems model to an area, the model components and the joined results are assessed (Gilbert and Terna 2000; Grimm et al. 2005; Wilensky and Rand 2007). This can be a challenging component of the modeling exercise. Individual components of the ecosystem model are typically vetted many times, such as methods to estimate potential evapotranspiration or soil surface temperature. Remotely sensed images are useful for assessment of the output from ecosystem models, such as comparison of greenness and biomass estimates from the model with normalized difference vegetation indices and net primary productivity from the Moderate Resolution Imaging Spectroradiometer (MODIS) platform (i.e., product MOD17). Such efforts can include comparisons with long-term average responses as well as month-to-month temporal and spatial responses. Assessing results from the household models (and the ecosystem model as well) involves comparisons with observed patterns from the system (Grimm et al. 2005). Agreement (or disagreement) between the simulated results and these patterns provides support for (or refutation of) the suitability of the application. These are sometimes novel patterns, and sometimes what may seem mundane patterns such as stability, but all provide evidence for or against model performance.

Once a model has been assessed, the pathways to discovery most often used in coupled human–natural systems are through experimentation and scenario analyses. In experimentation, simulations are viewed as including a suite of hypotheses that may be assessed (Peck 2004; Railsback and Grimm 2011). For example, one may hypothesize that the spatial heterogeneity of grazing resources influences grazing site selection by pastoralists (BurnSilver et al. 2003); that is, they favor areas where a diversity of forage patches makes it more probable that sufficient forage may be found for their animals. An analyst may build an agent-based model representing patch selection by pastoralists and include an option to enable or disable consideration of spatial heterogeneity in site selection. The two options may be simulated and then the results compared with observed site selection to support or refute the hypothesis.

The other means of discovery, scenario analyses, allow us to adopt the intuitive "what if" approach to simulation. In more structured approaches (van Notten et al. 2003), perhaps three to six alternative futures are described for a coupled human–natural system. A well-known example is the set of emission scenarios developed by the Intergovernmental Panel on Climate Change (2000). The scenarios (i.e., A1, A2, B1, and B2) include different rates of economic and population growth and technological development, each with its own implications for changes in the release of gases influencing climate change. Our work uses a less structured but still formally agreed on method of defining scenarios. Team members consider the internal and external drivers of change in a coupled system, often informed by participatory meetings where the interests and concerns of community members are gathered (Reid et al. 2009). Internal drivers of change are often incorporated into the processes and rules included in a simulation, and may or may not be the subjects of scenario analyses. External, slow drivers of change are typically accessed through scenario analyses (e.g., warming or changes in precipitation, changes in policies affecting access or land tenure) (Crépin 2007; Walker et al. 2012). In a creative and rewarding exercise (Axelrod 2006), team members discuss and agree on ways in which drivers of change may be reflected in the scenarios appropriate for the tools at hand. As examples, a scenario dealing with climate change may involve the replacement of observed temperature and precipitation data with data projected with use of a global circulation model; an increase in money spent on veterinary care may mean a change to household expenses in that regard and a small percentage increase in the average survival of livestock; and a change in access such as the addition of cultivated areas to a landscape that cannot be used by livestock may mean the editing of a spatial surface used in the model.

7.5 Samburu Coupled Human–Natural System

7.5.1 Samburu, Kenya

We conducted coupled human–natural system analyses among Samburu pastoralists in northern Kenya. The Samburu number about 200,000 and primarily reside in Samburu County in north central Kenya. They rely heavily on livestock for their livelihoods and herd cattle, sheep, goats, and in drier areas, camels. Most Samburu live in settlements comprising extended family households, often polygynous, usually from one male lineage. However, there is a trend toward smaller settlements, sometimes only one household, particularly in one of our research sites where land was privatized and households were required to move to their individual parcels. As noted earlier, like other pastoralist populations in East Africa, the Samburu are diversifying their livelihood strategies to include activities such as wage labor, trade in natural commodities (e.g., firewood, timber, milk, honey) and resale of purchased goods in small home-based shops. Ethnically, the Samburu are closely related to the Maasai, with whom they share many cultural traditions and social structures as well as language.

Our research sites are in southwest Samburu County, 42 km apart, but at different elevations (Fig. 7.1). The southern site, known as Mbaringon and at approximately 1790 m, is more typical Kenyan rangeland, with mixed grasses, dwarf shrubs (e.g., *Duosperma eremophilum*), and shrubs (e.g., *Vachellia reficiens*), The northern site, Siambu, is at approximately 2425 m and has a cool climate with moist grassland types, dense shrublands, and forest (Figs. 7.2 and 7.3). The western border of Siambu is the Rift Valley escarpment. In Mbaringon, we estimated that 370 families

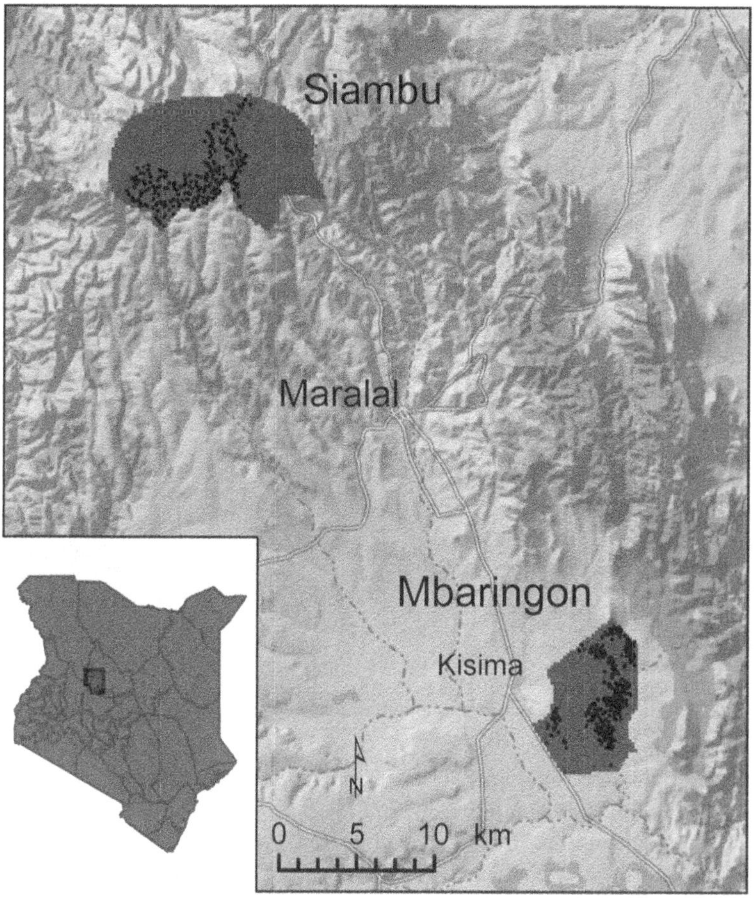

Fig. 7.1 The study areas, Siambu to the north and Mbaringon to the south. Households included in the models are shown as black dots. Underlying the study areas is the National Geographic base map (citing National Geographic, *ESRI* Environmental systems research institute, *NAVTEQ* Navteq, *UNEP-WCMC* United nations environment programme world conservation monitoring centre, *USGS US* geological survey, *NASA* National aeronautics and space administration, *ESA* European space agency, *METI* Japan ministry of economy, trade, and industry, *NRCAN* Natural resources canada, *GEBCO* General bathymetric chart of the oceans, *NOAA* National oceanic and atmospheric administration, IPC). The inset shows the areas in Kenya

occupied the area of interest, using most of the land communally but with some areas fenced for homesteads. In Siambu, a portion of the region has been formally subdivided into 240 individually owned parcels of about 9 ha each (one for each registered household) (Fig. 7.1). Some members of the community have sold part of their parcels, and others rent out their land for use in mechanized wheat cultivation. Both sites are located on the Lorroki plateau, which is the highest elevation and highest rainfall region within the 20,000-km^2 Samburu County. Although annual rainfall ranges from about 500 to 1200 mm on this plateau, the area still experiences

Fig. 7.2 A fenced and farmed parcel in Siambu, Kenya

periodic droughts. For example, serious droughts were experienced in this area in 1990–1993, 2000, 2006, and 2008–2009. Population growth is relatively high, whereas livestock numbers fluctuate, leading to reductions in livestock holdings per capita. According to our 2010 data, average livestock holdings were about two tropical livestock units (TLUs) per person and there is considerable stratification within the communities in terms of livestock holdings. Thus, increasing poverty among poorer pastoralists is a serious concern in this area, and our research also reveals high levels of undernutrition in the population (Iannotti and Lesorogol 2014). Although it is difficult to assess the degree to which climate change is affecting the environment in Samburu, shifts in land tenure, particularly the privatization of communal land, raise questions regarding the future of mobile pastoralism in the region as some owners restrict access to their private land. Understanding the relationships among household-level land use decisions in the light of privatization and falling livestock holdings motivated our scenario analysis using simulation modeling.

7.5.2 Simulation

Our primary research of coupled human–natural systems in East Africa began when we joined an ecosystem model called Savanna to a model representing households called PHEWS (for "pastoral household economic welfare simulator").

Fig. 7.3 A village in Siambu, Kenya

With those linked tools we assessed how cultivation in Ngorongoro Conservation Area, Tanzania, was contributing to the well-being of the resident Maasai (Thornton et al. 2003). We also used it when we were quantifying the effects of habitat fragmentation on Maasai livestock and Maasai households in Kajiado County (then Kajiado District), Kenya (Thornton et al. 2006). PHEWS was effective in answering the questions at hand, but had the limitation that households were represented as members of a population. In Ngorongoro Conservation Area, about 3800 households were collapsed into eight livelihood strategies, and each of those had up to three wealth levels (i.e., poor, moderate, and wealthy). That meant that, at most, the households were members of 24 populations. With households only in aggregate, there was no way to have a household occupy a specific place in space, or to own its own livestock, or be influenced by local environmental conditions. To address this limitation we created the DECUMA (for "decision making under conditions of uncertainty by modeled agents) model. That tool represents individual households as agents. As such, it allows households to have specific locations on the landscape, to own their own livestock herds, and for the people and their herds to interact with local environments so that their decisions are influenced by their environment and their management can alter the environment—a coupled human–natural system. The Savanna and DECUMA tools and the manner in which they join are expanded on in the following sections, with much more detail provided in Boone et al. (2011a).

7.5.3 Savanna

The Savanna ecosystem model is a series of Fortran modules written by Michael Coughenour and used originally more than 25 years ago in the Turkana region of Kenya, which is geographically near our study area. The model has since been used in many places around the world. Savanna is spatially explicit, with the spatial data it uses divided into a series of square cells that define the gridded landscape. Those spatial data include layers that describe elevation, slope, aspect, soils, and land cover. Climate data are used by the model to simulate weather as time passes, with precipitation and minimum and maximum temperature for a series of sites used. For a single weather station, additional information is used to simulate weekly weather and potential evapotranspiration, including wind speed, relative humidity, and Priestly–Taylor coefficients (Priestley and Taylor 1972). Plants are represented by a set of functional groups, and those groups compete for water, space, light, and nutrients (Fig. 7.4). Plant groups that compete successfully may expand their coverage at the expense of other functional groups, or the amount of bare ground may decrease. Quantities of photosynthate are estimated on the basis of the outcome of competition, and then distributed to roots, leaves, and stems by means of rules for plant allometrics. This provides estimates of primary production for the different functional groups, and plant populations are modeled depending in part of the proportion of production put to seeds and establishment and mortality rates.

Herbivores are represented in Savanna, again as functional groups, but often those groups are individual species. Animals are distributed on the landscape each week according to the suitability of the habitat, including forage quality and quantity, slope, elevation, cover, and the density of other herbivores. The herbivores feed on primary production and gain energy. They lose energy through basal metabolism, travel, thermal maintenance, reproduction, and lactation. An energy deficit leads to weight loss and an excess leads to weight gain. When joined to DECUMA, wildlife are represented in Savanna and livestock are represented in DECUMA to allow animals to be owned by individuals (Fig. 7.4). The results are summarized each month, with spatial and temporal data produced. See Ellis and Coughenour (1998), Boone (2000), and http://www.nrel.colostate.edu/projects/savanna/ for more details about Savanna.

7.5.4 DECUMA

The DECUMA model includes individual households placed appropriately on the landscape, with their initial attributes set on the basis of household surveys (Fig. 7.5). The set of surveys is smaller than the full set of household simulated, so surveys are used to initialize more than one household through random selection; these are initial conditions, and households may follow their own dynamics as a simulation progresses. Household heads own livestock herds, whose distributions are

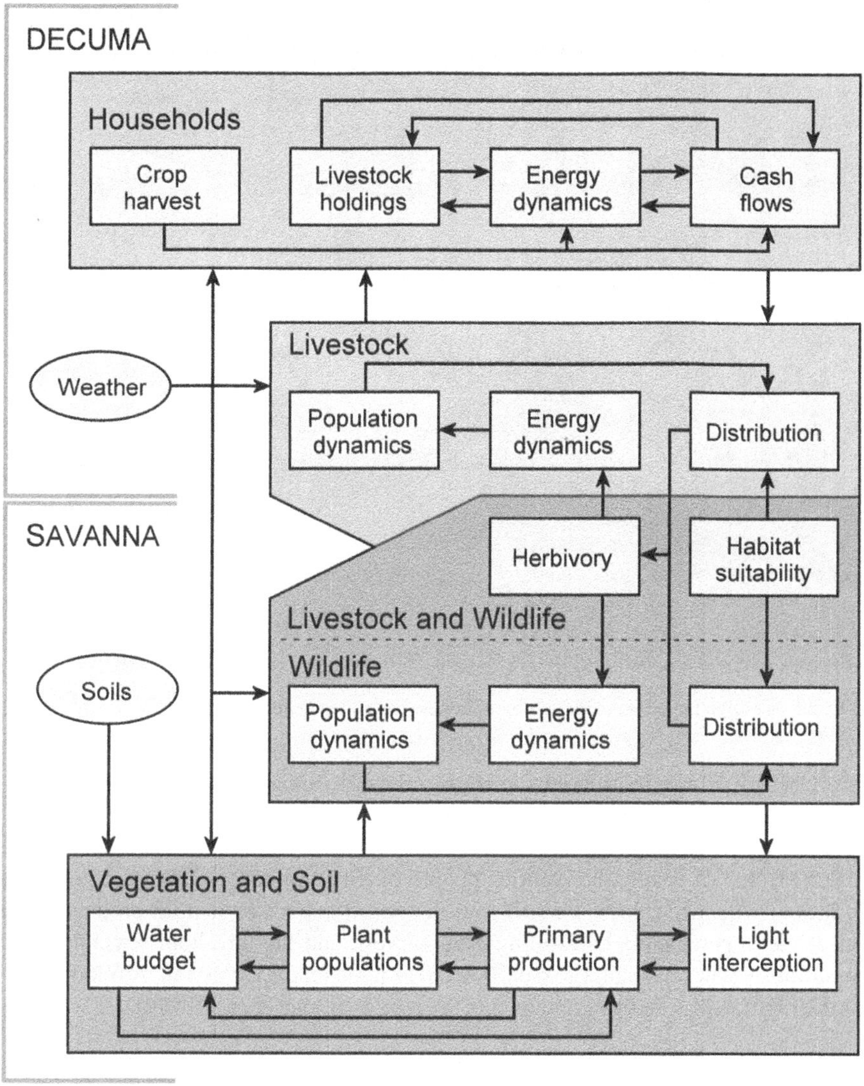

Fig. 7.4 Flow diagram showing the general connections between the Savanna ecosystem model and the household decision making model DECUMA, plus the main elements tracked in each modeling tool. (From Boone et al. 2011a)

simulated each week, and other household decisions are updated monthly (Fig. 7.4). Household family composition, the initial size and sex ratios of herds, and the areas of crops cultivated come from the surveys. The incomes of households, including wages, government leases or subsidies, livestock trading and remittances, and expenses, including food, other household supplies, and agricultural and veterinary inputs, are assigned from the household surveys as well. Another set of parameters

Fig. 7.5 Carolyn Lesorogol conducts an interview with a resident of Samburu

are used by all the households, such as values defining human adult equivalents (i.e., a means to summarize the numbers of people of different sex–age classes into a single value, with men equal to 1.0, women equal to 0.86, etc.; Boone et al. 2011a), caloric value of foods, the calories people of different sex–age class require, the prices at which livestock are sold and bought, the amount of milk produced and meat yield, parameters controlling crop harvest, and the threshold beyond which pastoralists will sell small (goat and sheep) or large (cattle) livestock. A full description of DECUMA and its components is given in Boone et al. (2011a).

7.5.5 The Coupled Savanna–DECUMA System

Intuitively, one may envision using a process-based ecosystem model and a household model separately, in a sequential way. The ecosystem model would provide spatial surfaces describing forage availability, and then those would be used as input into the agent-based model, for use by households in decision making. Such analyses can be instructive, but they do not allow human decision making to influence ecosystem function or the services provided. Instead, as cited, coupled systems modeling often includes the linking of an ecosystem model (or models, such as separate hydrology and vegetation models) with an agent-based model representing human activity. With the linked applications, ecosystem services can be estimated for a given time step (e.g., week, month), and then people may make decisions about their land use. Their land use may then alter the provision of services in succeeding time steps. These efforts can sometimes be challenging. The methods used to join Savanna and DECUMA provide an example of one approach, one where the main service of interest to people and at risk from overuse, and hence the main linkage, is the forage that an ecosystem provides.

Many options are available for the passing of information between coupled modeled. They may be tightly joined, where the models are essentially one large process and information flows transparently from one computer procedure to another. This approach can be computationally efficient and avoid errors. It can also marry the two models together more tightly than may be desirable; DECUMA, for example, has been joined to ecosystem models other than Savanna, and if it had been linked tightly to Savanna, untangling the connections may have been difficult.

Models may be more loosely joined, with information passed by use of computer dynamic linked libraries (DLLs) or compact and extendable data formats such as HDF (Hierarchical Data Format) and NetCDF (Network Common Data Form). After consideration, we decided that the most effective means of joining Savanna and DECUMA was through simple ASCII files, and that linkage has performed well. The spatial data were passed back and forth between models as rectangular blocks of values corresponding to the spatial landscapes used in modeling. This allowed Savanna and DECUMA to share information, but the loose connection means that DECUMA could be joined with any ecosystem model that could produce ASCII output, and the files themselves could easily be inspected to verify the information being passed between files.

The main pathway of information transfer used in Savanna–DECUMA is illustrated in Fig. 7.6. The image suggests paired cycles, corresponding to the passage of time in both Savanna and DECUMA. One may enter the cycles anywhere, but here we will consider the creation of habitat suitability indices as step 1 (Fig. 7.6). Savanna is aware of many attributes of each landscape cell. For each livestock species, Savanna produces a habitat suitability index (e.g., US Fish and Wildlife Service 1981) that indicates the suitability of a landscape patch for the species, spanning from 0 (unsuitable habitat) to 1 (most suitable). Parameters provide the relationship between some feature and the suitability of an area to a species [e.g., for slope, the parameter file may read, as a series of *x*–*y* pairs. 0, 1.0, 20, 0.3, meaning that slopes of 0 % are well suited to the species (1.0) but steep slopes (20 % or greater) are not (0.3), and the software performs linear interpolation for intermediate values.] Example elements that may help determine suitability include forage biomass and quality, distance to water, slope, elevation, the density of other grazers, snow depth, the density of standing dead biomass, daily temperature, and what are called force relationships. Force maps are used in the software to capture restrictions on animal movements that may not have an ecological basis, such as in delineating areas that are off-limits to grazing by policy. Each of these relationships may be used in the creation of a habitat suitability index or may be turned off in Savanna. Once the suitabilities of individual components have been determined on the basis of the attributes

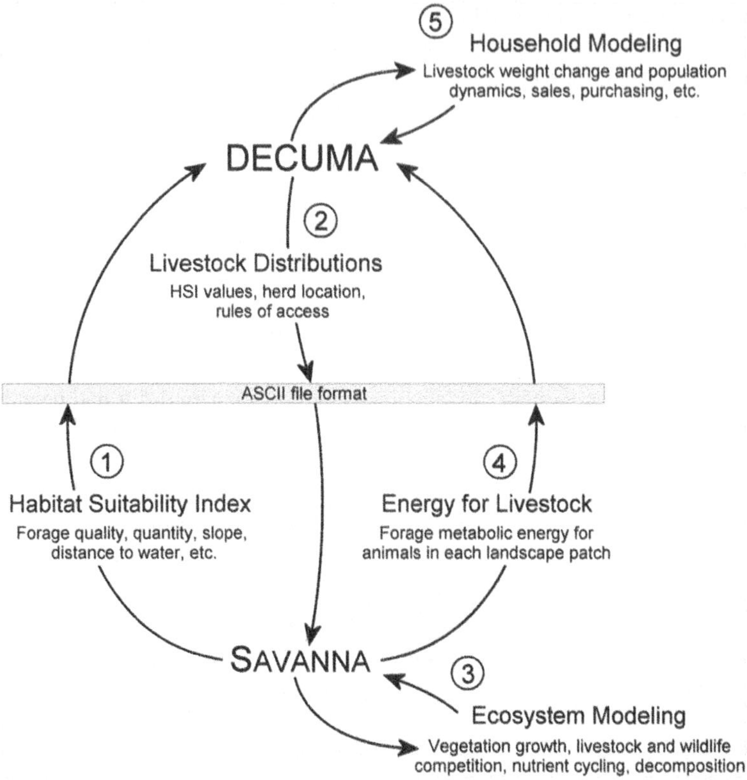

Fig. 7.6 Detailed information flows between the ecosystem model Savanna and the household decision making model DECUMA, with major processing steps shown associated with the stages of information transfer labeled. The *numbers* shown are cited in the text. *HSI* habitat suitability index. (From Boone et al. 2011a)

for a cell, they are multiplied together to yield a habitat suitability. Lastly, all cell suitabilities for the species, for that time step, are normalized to yield scores from 0.0 to 1.0. Here, suitabilities are produced for each species and placed in a single file, with each ASCII map preceded by a header that reports the species identifier and the month of simulation; the software double-checks to ensure that the two models remain in sync. After producing the habitat suitability indices, Savanna suspends programmatically, awaiting the creation of the next file produced by DECUMA, which in turn had been suspended while Savanna was doing these calculations.

After DECUMA has detected the availability of habitat suitability indices for the time step, it resumes execution and reads the suitabilities (step 2; Fig. 7.6). Household owners then decide, on the basis of the suitabilities plus their own movement rules, where to graze their animals. For example, homeowners have a grazing orbit assigned, the distance over which they may move their animals in a day before returning to the permanent or temporary household for the evening. DECUMA produces a file showing the number of animals of each species grazing in each landscape cell, and suspends.

After Savanna has detected that the file storing populations of distributed livestock is available, it resumes execution and reads the contents of that file. Programmatically, the material in Savanna that distributes, for example, livestock, was kept in place for simplicity and to avoid the introduction of potential errors. In practice, after Savanna has distributed wildlife and livestock, the distributions for livestock are overwritten with the information from DECUMA. Once that information is in place, Savanna has what is needed to do the bulk of its ecosystem modeling (step 3; Fig. 7.6), such as modeling vegetation growth, water dynamics and nutrient cycling, decomposition, and herbivory. That herbivory incorporates the diets of animals, which are known to Savanna through parameters, and the forage removed in each cell, by plant functional group, is calculated. From that, the model calculates the forage metabolic energy each species acquired in each cell. That information is written to a file, and Savanna suspends.

When DECUMA senses that the file containing forage metabolic energy per species is available, it resumes execution and digests that file (step 4; Fig. 7.6). DECUMA then has the information needed to model livestock dynamics. The forage metabolic energy acquired by animals of a given species in a herd is compared with the energy used by the animals (step 5; Fig. 7.6). The energy acquired may exceed the energy used by the animals, with the excess put to weight gain. If the acquired energy is less than that used, the animals will lose weight. The weights of animals are compared with an expected body weight given their sex and age, yielding a condition index. That condition index affects birth rates and, most importantly, death rates, which causes livestock herd dynamics. At this point in a simulation, DECUMA proceeds with its main computational activities, modeling livestock dynamics, household energy dynamics, livestock sales and purchasing, gifting, etc. (step 5; Fig. 7.6).

7.5.6 The Applications in Samburu

Two applications were created for the study region, one for Mbaringon and one for Siambu. In both, the vegetation was represented by eight functional groups: palatable grass, palatable forbs, unpalatable grass and forbs, palatable dwarf shrubs, unpalatable dwarf shrubs, palatable shrubs, unpalatable shrubs, and woods. Five types of animals were included, with cattle, goats, and sheep in each area, and zebras and antelope as wildlife species in Mbaringon. In Siambu, wildlife were rare and set to zero in the application. The landscape was divided into 200 m × 200 m cells.

There were insufficient data to include typical weather station data in the model application. Instead, we used the coarser-resolution Climatic Research Unit data (Mitchell and Jones 2005) to inform the model. Four cells were selected, and their centers were treated as weather stations. We adopted an approach included in Savanna where satellite images may be used to add spatial detail to interpolated precipitation. We used normalized difference vegetation indices from the MODIS sensor (Carroll et al. 2004) so that greener areas in the images could yield areas of higher interpolated precipitation. We were not interested in specific effects of observed climate histories, so we created five instances of randomized climate history for use in simulations.

Our household survey results were from 2010, but they were informed by more than a decade of annual visits to the area by Lesorogol and others thorough surveys of the same households in 2005 and 2000, as part of a longitudinal study. Spatial locations were gathered for some of the households surveyed (24 of 74 in Siambu and 33 of 83 in Mbaringon). The Siambu model included numbers of households equal to the parcels, 240, and the Mbaringon model included 370 households. The locations of households were digitized from high-resolution images available in Google Earth (Google, Mountain View, CA, USA). In Siambu, we ensured that each parcel contained a household, whether one was visible in the high-resolution imagery or not. Given that most household survey locations were unknown, we used ten representations of household initialization. In each instance, households with known locations were plotted as appropriate, and the remaining households were randomly selected from the locations and assigned attributes from randomly selected household interviews. In aggregate, households in Siambu were initialized with 1879 cattle, 1278 goats, and 3058 sheep. In Mbaringon, there were 3051 cattle, 3463 goats, and 8627 sheep, plus an estimated 100 zebras and 300 antelope of various types.

7.5.7 *Scenarios*

We defined eight scenarios for us to address with our modeling system; here we discuss two. The first scenario applies to both the high-elevation site, Siambu, and the more typical rangeland site, Mbaringon. The sites, especially Siambu, serve as a refuge for livestock and their herders during drought. Animals from other areas are brought into Siambu during regional droughts, to make use of pastures that remain productive. This sharing is typical given the reciprocal nature of the social system. But the high-elevation location of Siambu suggests that the residents are more often providing forage than relying on others for forage. Mbaringon too is a refuge for animals when areas in eastern Samburu are dry. We sought to quantify the degree to which immigration by herds from outside the areas affect the livelihoods of resident Samburu.

In modeling of the first scenario, we incorporated a drought in the seventh year of each simulation, which was defined as a moderate drought of one standard deviation below average annual rainfall. Additional animals were incorporated, in proportion to relative stocking of the different livestock species, into the Savanna model. Those animals used forage and were distributed in reasonable ways given the nature of the landscape and animal behavior, but were not owned by individual households. In simulations, we added 5 %, 10 %, 15 %, 20 %, and 25 % additional animals to the landscapes.

The second scenario dealt with wildlife populations, and included Mbaringon only (wildlife in Siambu are uncommon, and grazers may not be ecologically relevant). Over the long term, we estimated 100 zebras and 300 antelope occur in the Mbaringon study area, with large variations in their numbers. Residents have noted that there is less wildlife than in the past. The wildlife uses grazing resources also used by cattle [e.g., zebras (*Equus quagga* and *E. grevyi*) use grasses], sheep [e.g., Thomson's gazelle (*Eudorcas thomsonii*)], and goats [e.g., gerenuk (*Litocranius walleri*)]. Changes in one group of herbivores, wild or domestic, should cause commensurate changes in the other group. We quantified the trade-offs in the numbers of livestock and the numbers of wildlife that could be supported in Mbaringon. We may also expect increases in income for community members from the presence of wildlife if a community-based tourism opportunity develops.

In practice, the second scenario was simulated by our adding wildlife to the Savanna model, and the wildlife interacted with the livestock tracked in the DECUMA model. Zebra and antelope populations may be changed in Savanna, but they would quickly change themselves in response to livestock and forage availability, making comparisons difficult. Instead, wildlife populations were kept constant in Savanna, and livestock then responded to the presence of the wildlife. This may be thought of as residents having a goal of maintaining a given number of wildlife animals for tourism, and as us asking what effect that may have on livestock. Wildlife were taken from zero (no wildlife) to four times their estimated population (i.e., zebras at 0, 100, 200, 300, and 400 and antelope at 0, 300, 600, 900, and 1200). In some analyses, household incomes were not adjusted. In others, income was altered to represent increased income from tourism. The maximum payment pastoral landowners may expect for rental of their land for wildlife tourism is \$50 per hectare per year (Norton-Griffiths 2007), a value from the Maasai Mara National Reserve. An average value for renting land to a tour company is \$10 per hectare per year (Norton-Griffiths 2007). In the area, a herbivore that weighs 250 kg (i.e., termed a "large herbivore unit") required about 1.75 ha (Shaabani et al. 1992). With zebras weighing an average of 200 kg per animal and antelope weighing an average of 35 kg per animal, that equates to 320 large herbivore units for zebras and 168 large herbivore units for antelope at the maximum density of wildlife simulated. We estimate then that 854 ha [$(320+168)\times 1.75$] would be required to support wildlife at the maximum density used in the scenario. At \$10.2 per hectare, that equates to \$8711 per year for the entire area, and with 370 families and the exchange rate of 72 Kenyan shillings per dollar at the time, that equates to 141 Kenyan shillings paid to each family each month. At \$50 per hectare, the equivalent payment is 692 Kenyan shillings per month for each family. All unique settings for both scenarios were simulated 50 times, once for each combination of five weather histories generated and ten household distributions created. This allowed average responses to be calculated that were not dependent on the specific nature of a given weather history or household distribution.

7.5.8 Results

A moderate drought caused modest declines in the numbers of livestock the landscapes supported, but real-world TLUs per adult equivalent (Fig. 7.7, top) were less than 2.0 (1.88 TLUs per adult equivalent for Mbaringon, and 2.0 TLUs per adult

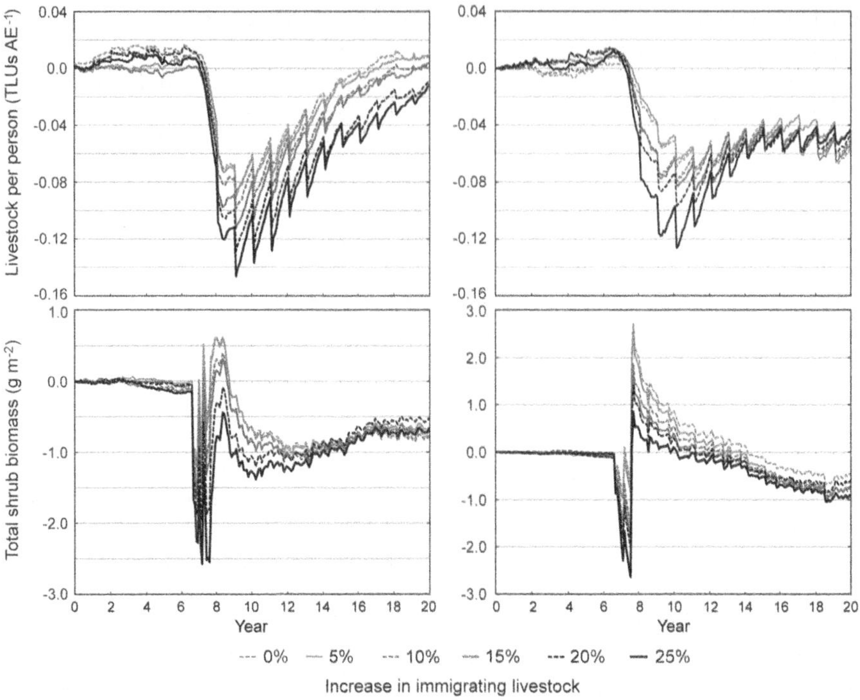

Fig. 7.7 The change in tropical livestock units per adult equivalent for Mbaringon (*top left*) and Siambu (*top right*) under a moderate drought in year 7 and different levels of livestock stocking from outside the study areas using forage during the drought. The change in total shrub biomass (g/m^2) in Mbaringon (*bottom left*) and Siambu (*bottom right*). Tropical livestock units allow livestock numbers of different species to be standardized, representing 250 kg of livestock biomass. Use of adult equivalents is a similar approach to standardize household family sizes, with a score of 1.0 assigned to men, a score of 0.86 assigned to women, etc. *AE* adult equivalent, *TLUs* tropical livestock units

equivalent for Siambu). As such, a decline of 0.1 TLUs per adult equivalent is a change that would have a significant effect on household well-being. In general, the effect of having up to 25 % additional livestock in the study areas during the drought caused up to a 0.5 decline in the number of TLUs per adult equivalent. Livestock populations approached their initial values by the end of the simulation in Mbaringon, but a threshold appeared to have been crossed in Siambu, and the populations only partially returned. For changes in total shrub biomass (Fig. 7.7, bottom), a modest decline was evident during the drought, but then shrubs rebounded and exceeded their baseline values, because of the release from some competition with herbaceous vegetation, as herbaceous vegetation was used most heavily during the drought. This rebound was most extreme in Siambu. The effects of the moderate drought in year °7 of the simulations on milk energy used by families are clear (Fig. 7.8, top), but differences caused by different levels of stocking of livestock from outside the study area were more modest. The amount of supplemental energy required by families increased

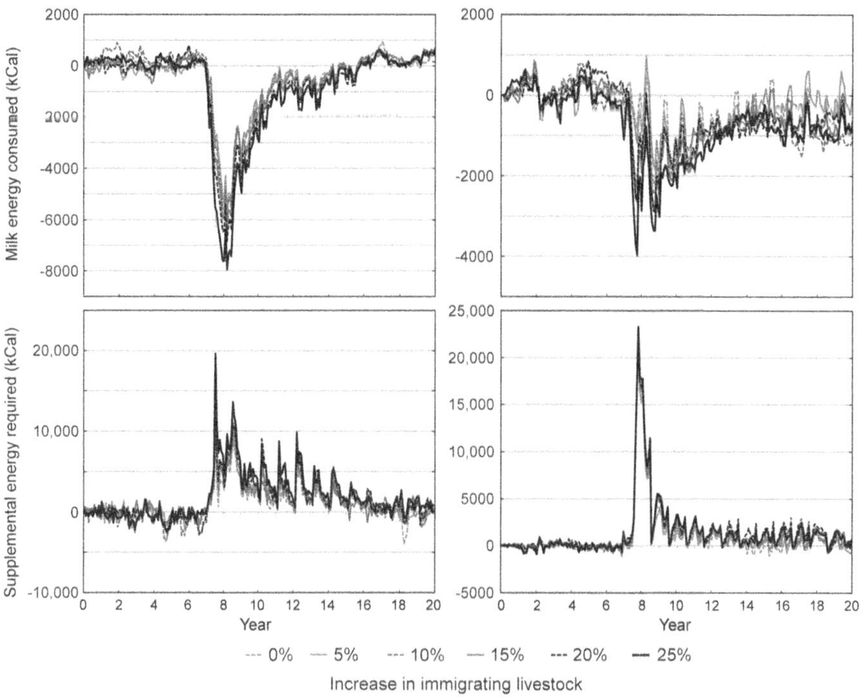

Fig. 7.8 The change in milk energy consumed in Mbaringon (*top left*) and Siambu (*top right*) under a moderate drought in year 7 and different levels of livestock stocking from outside the study areas using forage during the drought. Changes in supplemental energy required by families in Mbaringon (*bottom left*) and Siambu (*bottom right*)

during the drought, but returned to near baseline levels over some years (Fig. 7.8, bottom).

When wildlife numbers were changed in Mbaringon from their baseline numbers of 100 zebras and 300 antelope to other values (i.e., 0, 0; 200, 600; 300, 900; and 400, 1200), the changes in livestock numbers were in the direction expected (Fig. 7.9, top left). However, differences in payments to families led to very small differences in outcomes, such as small changes to livestock numbers (Fig. 7.9, top left). This may be foreseen, in that the payments of the equivalent of a few US dollars each month may seem insufficient to alter the well-being or behaviors of households. Relatively few hectares are required to support the numbers of wildlife posed in the scenario, and so payments that were divided between families were small. The amount of herbaceous biomass was somewhat higher when wildlife numbers were higher than the current state (Fig. 7.9, middle left), as livestock populations declined and wildlife were unable to show compensatory responses. Shrub biomass declined (Fig. 7.9, bottom left) likely associated with the mixed diets of the species included in the antelope group. Regarding household well-being, income (Fig. 7.9, top right) declined by up to about 6000 Kenyan shillings

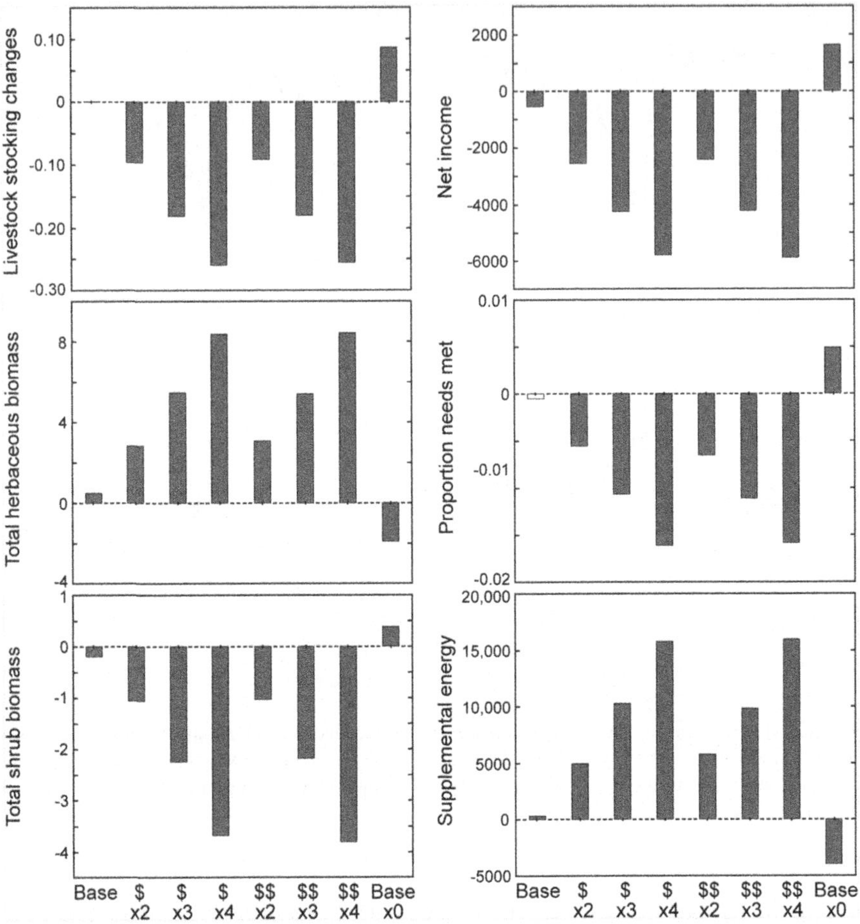

Fig. 7.9 Changes in a suite of Savanna–DECUMA-modeled responses averaged over the previous year simulated when wildlife numbers are altered (and kept constant in a given simulation) and payments to families are equivalent to $0 per hectare (i.e., *base*), $10 per hectare (*one dollar sign*), or $50 per hectare (*two dollar signs*). Changes in livestock stocking changes (tropical livestock units per adult equivalent; *top left*), total herbaceous biomass (g/m^2; *middle left*), total shrub biomass (g/m^2; *bottom left*), net income (Kenyan shillings; *top right*), proportion of caloric needs met by households' own production (*middle right*), and supplemental energy (Cal; *bottom right*)

when wildlife was most abundant. Loss of income was reduced by payments to families, but those payments were too small to offset losses associated with the families having fewer livestock. Families received less milk and meat from their livestock, such that they produced somewhat less of the food they needed (Fig. 7.9, middle right), and the amount of supplemental food required by families increased when wildlife reduced the numbers of livestock households owned (Fig. 7.9, bottom right). We have shared results from these scenarios, and others, with the communities in Siambu and Mbaringon (Figs. 7.10 and 7.11).

Fig. 7.10 The authors and others describe some of the simulation results to residents of Mbaringon, Samburu

7.6 Strengths and Challenges of Coupled Systems Modeling

A main strength of coupled systems modeling is implied by the term, the ability to simulate changes in linked ecological and social systems rather than having to infer how a change in one may influence the other. For example, in the pastoral systems in which we have worked, we have simulated ecosystem changes and how those led to changes in livestock numbers (e.g., Boone et al. 2002). There is a history of inferring pastoral well-being on the basis of the numbers of livestock per person (e.g., six to eight TLUs per adult equivalent are needed to lead a wholly pastoral lifestyle; Galvin et al. 2002; see the legend of Fig. 7.7 for definitions), and so conclusions about well-being could be made. But with the coupled systems modeling approaches we now use, we can speak of the number of livestock per person, their expected income, sources of energy acquired, cash flows, livestock trades, etc. That level of information richness in knowledge about the linkages between natural and social systems is valued.

The second main benefit we cite is related and again follows from the term "coupled systems." A coupled systems approach allows the elements in a system to interact and influence each other. Again, in our case, coupled models that allow the decisions of pastoral people to influence plant production through the distribution of their animals and to have that influence the people in turn is more satisfying than

Fig. 7.11 Carolyn Lesorogol shows information on a computer to Samburu residents

modeling the systems independently and inferring changes, or having one element (the ecosystem responses) be a static input into the other element (the social model). Also, a diverse set of inputs into the linked models increases the flexibility of the tool to address a variety of scenarios. Scenario analyses are most often conducted by the modification of inputs, and a diversity of inputs broadens the types of scenarios that may be addressed.

The challenges we have faced in coupled systems modeling are related to the complexity of the endeavor and of the sustainability questions being addressed. The teams involved in coupled systems modeling typically include several specialties, and as cited, communication among the team members can require a learning period. Some team members may be new to the idea of formalizing understanding of the systems they study to the degree that relationships may be programmed. Practical and technical difficulties often present themselves when one is joining two or more simulation models. Greater (and growing; see below) complexity of simulation models can make validation and assessment difficult. We cited the diversity of responses we now have for households, but assessing these can be challenging. Indeed, finding appropriate parameter settings to simply keep the many responses in a model stable through time can be challenging but rewarding (Boone et al. 2011a). Increasing the kinds of outputs from simulation models can be straightforward, but knowing whether those outputs are being simulated well remains nontrivial.

The next challenges we cite are again related. In projects that adopt a coupled systems viewpoint, team members often view the modeling efforts as the central tool for integrating and addressing core questions. It is rewarding for modelers to be central to project outcomes, of course, but having all outcomes flow from modeling is risky for investigators and stressful for modelers. To guard against that pathway, we encourage different subteams to produce products independently of the integrative modeling effort—to contribute to modeling efforts but to consider discipline-specific products as they have done in more traditional, noncoupled projects. That said, modeling is a good means of integrating findings across a project. As such, the work is typically dependent on a series of products from the project. That makes discovery using the tools dependent on the progress of others on the team, which can be challenging. Also, for this reason, products from modeling may come late in the life of a project, and funds to support outreach to stakeholders (e.g., Figs.7.10 and 7.11) may no longer be available.

The last challenges we cite follow from the nature of sustainability research itself—the problems are difficult, pressing, and imbedded in social and political settings. In graduate work, students may be trained in the most advanced means to reduce variability around some estimate, investing months to yield the best estimate for some attribute of an animal (e.g., an animals bite rate when foraging). But the areas in which we work in Africa can have a 15-year human population doubling time and hard-pressing ecological problems. Results from coupled systems modeling must be generated and shared in this environment, and the prospects of painstaking basic research may be viewed as a luxury, or even as misplaced.

Stakeholders often seek from researchers solutions to sweeping sustainability issues. Researchers should control expectations and make clear that modeling generally addresses the directions and magnitudes of change, with all conditions not specific to the scenario held constant, as cited. Researchers must also accept that rarely are solutions our goal. At some level, the solutions to sustainability questions are often clear, such as having less pollution, fewer people, or fewer grazing livestock. But that ignores the real challenge of sustainable research—the balancing of many diverse interest groups with different goals and levels of power. Rather than speaking of providing solutions, we prefer to speak of informing decision making. The stroke of a politician's pen may put in place policies that are counter to the pathways suggested by coupled systems modeling; researchers must accept that reality.

Finally, sustainability research focuses on pressing real-world problems, making the need to share results with land managers and other stakeholders even more relevant than in more typical research. Those guiding integrative modeling projects are often technically minded, and may struggle with communicating the relevance and results of a project to others. Those skills can be learned and communication improved, but for large projects, we encourage the inclusion of a communication specialist, someone who can convey the results of the integrative modeling in the clearest and most compelling way.

7.7 Future Coupled Systems Modeling

Perhaps the aspect of coupled human–natural systems simulation that has the most room for rapid growth is the modeling of the decision making and behaviors of humans. Researchers have had some success in establishing rules that emulate human behavior. But that work remains challenging. First, the history of simulating human systems is briefer than that of simulating ecological systems (Boone and Galvin 2014), and methods continue to develop. The underpinnings of behavior are less well known, and to strengthen that component can mean adding yet another expert to the team doing coupled human–natural system research, such as a cognitive anthropologist. Psychological research to understand the perceptions that inform cognition and determine behaviors (e.g., D'Andrade 1995) is improving modeling. The last challenge we cite is, put formally, the low spatial autocorrelation in human systems (Boone and Galvin 2014). Ecological modeling benefits from both the demonstrated similarity of the processes influencing a given tree species in a forest, say, and the (confounded) willingness of ecologists to treat those elements as similar in simulations—what we know about a tree can tell us a great deal about a tree 1 km away. Intuition and observation leads us to focus on the differences in humans. Both the differences we see in people within a community and the hesitation anthropologists have in treating them the same in models make simulation of human systems challenging—what we know about the behavior of a person may not tell us a great deal about the behavior of another person living just next door. This constraint is related to the cultural heterogeneity of an area, of course, and is less of a concern where people tend to live similarly. There is potential to gain a more complete understanding of individual and household behavior through ethnographic research done in conjunction with modeling, as we have done in this project. For example, discussions with Samburu research participants illuminate the choices they make about land use and the reasons for those choices (Lesorogol 2014; Figs. 7.5, 7.10, and 7.11). Such understanding is useful in defining the subject matter for simulation. A further challenge, however, is to translate local models or patterns of decision making into rules that can be inserted into an agent-based model such as DECUMA.

Simulations have become a normal approach to address questions of sustainability and resilience, both large and small. This is certain to expand as the availability of data and computational power expands dramatically (Hampton et al. 2013). The spatial and temporal grain of simulations is decreasing as well. For example, climate change surfaces projected with use of global circulation models or ensemble results have been downscaled to local areas and used in simulations (e.g., Jones and Thornton 2013). Enhanced and high-performance computing resources can make simulations at those scales efficient, and software development continues to make analyses more straightforward and available to wider audiences (e.g., Boone et al. 2011b, 2013). Primary processes underlying ecosystem functions have been explored for decades, and subtler relationships are being explored now. The shifting nature of what is considered timely to study means new processes are always being explored, such as our current focus on the effects of climate change. Also, our rapidly expanding ability to include many thousands of agents in a single simulation has the potential to identify novel emergent patterns across multiple scales. A simulation that depends on the bottom-up emergence of patterns through agent interactions may reach different end points when hundreds versus many thousands of agents are simulated.

Acknowledgments We thank the Samburu people who participated in our research, our research assistants, and the Kenyan government for research clearance. Support for preparation of this chapter was provided by US National Science Foundation grant BCS-0822752.

References

Axelrod R (2006) Agent-based modeling as a bridge between disciplines. In: Tesfatsion L, Judd KL (eds) Agent-based computational economics, vol 2, Handbook of computational economics. Elsevier, Amsterdam, pp 1565–1584
Baker PT, Little MA (1976) Man in the Andes: multidisciplinary study of high-altitude Quechua. Dowden, Hutchinson & Ross, Stroudsburg
Bateson G (1972) Steps to an ecology of mind. Jason Aronson, Northvale
Benson MH, Garmestani AS (2011) Can we manage for resilience? The integration of resilience thinking into natural resource management in the United States. Ecol Manage 48:392–399
Boone RB (2000) Integrated management and assessment system training manual. University of California, Davis
Boone RB, Coughenour MB, Galvin KA, Ellis JE (2002) Addressing management questions for Ngorongoro Conservation Area using the Savanna modeling system. Afr J Ecol 40:138–150
Boone RB, Galvin KA, BurnSilver SB, Thornton PK, Ojima DS, Jawson JR (2011a) Using coupled simulation models to link pastoral decision making and ecosystem services. Ecol Soc 16(2):6, http://www.ecologyandsociety.org/vol16/iss2/art6/
Boone RB, Conant RT, Hilinski TE (2011b) G-Range: development and use of a beta global rangeland model. Final report to the International Livestock Research Institute, Nairobi, Kenya
Boone RB, Conant RT, Sircely J (2013) Adjustment and sensitivity analyses of a beta global rangeland model. Final report to the International Livestock Research Institute, Nairobi, Kenya
Boone RB, Galvin KA (2014) Simulation as an approach to social-ecological integration, with an emphasis on agent-based modeling. In: Manfredo M, Vaske JJ, Rechkemmer A, Duke EA (eds) Understanding society and natural resources: forging new strands of integration across the social sciences. Springer, Dordrecht, pp 179–202
Boyd R, Richerson PJ (2005) The origin and evolution of culture. Oxford University Press, Oxford
Brenner T (ed) (1999) Computational techniques for modeling learning in economics. Kluwer, Norwell
BurnSilver S, Boone RB, Galvin KA (2003) Linking pastoralists to a heterogeneous landscape: the case of four Maasai group ranches in Kajiado District, Kenya. In: Fox J, Mishra V, Rindfuss R, Walsh S (eds) Linking household and remotely sensed data: methodological and practical problems. Kluwer, Boston, pp 173–199
Carroll ML, DiMiceli CM, Sohlberg RA, Townshend JRG (2004) 250m MODIS normalized difference vegetation index, collection 4. University of Maryland, College Park
Case M (2006) Climate change impacts on East Africa: a review of the scientific literature. World Wildlife Fund for Nature, Gland
Coughenour MB, Ellis JE, Swift DM, Coppock DL, Galvin K, McCabe JT, Hart TC (1985) Energy extraction and use in a nomadic pastoral population. Science 230:619–625

Crépin A-S (2007) Using fast and slow processes to manage resources with thresholds. Environ Resour Econ 36:191–213

D'Andrade RG (1995) The development of cognitive anthropology. Cambridge University Press, Cambridge

Ellis JE, Swift DM (1988) Stability of African pastoral ecosystems: alternative paradigms and implications for development. J Range Manage 41:450–459

Ellis JE, Galvin KA (1994) Climate patterns and land-use practices in the dry zones of Africa. Bioscience 44:340–349

Ellis J, Coughenour MB (1998) The SAVANNA integrated modeling system. In: Squires V, Sidahmed A (eds) Drylands: sustainable use of rangelands in the twenty-first century. International Fund for Agricultural Development, Rome, pp 97–106

Epstein JM (2008) Why model? J Artif Soc Soc Simulat 11(4):12, http://jasss.soc.surrey.ac.uk/11/4/12.html

Eyakuze A, Salim AI (2013) The state of East Africa 2013: one people, one destiny?: the future of inequality in East Africa. Society for International Development, Regional Office for Eastern Africa, Nairobi

Galvin KA, Ellis J, Boone RB, Magennis AL, Smith NM, Lynn SJ, Thornton PK (2002) Compatibility of pastoralism and conservation? A test case using integrated assessment in the Ngorongoro Conservation Area, Tanzania. In: Chatty D, Colester M (eds) Displacement, forced settlement, and conservation. Berghahn, Oxford, pp 36–60

Galvin KA, Hobbs NT, Reid RS, Behnke RH (eds) (2008) Fragmentation of pastoral rangelands. Consequences for human and natural systems. Springer, Dordrecht

Gilbert N, Terna P (2000) How to build and use agent-based models in social science. Mind Soc 1:57–72

Griffin WA (2007) Agent-based modeling for the theoretical biologist. Biol Theory 1:404–409

Grimm V (1999) Ten years of individual-based modelling in ecology: what have we learned and what could we learn in the future? Ecol Model 115:129–148

Grimm V, Revilla E, Berger U, Jeltsch F, Mooij WM, Railsback SF, Thulke H-H, Weiner J, Wiegand T, DeAngelis DL (2005) Pattern-oriented modeling of agent-based complex systems: lessons from ecology. Science 310:987–991

Gunderson LH, Holling CS (eds) (2002) Panarchy: understanding transformations in human and natural systems. Island, Washington

Hampton SE, Strasser CA, Tewksbury JJ, Gram WK, Budden AE, Batcheller AL, Duke CS, Porter JH (2013) Big data and the future of ecology. Front Ecol Environ 11:156–162

Hobbs NT, Galvin KA, Stokes CJ, Lackett JM, Ash AJ, Boone RB, Reid RS, Thornton PK (2008) Fragmentation of rangelands: implications for humans, animals, and landscapes. Glob Environ Chang 18:776–785

Hoffman T, Vogel C (2008) Climate change impacts on African rangelands. Rangelands 30:12–17

Huston M, DeAngelis D, Post W (1988) New computer models unify ecological theory. Bioscience 38:682–691

Iannotti L, Lesorogol C (2014) Animal milk sustains micronutrient nutrition and child anthropometry among pastoralists in Samburu, Kenya. Am J Phys Anthropol 155:66–76

Intergovernmental Panel on Climate Change (2000) Emissions scenarios. Summary for policymakers. A special report of Working Group III of the Intergovernmental Panel on Climate Change. Intergovernmental Panel on Climate Change, Geneva

Jones PG, Thornton PK (2013) Generating downscaled weather data from a suite of climate models for agricultural modelling applications. Agric Syst 114:1–5

Kareiva PM, Andersen M (1988) Spatial aspects of species interactions: the wedding of models and experiments. In: Hastings A (ed) Community ecology. Springer, New York, pp 38–54

Kates RW, Clark WC, Corell R, Hall JM, Jaeger CC, Lowe I, McCarthy JJ, Schellnhuber HJ, Bolin B, Dickson NM, Faucheux S, Gallopin GC, Grübler A, Huntley B, Jäger J, Jodha NS, Kasperson RE, Mabogunje A, Matson P, Mooney H, Moore B III, O'Riordan T, Svedin U (2001) Environment and development: sustainability science. Science 292:641–642

Kohler TA, Gumerman GJ (eds) (1999) Dynamics in human and primate societies: agent based modeling of social and spatial processes. Santa Fe Institute studies in the sciences of complexity. Oxford University Press, Oxford

Lesorogol C (2008) Land privatization and pastoralist well-being in Kenya. Dev Change 39:309–31

Carolyn K. Lesorogol (2014), Redefining the Meaning of Land: Property Rights and Land Use in a Privatized Commons in Kenya, in Donald C. Wood (ed.) Production, Consumption, Business and the Economy: Structural Ideals and Moral Realities (Research in Economic Anthropology, Volume 34) Emerald Group Publishing Limited, pp. 211–233

McCabe JT (2004) Cattle bring us to our enemies: Turkana ecology, politics, and raiding in a disequilibrium system. University of Michigan Press, Ann Arbor

Mitchell TD, Jones PD (2005) An improved method of constructing a database of monthly climate observations and associated high-resolution grids. Int J Climatol 25:693–712

National Science Foundation (2014) Dynamics of coupled natural and human systems (CNH) program solicitation, NSF 14-601. US National Science Foundation, Arlington

Norton-Griffiths M (2007) How many wildebeest do you need? World Econ 8:41–64

Ogutu JO, Piepho H-P, Reid RS, Rainy ME, Kruska RL, Worden JS, Nyabenge M, Hobbs NT (2010) Large herbivore responses to water and settlements in savannas. Ecol Monogr 80:241–266

Orlove BS (1980) Ecological anthropology. Annu Rev Anthropol 9:235–273

Peck SL (2004) Simulation as experiment: a philosophical reassessment for biological modeling. Trends Ecol Evol 19:530–534

Pricope NG, Husak G, Lopez-Carr D, Funk C, Michaelsen J (2013) The climate-population nexus in the East African Horn: emerging degradation trends in rangeland and pastoral livelihood zones. Glob Environ Chang 23:1525–1541

Priestley CHB, Taylor RJ (1972) On the assessment of surface heat flux and evaporation using large-scale parameters. Mon Weather Rev 100:81–92

Railsback SF, Grimm V (2011) Agent-based and individual-based modeling: a practical introduction. Princeton University Press, Princeton

Reid R, Nkedianye D, Said M, Kaelo D, Neselle M, Makui O, Onetu L, Kiruswa S, Ole Kamuaro N, Kristjanson P, BurnSilver S, Goldman M, Boone R, Galvin KA, Dickson N, Clark W (2009) Evolving models to support communities and policy makers with science: balancing pastoralism and wildlife conservation in East Africa. Proc Natl Acad Sci U S A. doi:10.1073/pnas.0900313106

Reid RS (2012) Savannas of our birth: people, wildlife, and change in East Africa. University of California Press, Oakland

Shaabani SB, Walsh M, Herlocker DJ, Walther D (1992) Range management handbook of Kenya, vol II, 2: Samburu District. Republic of Kenya Ministry of Livestock Development, Nairobi

Steward JH (1955) Theory of culture change: the methodology of multilinear evolution. University of Illinois Press, Urbana

Thornton PK, Galvin KA, Boone RB (2003) An agro-pastoral household model for the rangelands of East Africa. Agric Syst 76:601–622

Thornton PK, BurnSilver SB, Boone RB, Galvin KA (2006) Modelling the impacts of group ranch subdivision on agro-pastoral households in Kajiado, Kenya. Agric Syst 87:331–356

US Fish and Wildlife Service (1981) Standards for the development of habitat suitability index models. 103ESM. US Department of the Interior, Fish and Wildlife Service, Division of Ecological Services

van Notten PWF, Rotmans J, van Asselt MBA, Rothman DS (2003) An updated scenario typology. Futures 35:423–443

Walker BH, Carpenter SR, Rockstrom J, Crépin A-S, Peterson GD (2012) Drivers, "slow" variables, "fast" variables, shocks, and resilience. Ecol Soc 17(3):30. doi:10.5751/ES-05063-170330

Walker B, Carpenter S, Anderies J, Abel N, Cumming G, Janssen M, Lebel L, Norberg J, Peterson GD, Pritchard R (2002) Resilience management in social-ecological systems: a working

hypothesis for a participatory approach. Conserv Ecol 6(1):14, http://www.consecol.org/vol6/art14
Walker B, Holling CS, Carpenter SR, Kinzig A (2004) Resilience, adaptability and transformability in social-ecological systems. Ecol Soc 9(2):5, http://www.ecologyandsociety.org/vol9/
Wilensky U, Rand W (2007) Making models match: replicating an agent-based model. J Artif Soc Soc Simulat 10(4):2, http://jasss.soc.surrey.ac.uk/10/4/2.htmliss2/art5
Williams AP, Funk C (2011) A westward extension of the warm pool leads to a westward extension of the Walker circulation, drying eastern Africa. Clim Dyn 37:2417–2435

Chapter 8
Conclusion

Karim-Aly S. Kassam

This work has been more than a survey of pastoralism across the globe, covering all continents except Antarctica (see Fig. 8.1). After a global overview in Chap. 1, subsequent chapters examined specific case studies from North and sub-Saharan Africa, South America, and Central and Inner Asia. The case studies reveal key insights that are both unique to the context and shared across pastoral societies worldwide. In this sense, the chapters collectively provide a window into the unique and diverse contexts in which pastoralism manifests itself while also hinting at a broader understanding of the state of pastoralism internationally.

Although the cases of pastoralism have been presented as coupled human and natural systems, it is clear that this coupling is merely an analytical construct. The fact is that that human systems are embedded within ecological systems. Human systems cannot exist outside of their ecological context. The survey of pastoralism in Chap. 1 makes that imminently clear. Similarly, pastoralism is a sociocultural system embedded within an ecological context. Pastoralism has multiple centers of origin. It predates any of the political and economic ideologies that inform our modern global economic system, from trade liberalization and capitalism to communism. Arguably, over the course of human history, in very practical terms, pastoralism has withstood the test of time and continues to represent an economically valid and ecologically sound livelihood strategy used by humans. Because of human and animal mobility, pastoralism is a highly adaptive approach to environmental change. Pastoralists negotiate a complex set of political, economic, cultural, and environmental factors as they seek to sustain a livelihood. As all the chapters indicate, there is diversity in pastoral approaches, including a variety of animals, depending on the ecological context and economic circumstances. These unique conditions have generated complex and dynamic sociocultural systems.

K.-A.S. Kassam (✉)
Department of Natural Resources, Cornell University, Ithaca, NY, USA
e-mail: ksk28@cornell.edu

S. Dong et al. (eds.), *Building Resilience of Human-Natural Systems of Pastoralism in the Developing World: Interdisciplinary Perspectives*,
DOI 10.1007/978-3-319-30732-9_8

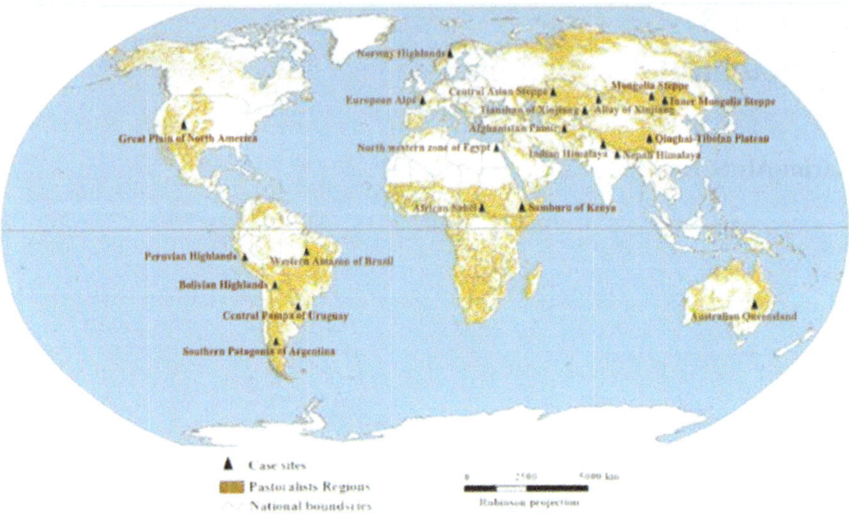

Fig. 8.1 Pastoral regions studied in this book

To address pastoral vulnerability and resilience, Chap. 1 asserts, and the remaining chapters illustrate, that pastoralism must be viewed as a sociocultural and ecological system. Chapter 2 discusses pastoralism on a global scale. In the twenty-first century, pastoralism as a way of life is facing multiple and complex pressures, such as population growth, economic policies geared toward rapid urbanization, resulting land use changes, ill-informed government policy, and anthropogenic climate change. Yet pastoralism is not only key to subsistence for the diverse ethnic communities in a variety of ecological zones, it is necessary to the food systems of lowland communities. The authors argue that social–ecological learning, technical and management innovations, ecological system renewal, and reorganization of institutions are pathways to mitigate the negative causes and effects of pastoralism's vulnerability.

Chapter 3 examines three cases from Nepal, India, and China to illustrate that Himalayan pastoralism is not bound to a particular economic system. Furthermore, these cases show that multiple land tenure types exist, where some pastoral communities have ownership of rangelands for mobile livestock grazing, whereas others have to graze their livestock with a formally or informally contracted migration routine on public or private lands for which they do not have political or legal ownership. In addition, pastoral groups are also very diverse in their sociopolitical structure, ranging from state-controlled groups and community-based collectives to individualized households. The pastoral system in the Himalayas is marked by ecological and cultural diversity. The authors maintain that the interactions and feedback between human and natural components of pastoralism in coping with the stresses and the integration of various tools and strategies from the ecological and social sciences as well as other disciplines can promote sustainable pastoral development in the Himalayas.

Chapter 4 maintains that variation and difference are the hallmarks of pastoralism. Consequently, pastoralism is not merely a livelihood strategy but is a way of life that is fundamentally based on adaptation to changing seasonal and climatic conditions within wide-ranging ecological contexts. Mobility through pastoral activities and the subsequent food security arising from those undertakings are not only a necessity but a recognized behavioral norm with sociocultural significance. Using three Inner cases, two from Xinjiang and Inner Mongolia, China, and one from Badakhshan, Asia, Afghanistan, the authors show pastoralism is not only an ecological profession strategic to securing human survival, but in turn generates a mutually reinforcing sociocultural identity that draws primarily from connectivity with the ecosystems in which humans seasonally dwell. The key argument put forward in Chap. 4 is that pastoralism itself is a manifestation of livelihood diversity and continues to be relevant in the third millennium in Inner Asia despite governmental pressures to sedentarize and homogenize the livelihood structures of different ethnic communities. Livelihood variation once sedentarization has occurred is really a euphemism for reducing genuine diversity in ecological professions. Building on historical pastoral cultural institutions under these conditions continues to offer hope for effective economic stability of livelihoods and food security while not compromising on ecological conservation of rangelands.

On the basis of three case studies from different ecological regions (i.e., southern Patagonia, Argentina, the central Pampas, Uruguay, and western Amazonia, Brazil), Chap. 5 argues that colonialism accounts for the arrival of an intensive form of pastoralism in South America. In essence, the European settlers and the ruminants that accompanied them were invasive species that had sociocultural and ecological impacts on the habitat they colonized. Land tenure and land use relationships were transplanted from Europe and were driven by the concept of private ownership. The absence of endogenous sociocultural and ecological norms of pastoralism makes the character of herding by European settlers in South America a different ecological and economic type compared with that in North Africa, sub-Saharan Africa, and Central Asia. Within a century, European settlers were able to transform ecological regions. The aims of European settlers with respect to agropastoralism and the resultant products were beyond meeting basic local and regional needs and were intended for distant markets and therefore agropastoralism was intensive in character from the beginning. Nonetheless, the authors are guardedly optimistic about sustainable rangeland management, describing recent efforts in their case study sites which require the support of long-term oriented policies as well as strong enforcement monitoring by governments.

Chapter 6 on the North African Bedouin in the northwestern coastal zone illustrates how tribal sociocultural institutions remain central for effective implementation of economic and ecological policy related to pastoralism. The northwestern coastal zone in Egypt is located in the southeastern Mediterranean, between the Nile delta and the Libyan border. In this area, land tenure and land use are key factors affecting pastoralism with respect to tourism development and agricultural expansion in the Mediterranean. The Bedouin tribe, historically and presently, is the operational network that affects the sociocultural as well as the ecological context of pastoralists by directly influencing economic and political decisions in times of hardship as well as relative stability. Therefore, the key implication of this chapter is that the tribe is the central institution from which any adaptation strategies for policy formulation must originate in order to be meaningful ecologically and economically for pastoralists in this region of North Africa.

Chapter 7 seeks to model resilience or nonresilience of human–ecological systems in East Africa with respect to changes such as rapid population growth, poverty, degraded rangelands, declining wildlife, land fragmentation, reduced mobility, and more frequent drought associated with climate change. By simulating short-term and long-term perturbations in combining ecosystem and human decision making models, the authors seek to reveal vital relationships and insights within a complex system. As a result of analyzing the scenarios emerging from various perturbations noted above, for direction and magnitude of change, researchers can inform policymaking.

All the chapters in this publication, in some form or other, address policy toward pastoralism. This suggests that pastoralism continues to be a viable ecological profession and economic livelihood strategy for humans globally in the third millennium. These chapters indicate that pastoralism is dynamic and adaptive. Rather than viewing mobility as both an ecological and an economic asset, some governments and policymakers frame pastoralism as an anachronistic ecological profession warranting forced sedentarization. Governments, to strengthen their control over pastoral societies in the name of "progress" and "ecological conservation" but more insidiously to generate a low-wage labor force, insist on sedentarizing pastoralists. Nonetheless, evidence from these chapters indicates such externally imposed policies, in fact, provide limited ecological and economic benefits. It behooves applied researchers and government policymakers to acknowledge pastoralism as a genuinely time-tested human endeavor that has demonstrated ecological and economic resilience. We should therefore investigate what strategies are required to truly address rangeland degradation, food insecurity, growing populations, and climate change in the twenty-first century.

Index

S. Dong et al. (eds.), *Building Resilience of Human-Natural Systems of Pastoralism in the Developing World: Interdisciplinary Perspectives*, DOI 10.1007/978-3-319-30732-9

Q

About the Editors

Dong Shikui is a full Professor at School of Environment, Beijing Normal University and has been appointed as an Adjunct Professor at Natural Resource Department of Cornell University (USA), as well as a fellow of India China Institute, the New School (USA). He also currently serves as an Invited Scientist for Environment Impact Assessment Center for China's Ministry of Environmental Protection and Vice Secretary for Chinese Grassland Society. He got his MSc and PhD in Grassland Ecology from Gansu Agricultural University in 1998 and 2001, respectively. He has conducted over 10 projects from National Science Foundation of China (NSFC), Ministry of Science and Technology (MOST) of China, Ministry of Agriculture (MOA) of China (MAC) and international funding resources such as Asian Scholarship Foundation (ASF) to examine the problems and solutions in sustainable environments in rural Western china and fragile areas of Hindu-Kush Himalayan regions since 1998. He has authored over 220 peer-reviewed papers, over 30 symposium proceedings and abstracts, and 8 books in recent years on Ecological Restoration, Natural Resources Management, as well as Strategies for Sustainable Development.

S. Dong et al. (eds.), *Building Resilience of Coupled Human-Natural Systems of Pastoralism in the Developing World*, DOI 10.1007/978-3-319-30732-9

Karim-Aly S. Kassam is International Professor of Environmental and Indigenous Studies in the Department of Natural Resources and the American Indian and Indigenous Studies Program at the College of Agriculture and Life Sciences, Cornell University. In 2014, he was elected to the Board of Directors of the International Society of Ethnobiology. Since 2013, he serves as Associate Editor of the Journal *Action Research*. In 2013, Dr. Kassam received the College of Agriculture and Life Sciences Diversity Award for creating and fostering diversity in research and teaching. In 2003, he was the first Canadian to receive the Organization of American States—Fulbright Ecology Fellowship. Dr. Kassam's objective is to seamlessly merge teaching with applied research in the service of communities. His research focuses on the complex connectivity of human and environmental relations, addressing indigenous ways of knowing, food sovereignty, sustainable livelihoods, and climate change in the Alaskan, Canadian, and Russian Arctic and Sub-Arctic; the Pamir Mountains of Central Asia; and the forests in the south of India. Currently, Dr. Kassam is Principal Investigator on three participatory research projects which include: the St. Regis Mohawk Tribe and the Standing Rock Sioux Tribe in North America as well as several indigenous communities in the Pamir Mountains of Central Asia (Afghanistan, China, Kyrgyzstan, and Tajikistan).

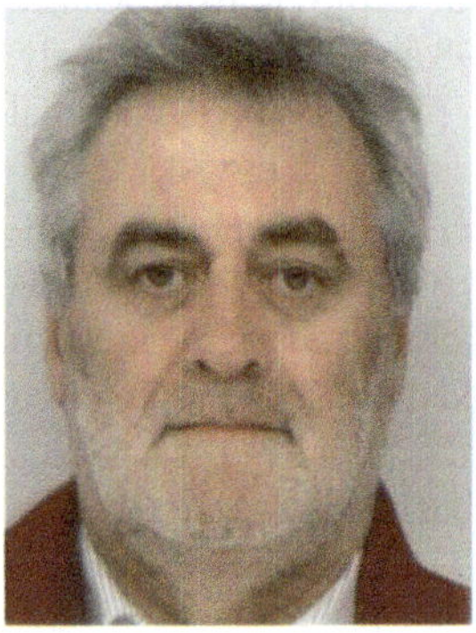

Jeans F. Tourrand, 59, is researcher at Cirad (*Centre for International Research and Cooperation in Agriculture for Development*)—Green Research Unit. He graduated in Veterinary Medicine (Toulouse, France, 1981), with Master and PhD in Animal Science (*University Paris XII*, respectively in 1983 and 1993). He is General Inspector of Veterinary Public Health at the *French Ministry of Agriculture*. During past three decades, he coordinated several interdisciplinary research projects and post-graduate courses focusing on rural development and the interactions between society and environment. He was successively visiting professor in West Africa, North and South America, China, and more recently in Mediterranean. His research is based on the comparative analysis in contrasted livestock regions, especially in South and North America, North and West Africa, Europe and China. He has supervised around 30 PhD and 30 MSc students in different post-graduate courses in Europe, America, Africa and China. He has participated to build scientific networks as Transamazonia at Amazon basin scale, SMART—*Strategic monitoring of South American regional Transformations*, and more recently LiFLoD—*Livestock Farming Systems and Local Development*, www.liflod.org). He has published approximately 150 papers and 10 books.

Randall B. Boone is a Research Scientist within the Natural Resource Ecology Laboratory, Professor in the Department of Ecosystem Science and Sustainability, and affiliate faculty member of the Department of Anthropology, Colorado State University. Randall is a wildlife ecologist, but his experience is diverse, including research in ecological modeling, spatial analyses, landscape ecology, biogeographical relationships, species/habitat relationships, coupled systems modeling, and wildlife and livestock mobility, reported in more than 50 publications and 18 book chapters. Randall's current research analyzes how landscape fragmentation can alter the numbers of livestock and wildlife that can be supported on an area, global rangeland modeling, methods of using computer simulation in education and the inclusion of underrepresented groups, agent-based approaches to household modeling, and as a means of including competition in niche dimension models assessing effects of climate change. He teaches courses in ecosystem science, spatial analyses and geographic information systems, and agent-based modeling of ecological and social systems.

Book Review Commentaries

"This is a fantastic book, which provides an album of rich information and deep insights for both researchers and policymakers in framing research/monitoring programs and designing policy actions in vast pastoral regions worldwide. I highly recommend it to scientists, planners, educators, students and many more who are involved in the actions for promoting the sustainability of worldwide pastoralism. This is a book that attracts you to read, think, and act right away."

Xinquan Zhao, PhD, Professor, Director,
Chengdu Institute of Biology, Chinese Academy of Sciences

This book represents an important contribution to understanding and protecting the world's extensive grasslands that convey critical ecosystem services to millions of people in developing countries. Challenges to maintaining pastoralism in the face of global change have led to continuing land degradation and the concurrent loss of rural livelihoods, biodiversity, and critical watershed functioning. Professor Dong and his colleagues have successfully addressed the complexity of this wicked environmental problem by coupling the dynamics of human behavior and ecosystem processes and providing strategies for promoting sustainable pastoralism worldwide. Their comprehensive assessment will provide guidance for scientists, managers, and government agencies seeking the sustainable development of the world's valuable grassland ecosystems.

James P. Lassoie, Ph.D.
International Professor of Conservation
Cornell University

S. Dong et al. (eds.), *Building Resilience of Coupled Human-Natural Systems of Pastoralism in the Developing World*, DOI 10.1007/978-3-319-30732-9

Pastoral systems are good examples of the coupled linkages between human and natural systems.

Current impacts of globalization, global environmental changes and local social-economic changes are affecting the adaptive capacity of the systems beyond the historic ranges of the pastoral communities to cope.

Exploring how to build enhanced resilience in these systems are critical to the survival of these communities in their current situations

The book provides an excellent insights in how to harness available ecosystem services, enhance land management practices, and to better align policy and socio-economic instruments to support these pastoral systems in the developing world.

Dennis Ojima
Professor
Natural Resource Ecology Laboratory,
Colorado State University

The manufacturer's authorised representative in the EU is Springer Nature Customer Service Centre GmbH, Europaplatz 3, 69115 Heidelberg, Germany. If you have any concerns regarding our products, please contact ProductSafety@springernature.com

Printed and bound by CPI Group (UK) Ltd, Croydon, CR0 4YY
30/09/2026
02225230-0007